Agriculture and the Citizen

Colin R.W. Spedding

Chairman Farm Animal Welfare Council and
Consultant Director of the Centre for Agricultural
Strategy, University of Reading, Reading, UK

Published by Chapman & Hall, 2–6 Boundary Row, London SE1 8HN

Chapman & Hall, 2–6 Boundary Row, London SE1 8HN, UK

Chapman & Hall GmbH, Pappelallee 3, 69469 Weinheim, Germany

Chapman & Hall USA, 115 Fifth Avenue, New York, NY 10003, USA

Chapman & Hall Japan, ITP-Japan, Kyowa Building, 3F, 2-2-1 Hirakawacho, Chiyoda-ku, Tokyo 102, Japan

Chapman & Hall Australia, 102 Dodds Street, South Melbourne, Victoria 3205, Australia

Chapman & Hall India, R. Seshadri, 32 Second Main Road, CIT East, Madras 600 035, India.

First edition 1996

© 1996 Professor Sir Colin Spedding

Typeset in 10½/12½ Sabon by Westkey Ltd, Falmouth, Cornwall
Printed in Great Britain at the Alden Press, Oxford

ISBN 0 412 71520 1

A catalogue record for this book is available from the British Library

Library of Congress Catalog Card Number: 96–83045

∞ Printed on permanent acid-free text paper, manufactured in accordance with ANSI/NISO Z39.48-1992 and ANSI/NISO Z39.48-1984 (Permanence of Paper).

Agriculture and the Citizen

To my mother
Mrs Ilynn Spedding
in her
102nd year

Contents

Preface

The present book arises out of the first two editions of *An Introduction to Agricultural Systems*, recognizing that interest in agriculture is not now confined to those engaged in the industry, or training to be so.

Agriculture has always mattered to people in general, because its products are essential, but for many years, in the UK, it has been possible to regard the supply of food as guaranteed and no longer under threat. In an atmosphere of food mountains and concern over the cost of the Common Agriculture Policy, the public have seen no reason to understand how their food was produced. In recent years, however, the situation has changed markedly. The effects of agriculture on the environment, on conservation of wildlife, on landscape, on food safety and human health, have all become major matters of public concern.

The latter also extends now to what happens within agriculture, the most high profile issue being animal welfare – and not only in this country. As part of the European Union (EU), and importing and exporting food, we are all exposed to the consequences of trade agreements (such as the General Agreement on Tariffs and Trade) and, for example, their effects on developing countries.

This book attempts to cover all these issues. The field covered is therefore very large and presents problems in condensing the whole subject to manageable proportions. Almost every chapter could form a book and, indeed, books exist on many of them. The aim is still to provide an overall view but it cannot hope to be comprehensive in detail.

A somewhat different approach (explained in Chapter 1) has therefore been adopted, with the intention of encouraging debate on the important issues raised. It is hoped that the book will therefore be of interest to a wider range of readers than those specifically concerned with agriculture.

Acknowledgements

I wish to acknowledge the enthusiastic support and encouragement of Nigel Balmforth, of Chapman & Hall, from the original idea of this book until its completion.

It is a pleasure also to acknowledge the invaluable assistance of my secretary, Mrs Mary Jones, not only for her efficient typing of successive drafts but for her help throughout the preparation of this book.

I wish to thank Brian Ford for supplying Table 16.1.

Abbreviations

ADI	Acceptable daily intake
AFRC	Agriculture and Food Research Council
AI	Artificial insemination
BBSRC	Biotechnology and Biological Sciences Research Council
BSE	Bovine spongiform encephalopathy
BST	Bovine somatotropin
CAP	Common Agricultural Policy
CAS	Centre for Agricultural Strategy
CC	Consensus conference
CFCs	Chlorofluorocarbons
CO_2	Carbon dioxide
CSTI	Council of Science and Technology Institutes
C_3	Plants that form stable compounds containing three carbon atoms
C_4	Plants that form stable compounds containing four carbon atoms
DDT	An organo-chlorine insecticide
DM	Dry matter
DNA	(rDNA) recombinant deoxyribonucleic acid
DO%	(or KO%) Dressing-out or killing-out percentage, calculated as the weight of carcass per 100 units of live weight
DOM	Digestible organic matter – a measure of the nutritional value of food to an animal
EU	European Union
FSP	Farming systems perspective
FSR	Farming systems research
GATT	General Agreement on Tariffs and Trade
GMO	Genetically modified organism
HMSO	Her Majesty's Stationery Office
HYVs	High yielding varieties
IBP	International Biological Programme
IFOAM	International Federation of Organic Agriculture Movements
IoB	Institute of Biology
IPM	Integrated pest management
IPR	Intellectual property rights
ITDG	Intermediate Technology Development Group
LU	Livestock unit
MJ	Megajoule
MRI	Magnetic resonance imager
MRLs	Maximum residue levels
NFSD	New Farming Systems Development

NMR	Nuclear magnetic resonance
NSAs	Nitrogen sensitive areas
ODA	Overseas Development Administration
OFR	On-farm research
ppm	Parts per million
RASE	Royal Agricultural Society of England
RSA	Royal Society of Arts (The Royal Society for the encouragement of Arts, Manufactures and Commerce)
UKROFS	United Kingdom Register of Organic Food Standards
WHO	World Health Organization

List of tables

List of figures

List of boxes

Chapter one

Introduction

> The human mind, like the parachute, is best kept open when in use.
>
> *Sir Crispin Tickell**

This book starts with the cover and the questions posed on it. These questions are surely of concern to most people but they are complex and not susceptible to glib answers. Answering them requires knowledge, on a very wide range of topics, especially agriculture, but knowledge is not all that is needed.

The questions involve moral judgments and thus some moral standpoint, but answering them also needs an understanding of people, economics and international relations, including trade, as well as the underlying sciences, especially biology.

It is assumed that the reader would like to be in a position to say 'this is what I think about that', rather than necessarily to be able to answer each question, recognizing that the latter would involve oversimplification of the world as it is.

Certainly, this book is not intended to suggest how readers should answer the questions: it is primarily aimed at helping them to reach their own conclusions. Where moral judgments are involved this must surely be right: no one is entitled to tell you what to think.

Of course, some questions can be answered with some degree of confidence and, where this is so, answers are provided. But even the factual knowledge that is needed is subject to all kinds of reservations. Chapter 2, for example, describes 'world agriculture' but has to recognize: (1) that any such

picture has to be selective and (2) that world data are subject to all kinds of inaccuracies. Just ask yourself, when confronted by statistics about world agriculture, 'how do we know that?'

Other chapters (e.g. Chapter 5 – The role of plants, Chapter 6 – The role of animals) are also factual, but not every species can be mentioned, so there is selection, usually based on relative importance (now?, everywhere?, to whom?).

Feeding the world (Chapter 3), Helping developing countries (Chapter 4), Agriculture and the environment (Chapter 10), Animal welfare (Chapter 12) and Agriculture and human health (Chapter 13), address some of the major concerns of any citizen – all complex issues where one needs to have a considered view in order to contribute to the democratic process.

The word 'citizen' (see Chapter 14) is used (with some worries that it may sound pompous or pretentious) because it expresses the combination of rights and responsibilities involved in these issues.

One of the citizen's natural concerns relates to the safe and responsible use of resources, such as water and energy (Chapter 8), and their impact on people (Chapter 13) and the environment (Chapter 10).

In order to comprehend the mass of information available to us, we need to understand certain important concepts, such as efficiency (Chapter 9), sustainability (Chapter 11) and a systems approach (Chapter 7).

*For further details see Appendix A.

However, it is not enough to pose questions and discuss what you should take into account in answering them. Most of the questions need answers in order to get something done about the problems reflected in them. Getting something done is not, of course, within the power of many of us, but the first step is to try and identify what should be done, within a realistic framework of current conditions, now and in the future.

Sometimes, education represents a way forward (Chapter 15): where knowledge is lacking, relevant research may be needed (Chapter 16). Since both involve public funding, we have to work out how best to conduct our educational and research activities, in ways that encourage the enormous contributions that they can make to the improvement of the world. Not many would find the current world satisfactory and most would wish to press for improvement.

But what constitutes an improvement may look different to different people and may genuinely be so: it cannot be a question, therefore, of deciding what should happen and imposing it on others. Thus, the whole question of how to bring about change for the better is infinitely complicated and has to recognize that the future will almost certainly differ from what we can foresee.

The present book arose because a third edition to my book on *An Introduction to Agricultural Systems* (Spedding, 1979, 1988) was proposed. On thinking about this, it became obvious that the original aim, of establishing the value of a systems approach to agriculture, had been largely achieved. However, in developed countries, the development of a sophisticated food industry meant that it was now necessary to consider the whole food chain (Spedding, 1989) and not just the farm.

Further, it had also become clear that the farm could no longer be usefully viewed as a self-contained system. The interactions between the farm, the food chain, the public and the whole (even global) environment, now required a much wider view (Spedding, 1994a).

It is also clear that public concerns about agriculture now include not only the quality and safety of its products but the acceptability of its methods – especially in relation to animal welfare and environmental impact. Thus, the idea of focusing on the concerns of the citizen (Spedding, 1994b) was born.

Some of these concerns (e.g. with modern biotechnology) are of relatively recent origin but are tied to what may happen in the future. The final chapter (Chapter 17) therefore discusses how these and other developments may shape the future and the importance of allowing citizens to play their full and proper parts in that process.

But we must not view agriculture in a wholly negative fashion, as if it simply gives rise to concerns. The world is fed as a result of agricultural activity and, as Blundell (1995) has pointed out, 'British industries that depend on biological systems – pharmaceuticals, food, healthcare and agriculture – are among the most productive and wealth creating.'

The book draws heavily on the work of many specialists, who know a vast amount about their areas of special interest. It is, of course, inevitable that deep and detailed knowledge is the preserve of specialists: none of us can know all about everything. But the problem is that activities and subjects such as agriculture do not exist in isolation and cannot be fully understood by looking within their boundaries. This idea was expressed, many years ago, in a poem entitled *Harbours*:

> Oh what know they of harbours
> Who toss not on the sea?

This recognizes that, however detailed one's knowledge of a harbour, it is extremely difficult to appreciate its main function – that of conferring protection from the unbridled sea – without experiencing: (1) what it is like outside and (2) what it is like to come inside. The last lines of the poem are:

> Oh what know they of harbours
> Who only harbours know?

Questions

At the end of each chapter there is a short list of questions. These are not intended as a trivial test of knowledge or understanding but represent the kind of everyday questions which force one to think hard about the issues raised. They are a recognition that each chapter is only opening up a whole area of important debate; it is hoped that the questions will lead you further into it.

Chapter two

World agriculture

Never stand between a hippopotamus and the water.

African proverb

2.1 What is agriculture?

The definition that I have used before (Spedding, 1988) is:

> Agriculture is an activity (of Man), carried out primarily to produce food and fibre (and fuel, as well as many other materials) by the deliberate and controlled use of (mainly terrestrial) plants and animals.

This would exclude gardening and landscaping unless products could be described for them (such as money?), but forestry, fish farming and a number of industrial processes would be included.

The word 'primarily' implies that there are other important products and this is indeed so: the main crop and animal products are listed in Tables 2.1 and 2.2. Of course, not all crop – or, indeed, animal – products are directly used by humans, many are grown to feed to animals, so the 'production of food' has to include food for animals as well. Alternatively, feed for animals can be regarded as an intermediate product, or feed production as a process that takes place within an agricultural system. (The word 'feed' is sometimes used to distinguish 'feed' for animals from 'food' for humans.)

Definitions are never as permanent as they sound and there is no reason to retain one (including that proposed here) if a more useful version can be found. Furthermore, we must always be careful not to accept too readily what seems obvious.

This brings us to a common difficulty of definitions. Although a cow is obviously an agricultural animal, it is really only so when it is being used agriculturally (i.e. for food production, for example). Even accepting that animals do have other agricultural functions and roles other than just food production, a pet cow is clearly not an agricultural animal. Similarly, there is no difficulty at all in accepting that the horse may or may not be an agricultural animal, just as is the case for land, people, plants and water.

These differences and distinctions are of great importance where subsidies, laws or regulations, for example, may only apply to activities regarded as 'agricultural'. Furthermore, these distinctions become increasingly difficult to make in those countries (such as the UK) where farming is becoming integrated with non-farming enterprises. Recreational activities on farms may use horses; fish production may be undertaken to provide sport for anglers; rare breeds may be kept because visitors like to look at them; and crops may be grown as cover for game birds. For these reasons, the range of crops and animals that are regarded as 'agricultural' may increase.

So particular animals (species, breeds or individuals) and particular plants (species, varieties or individuals) may serve as illustrations of agricultural organisms but whether they are themselves agricultural or not depends entirely on whether they are embedded in agricultural systems (or processes) or not: and processes are only agricultural

TABLE 2.1 The main crop products

Food (for humans)	Cereals
	Starchy roots
	Sugar
	Seeds, especially pulses
	Nuts
	Oils
	Vegetables
	Fruits
	Beverages
	Flavourings
Medicines	Quinine, opium, cocaine
Fumitories	Tobacco
Masticatories	Betel nuts
Feed (for animals)	Fresh green feed (grass and forages)
	Conserved feeds (hay and silage)
	Roots
	By-products (bagasse, straw, beet pulp)
	Concentrates (cereals, oil seed cake)
Materials	
Construction	Timber
Paper	
Textiles	Cotton, hemp
Rubber	
Household goods	Cork, woven utensils
Fertilizers	Green crops
	Crop wastes
Fuel	Wood, charcoal, alcohol methane
Industrial oils	Linseed, cottonseed, corn oil
Essential oils	Perfumes, camphor
Gums	Gum arabic, gum tragacanth
Resins	Lacquer, turpentine
Dyes	Logwood, indigo, woad
Tannins	Hemlock, oak, mangrove, wattle
Insecticides	Roterone, pyrethrum

TABLE 2.2 The main animal products

Food	Meat
	Milk
	Eggs
	Fish
	Honey
	Blood
Fibre	Wool
	Hair
	Fur
	Silk
Skins	Leather
Fertilizer	Faeces
	Bone
	Feathers
	Horn
Work	Transport
	Traction

of agriculture is generally called farming and occurs most usually in a land-use context.

The term 'agricultural science' is sometimes used to describe the scientific study of agriculture but this can be rather misleading because agriculture embraces more than the contributing sciences and cannot, therefore, necessarily be studied scientifically in all its aspects. (A similar difficulty applies, for example, to 'social science'). 'Agricultural science', on the other hand, can be used to group those sciences (or parts of them) that are most relevant to agriculture or underpin its processes and operations.

Since agriculture is an activity undertaken by, and involving, people, however, and is carried out for productive (and profitable) purposes, it must necessarily include aspects of economics and management, as well as biology. The multi-disciplinary nature of the subject is, to many, one of its major attractions, partly because it is not confined by arbitrary boundaries and partly because it deals with real-life situations, characterized, as they so often are, by being mixtures of different disciplines.

A simple picture (or model) of agriculture is shown in Figure 2.1 but, as can be seen from the tables for this chapter, a great many quite different versions can be derived just by inserting different plants and animals. Varying products, methods, inputs and waste disposal systems produce an even greater number of versions in any one country.

when embedded in such systems. Chapter 7 deals with agricultural systems and the problems of what constitutes a system and how we know when we are only looking at a part of one.

The agricultural activity itself, of course, may only be a part of, for example, national life. Similarly, the activity that we have defined agriculture to be can also be studied and taught, as well as practised. The subject studied would also be called agriculture – the study of it does not have a special word, equivalent to zoology for the study of animals or botany for the study of plants. The practice

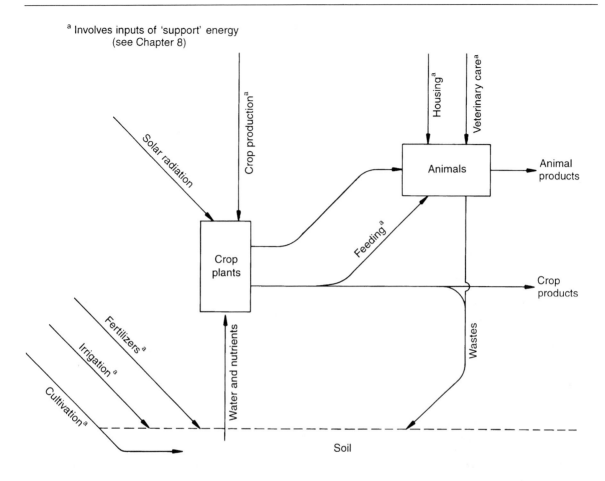

a Involves inputs of 'support' energy
(see Chapter 8)

FIGURE 2.1 A model of agriculture.

Worldwide, the possible variation is enormous. Thus there is really no such thing as **world agriculture** and one must be very careful in generalizing about it. Some picture of agriculture worldwide is, however, essential.

2.2 World agriculture

Agriculture is often described as the world's largest industry, in terms of land use and employment of the population, but, like so many industrial groupings, it is a somewhat artificial aggregation.

Within a country, the importance of the agricultural industry can be more readily assessed. For example, in the UK, the turnover of the industry is some £60bn (4% of gross domestic product (GDP), 13% of manufacturing GDP), it employs about one million people over the whole food chain and the 'value added' by farmers is c. £8bn, by manufacturers c. £11bn and, by retailers and caterers, c. £13bn.

In world terms, the situation is less clear because of the enormous variation between countries. Certainly, farm-land constitutes about 35% of the total land area (of c. 13bn ha), compared with forest and woodland at c. 31%. It employs about 46% of the 'economically-active' population (c. 2285m in 1988), a proportion that fell by c. 4% between 1975 and 1988, although the total population has increased substantially.

It is worth looking, in turn, at these major resources in a little more detail.

2.3 Land

The total land area is about 13 077 million ha, of which some 4565 million ha are farmed. Table 2.3 illustrates the distribution of this land amongst different usages. The areas devoted to the main crops are shown in Table 2.4. It is more difficult to relate livestock to the areas used, since some are on grassland and some fed on the crops already listed. Livestock numbers are given in Table 2.5. However, these animals differ greatly in size and are far from uniformly distributed between regions of the world.

Thus, in 1988, 236.5m cattle produced 48.1m tonnes of meat, 442m sheep produced 6.6m tonnes and 870m pigs produced some 64.4m tonnes.

The variation in numbers across the world and the production of an important non-food commodity are shown in Table 2.6, for sheep.

It will be noticed that by far the most numerous large animals are grazers. Grassland is therefore a rather special form of land use.

2.4 Grassland

It has been suggested that, prior to major agricultural and industrial development, 40–45% of the land surface of the world was grassland (Coupland, 1974). Grassland meant land on which grasses and grass-like plants were dominant and trees were largely absent. Much of this land has since been used for arable cropping and produces a high proportion of the world's food.

There is no doubt, however, that the world area of grassland is enormous (3×10^9 ha) by comparison with that of Britain (14.4×10^6 ha), and it is claimed that the potential grazing areas are probably much greater.

Furthermore, whereas arable land in different parts of the world often produces quite different crops, grassland tends to produce quite similar products even in diverse climatic regions.

TABLE 2.3 Global land use (1987)

	Area (m ha)
Total land area	13 077
Arable	1 373
Permanent crops	100
Permanent pasture	3 214
Forests and woodland	4 069
Other	4 320

TABLE 2.4 Areas of main crops (1987)

	Area ('000 ha)	Area ('000 ha)
All cereals	694 000	
Wheat		221 600
Rice paddy		143 750
Maize		126 100
Barley		81 000
Sorghum		44 100
Millet		40 100
Oats		25 700
Roots and tubers	46 200	
Potatoes		19 000
Cassava		13 800
Sweet potatoes		8 900
Pulses	67 645	
Beans		27 983
Chick-peas		9 530
Oil-seeds	143 700	
Soya beans		52 500
Cotton seed		30 200
Groundnuts		18 500
Sunflower		14 500
Rape-seed		15 800
Sugar cane	16 300	
Coffee	11 000	
Cocoa	5 600	

TABLE 2.5 Global livestock numbers

	Number (m head)
Cattle	1 250
Sheep	1 200
Buffaloes	137
Camels	19
Goats	520
Pigs	823
Poultry	10 000

What happens on the grassland of the rest of the world is therefore of some consequence to the UK.

2.5 The grazing resource

The area devoted to permanent meadows and pastures was given by Pendleton *et al.* (1979) as 3 058 083 000 ha (Table 2.7) and carried a total of livestock equivalent to 1367×10^6 LUs (Table 2.8). Most of these animals were on permanent pasture, which represents a substantial proportion of the total land area in Africa, Latin America, the Near East, the Far East, what used to be called the centrally-planned economies, Europe and the USSR.

Much of the world's meat, milk, wool and hides comes from these areas. Indeed, it was estimated by Van Dyne *et al.* (1978) that over 90% of the feed for livestock on a worldwide basis came from forages.

Although grassland is characterized by graminoids and non-grass-like herbs (forbs), the number of species is extremely large. Natural grassland is usually too arid for closed forest and is therefore usually found in semi-arid to subhumid zones with marked variation in precipitation. Rainfall is in the range of 250–750mm year – for temperate grasslands and 600–1500mm for tropical and subtropical regions.

The growing season is usually limited, by water supply in the tropics and subtropics (to 120–190 days) and by low temperature in temperate regions (to *c.* 165 days). Grasslands tend to have higher than average wind speeds and fire is a common feature, often preventing the extension of forests. The topography is generally undulating and grassland is often important in the control of erosion, the encouragement of soil fauna and the improvement of soil structure.

The desert grasslands and short-grass prairies are thought to be the least likely to be used for cultivation. Thus, it can be argued that much may depend upon learning how best to use the low-rainfall temperate and tropical grasslands. A surprisingly large area of the tropics is thought to be potentially available for grazing but many of the tropical and subtropical grasslands are currently degraded by overgrazing.

TABLE 2.6 Global sheep numbers and greasy wool production (after Pickering, 1992)

	Number (m)	Wool ('000 tonnes)
Africa	200	224
North and Central America	19	46
South America	110	310
Asia	332	486
Europe	142	311
Oceania	229	1270
Russia	141	476

TABLE 2.7 Permanent meadows and pastures for continental areas (1976) (source: Spedding, 1981, after Pendleton, Van Dyne and Whitehouse, 1979)

Continental area	Area (kha)
Africa	800 437
North and Central America	346 735
South America	441 834
Asia	538 310
Europe	87 606
Oceania	469 761
USSR	373 400
World total	3 058 083

TABLE 2.8 Livestock units (LUs) for continental areas (1977) (source: Spedding, 1981, after Pendleton, Van Dyne and Whitehouse, 1979)

Continental area	Total LUs[a] ('000)
Africa	184 983
North and Central America	177 741
South America	208 306
Asia	503 299
Europe	129 325
Oceania	53 447
USSR	109 823
World total	1 366 924

[a] Standard LUs: horses, mules and buffalo, 1.0; cattle and asses, 0.8; sheep and goats, 0.1; camels, 1.1.

2.6 The productivity of grasslands

The productivity of natural grasslands is generally low. For West African savanna, Kowal and Kassam (1978) concluded that the yield of range-land vegetation was very low, irrespective of species composition, ranging from 0.1 to 4.5 tonnes DM ha^{-1} yr^{-1}.

For East Africa, Pratt and Gwynne (1977) expressed productivity in terms of annual carrying capacity (i.e. the capacity to support animals), mainly within subsistence pastoral systems, which ranged from 0.8 to 42.0 ha LU^{-1} varying with the ecological zone. This means that the number of ha required to support one head of population varied from 2.0 to 189 over the same range of zones. Coupland (1979) assembled all the data from the IBP studies on natural grassland and found estimates of total annual net primary production of biomass to range from 239 g DM m^{-2} in semi-arid tropical grassland to 4557 g DM m^{-2} in subhumid tropical areas. The estimates for natural and semi-natural temperate grasslands showed a narrower range (702–3470 g DM m^{-2}). Above-ground biomass, however, ranged from 82 to 3396 g DM m^{-2} in the tropics to 98–2430 g DM m^{-2} for temperate grasslands.

All this illustrates two characteristic features of grassland. First, a high proportion, often more than 50%, of the biomass is underground; secondly, the animal products harvested are always a small proportion of the nutrients and energy in the system.

None of the foregoing gives much of an indication of the potential world output, however, and indeed this would be very difficult to estimate. Two features stand out from any analysis of the world's grassland, which have a considerable bearing on what that potential might be. The first is the amount of potential grazing land in desert areas (904m ha) and in the tropics (Table 2.9). The second is the extremely high proportion of the total grazing area where either moisture or temperature (or both) limit production (Table 2.10).

If these limitations could be lifted, or signifi-

TABLE 2.9 Grazing areas in different climatic zones (source: Spedding, 1981)

Climatic zone	Grazing area (m ha)
Polar and sub-polar	0
Cold-temperate boreal	188
Cool temperate	984
Warm-temperate sub tropical	832
Tropical	1 608
Total	3 612

TABLE 2.10 World grazing areas limited by temperature or moisture (source: Spedding, 1981)

Limitation	Grazing area (m ha)
Temperature	928
Moisture	1 456
Temperature and moisture	912
Neither temperature nor moisture	324
Total	3 620

cantly reduced, output could be raised enormously. Even without such developments, the use of fertilizers and improved plant varieties could raise yields to nearly four times the level otherwise achievable: legumes could also be used where fertilizers are scarce or expensive.

2.7 Implications for British grasslands

There is no doubt that, however important British grasslands may be to the British economy, the area involved is a very small proportion of the world's grassland (0.2–0.48%, according to what is included in each category).

It would be idle to suppose that we can predict with any confidence what levels of productivity will be achievable or achieved in the future, having regard to the problems associated with the use of fossil fuels but bearing in mind possible technical developments, both foreseeable and quite unforeseeable.

Since food is likely to be scarce in many parts of the world and the cost of transport is likely to increase, it might be expected that more food will

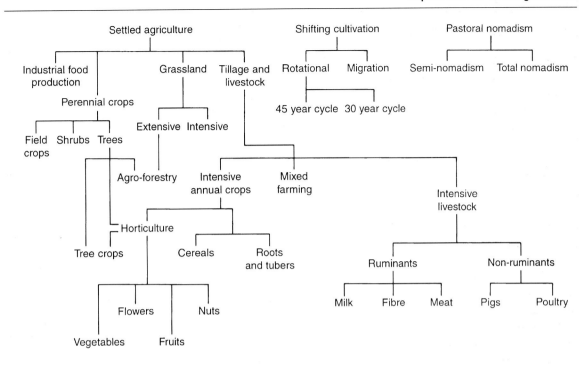

FIGURE 2.2 Example of a classification of world farming systems.

be consumed close to where it is produced. The existence of a relatively vast reservoir of grassland outside Britain may therefore make less difference than might otherwise be the case.

Perhaps greater significance should be attached to the opportunity that our grasslands present, to develop ways of utilizing them that are relevant to less developed countries and thus to make a contribution to the capacity of poor countries to feed themselves (see Chapter 4).

We start from an advanced technical base and it would be a pity, and a poor use of our resources, if our research efforts focused only on our own needs. The world's grasslands represent an enor-

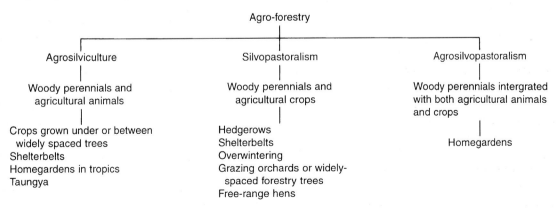

FIGURE 2.3 Classification of agro-forestry systems.

TABLE 2.11 Employment in agriculture for selected countries[a]

Country	Population m (year)	Economically-active population m (A) (year)	Nos employed in agriculture (B) m (year)	$\frac{B}{A} \times 100$ %	Source
UK	57 (1991)	27.6 (1991)	0.552 (1991)	2.1	Marks (1992)
EC-12		130 (1989)	9.2 (1989)	7.0	Marks (1992)
China	844 (1985)	505 (1985)	304 (1985)	60	Xu and Peel (1991)
Sudan		5.2 (1979–1980)	3.4 (1979–1980) (74% men)	66	Craig (1991)
Egypt	52.43 (1990)	14.5 (1990)	5.9 (1990)	40	Craig (1993)
Malaysia	16.6 (1988)	6.6 (1988)	2.3 (1988)	35	Pickering (1992)
India				71	Glantz (1985)
Africa (for different countries)	1.6–33 (1984)			60–91	Glantz (1985)

[a] It is not always easy to obtain up-to-date figures or, sometimes, to be quite sure what they refer to.

mous opportunity to help in the problem of reducing world hunger and this combines the attributes of being fundamentally right with being, in the long run, in our own interest.

2.8 Other land-use systems

Grassland represents only one of the major land-use systems of the world: the others are identified in Figure 2.2. Any part of Figure 2.2 could be expanded to show more detail, as in Figure 2.3 for agro-forestry.

2.9 Capital

African subsistence farmers will have hardly any capital of any kind. They may be entitled to clear and work a couple of hectares but their only capital may be their few tools. At the other extreme, industrialized countries have individual farmers and companies with capital investments of many millions of pounds.

Access to capital may be very limited and the opportunity to use it may be rare: the possibility of obtaining an adequate return on it may be entirely at the mercy of quite unpredictable prices.

Today, people farm for a variety of reasons. Agriculture may be a business, a way of life or a means of sustaining life.

2.10 People

Historically, most people will have been engaged in agriculture (in order to feed themselves) and it will have involved whole families. The world trend is for less involvement and, in industrialized countries, the proportion of the population now employed in farming is almost negligible. However, in precisely these same countries, an elaborate food industry has developed and the number of people employed in this usually greatly exceeds the numbers in farming.

Since the countries of the world are at very different stages of development, there is enormous variation in the proportion of the population engaged in agriculture (see Table 2.11), but the world average (see Table 2.12) still shows that nearly half the economically-active population is so engaged. The proportion appears to be related to the level of income characteristic of the economy (see Box 2.1). This diversity in the proportion of the people involved in agriculture is characteristic of world patterns for almost anything.

TABLE 2.12 Global population (millions) (1988) (after Pickering, 1992)

Total	Economically active (A)	Economically active in agriculture (B)	$\frac{B}{A} \times 100$ %
5.1	2.3	1.1	48

Box 2.1 Type of economy and labour in agriculture (%) (based on Marsh, 1992)

Economic category	Population (m)	Labour force in agriculture (%)	GNP per head ($)	Area ('000 km²)
Lower income	2 493	71	270	33 608
Lower-middle income	691	59	750	15 029
Upper-middle income	577	29	1 890	22 238
Industrial market economies	742	7	12 960	30 935

Clearly, climate, soil and topography show enormous variation across the world, but it is also the case for people (ethnicity, size, appearance) and, more importantly for this chapter, for the human condition (see Box 2.2).

Although some of these features may appear to have little to do with agriculture, they greatly affect the ability of people to operate. The availability of water and fuel determine how long people have to spend collecting them and thus how much time they have available for weeding etc.

Their physical fitness and nutrition affect how hard people can work and the proportions of children and old affect the needs of the family.

But none of these figures, or the national averages, reveal the large discrepancies in wealth and condition within each country – a variation which applies to virtually all countries, however rich or poor.

2.11 The food chain

One of the very big differences between countries is the extent to which agriculture is represented largely by farming and the extent to which the output of farming is processed by a well-developed food industry. The general shape of this food chain is illustrated in Box 2.3, but in many countries some stages will be omitted, at least, for some sections of the population, and others will be much simplified.

In fact, the structure of the chain may be different. For example, in subsistence farming, it is not only the food preparation and cooking that may be done in the 'kitchen' or the homestead, but also the storage and processing.

At the other end of the scale, agriculture is a big industry and in the UK, for example, it

Box 2.2 Attributes of developed and developing countries [north versus south (see Chapter 4)]

Developed countries North	Developing countries South
Overproduction	Underproduction
High yields	Low yields
Overeating	Starvation
Obesity	Low body weight
Affluence	Poverty
Meat consumption	Crop product consumption
Oil powered	Animal powered
High percentage of old people	High percentage of young people
22% of population rural	79% of population rural

Box 2.3 The human food chain
In developed countries the human food chain is now very elaborate, partly because of the development of a complicated food industry and partly because of the network of input industries that feed into it at all points. These are only briefly indicated in figure (a) below (the core of which is based on Nursten, 1992).

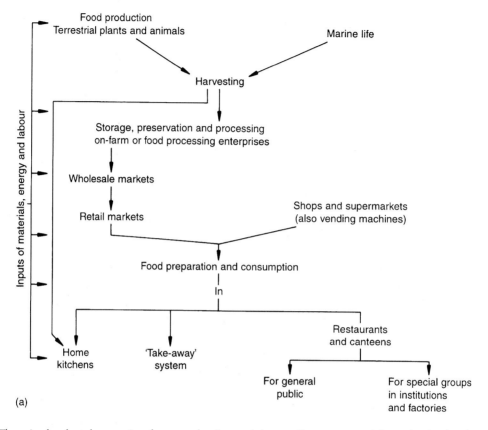

(a)

Thus, in developed countries, farm production mainly supplies raw materials to the food industry as illustrated in figure (b).

In developing countries, the food industry part is usually much simpler but the labour inputs may be complex, involving members of the family who undertake other activities as well, such as wood collection to supply the cooking.

contributed £13.8bn (in 1991) to the economy, although this only represents c. 1.4% of the GDP. Of course, few industries contribute very large percentages unless they are aggregated in a major way.

The value of the food produced on UK farms in 1991 amounted to £9.2bn, with a further £8.6bn being imported. Since consumers spent about £52.9bn in the same year, it can be calculated (Marks, 1992) that food manufacturers, wholesalers, retailers and caterers added value to the extent of c. £38bn. The importance of catering is illustrated by the fact that about one-third of the food is now eaten outside the home (in the USA it is about 50%).

The dependence of the UK on food imports has led to considerable concern about the 'food trade gap' (see Box 2.4).

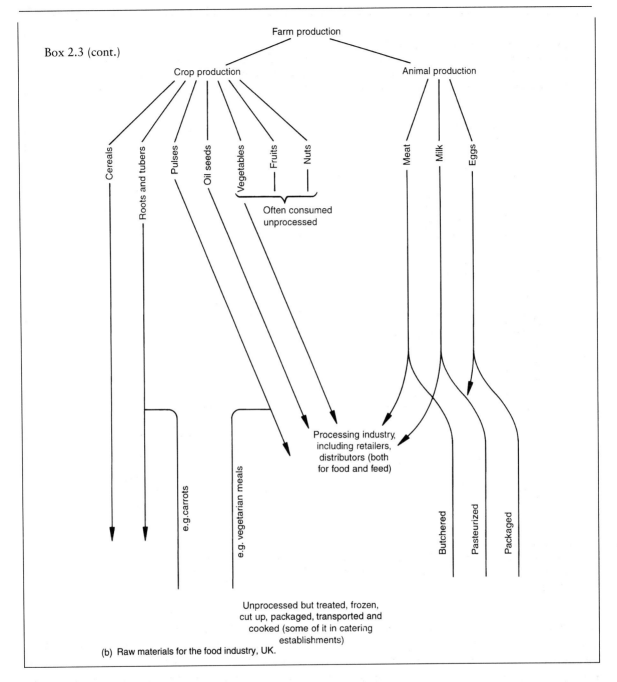

Box 2.3 (cont.)

(b) Raw materials for the food industry, UK.

2.12 International trade

Preoccupation with 'self-sufficiency' in food, understandable in countries like the UK that have been vulnerable in the past two wars threatening their vital supplies, has resulted in many countries protecting their agricultural industries by support of various kinds (as with the Common Agricultural Policy (CAP) – (Box 2.5).

This, by and large, has not succeeded in protecting

Box 2.4 The UK food and drink trade gap

Since the UK cannot produce all the food commodities its consumers want, and even those it can produce may be only seasonally available, imports are needed. In some cases (e.g. coffee), food raw materials are imported, processed and re-exported.

Food will be imported if it is cheaper, even when it could be home grown: equally, food can only be exported if it can be produced at prices that other countries are willing to pay.

It is easy to understand a national view that reducing its trade deficit represents a worthwhile goal, but the political issue should not be confused with the economic problem or the geographical limitations. There are dangers in focusing just on one sector, such as food, and it has to be recognized that there are alternative uses for resources required to produce more food. For example, land could be used to grow trees and thus to reduce the even larger (in percentage terms) trade gap in timber and wood pulp.

Nevertheless, food is focused on because of its prominence as part of the overall trade gap, see below.

Total trade balance (£bn)

Year	Current values	Inflation adjusted to 1991
1981	+6.7	+12.0
1990	−15.4	−16.3
1991	−4.4	−4.4

Source: Customs & Exercise, FFB (1992).

Within the total, food and drink is an important segment being consistently in deficit by *c.* £5.5–6.5bn after adjusting for inflation. In proportion to the total, however, the contribution increased dramatically over this period (1981–1991) so that food and drink is now the largest element of the trade deficit.

Food and drink sector trade balance (£bn)

	Current values	Inflation adjusted to 1991
1981	−3.5	−6.3
1990	−6.0	−6.3
1991	−5.5	−5.5

Source: Customs & Excise, FFB 1992.

Particular sectors can be identified, such as fruit and vegetables, where the deficit is very high.

Box 2.4 (cont.)

The UK food and drink trade balance with the rest of the world 1992

Sector	Exports	Imports (£m)	Balance[a]
Meat	827	2 033	(1 206)
Dairy	534	1 111	(577)
Fish	568	996	(428)
Cereals	1 203	1 004	199
Fruit and vegetables	331	3 118	(2 788)
Sugar	301	762	(461)
Tea, coffee and cocoa	509	890	(381)
Other groceries	341	559	(217)
Oil seeds and nuts	37	229	(192)
Fats and oils	86	423	(337)
Animal feed	337	717	(379)
Total food	5 074	11 842	(6 767)
Drinks	2 448	1 565	883
Total food and drink	7 522	13 407	(5 885)

[a] Adverse trade balances are shown in brackets.

Of the £13.4bn imports, it has been estimated that about £7.5bn are 'unavoidable' because of climate, seasonality, special access for third countries and consumer demand for 'authentic' foreign products. Development in some sectors is limited by CAP constraints. Thus, the opportunity for import substitution is c. £6bn, although, in practice, improvements in the trade gap are at least as likely to come about through increased exports as through import substitution.

The food trade gap is of political importance but the economic significance of a specific deficit or surplus can only be determined by detailed analysis of its causes, the use made of imported or exported capital and domestic opportunities. The crucial need is for increased competitiveness and economic efficiency.

rural economies, is very costly and has resulted in the production of surpluses that are costly to produce, store and dispose of. Most economists now favour liberalization of trade and the General Agreement on Tariffs and Trade (GATT) aims to do this (see Box 2.6).

In the UK, it is now thought to be more sensible to support rural policies and environmental protection and improvement more directly, in order to minimize the distorting effect on agriculture.

As Marsh (1992) has pointed out: 'Among the developing countries of the world, the political economy of agriculture is closely linked to the overall problem of development and poverty.' This will be pursued in Chapter 4.

The fact is that the world appears divided into regions, not only of soil, climate and topography, but of affluence, and this last difference is so great as to demand action. That is why Chapter 4 concentrates on how the affluent, developed part of the world can help those suffering from the ills of powerlessness and poverty.

Starvation and malnutrition are only part of the problem but providing food for the whole world is

Box 2.5 The CAP

The Common Agricultural Policy (CAP) was established in 1957 as part of the Treaty of Rome and set out to provide a reasonable standard of living for Europe's farmers, to keep farmers on the land and to ensure that consumers were guaranteed food at reasonable prices.

Vast amounts of money have been paid to farmers (as subsidies) in order to achieve these aims. But much of the money went to input suppliers – not to the farmers – and much of it became capitalized in the greatly increased values of land [see figures (a) and (b) for the UK picture]. (Other countries, e.g. the USA, also subsidized farmers to a massive extent.)

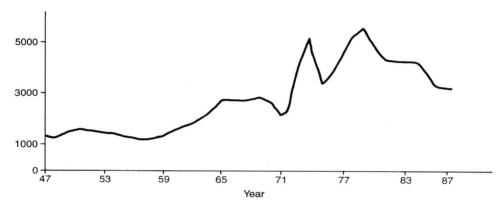

(a) Changes in the value of land ($£\ ha^{-1}$ in real terms, 1985). Smoothed curve based on data from Harvey (1991).

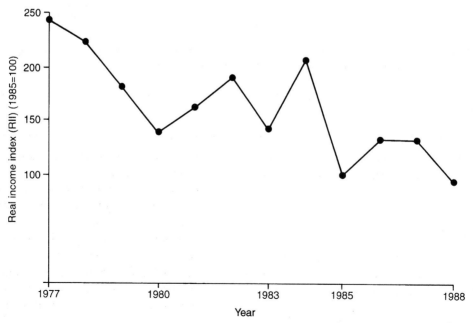

(b) Change in real farm incomes 1977–1986 (based on Byrne and Ravenscroft, 1990).

Box 2.5 (cont.)

The cost of this support has risen dramatically over the years and the cost of the CAP to the European taxpayer has been variously estimated (because there are different ways of calculating these things) but is expected to reach £26 288m in 1992 (Rosen, 1991).

The costs incurred [see figure (c)] consist of subsidies, the 'intervention' cost of buying unwanted products and the cost and disposal of these surpluses (often referred to as 'mountains' and 'lakes'). They have represented a high proportion of the EC budget: up to 1981, 65–75% (Body, 1982), and in 1991, 59% (Rosen, 1991).

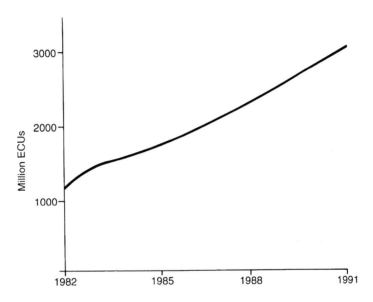

(c) CAP support from EC budget: mainly on oil seeds, milk products, cereals, beef and veal, sugar, sheep meat, fruit and vegetables (in descending order of magnitude). After Harvey (1991).

CAP reform has been a major subject of discussion for many years and now has to be seen within the context of the GATT (see Box 2.6).

the obvious starting point for those concerned with agriculture. The next chapter therefore focuses on Feeding the world.

Questions

1. Why do we talk about world agriculture? (We don't talk about world retailing, medicine, flower arranging, electricity generating, shoe-making, etc!)
2. Who is responsible for world agriculture?
3. Why don't we farm the sea? (It covers 90% of the earth's surface.)
4. Is it sensible to transport food (or live animals) from one country to another? Or from one part of a country to another?
5. Is it better for more people to be engaged in farming? Or is it efficient to employ the minimum necessary to produce the food required?
6. Why is it necessary to employ so many people in processing, packaging and distributing food?

Box 2.6 The GATT

It took seven years to complete the most recent ('Uruguay') round of negotiations of the General Agreement on Tariffs and Trade (GATT). The agreement is based on the principle that trade should be non-discriminatory and that national legislation should not aim to give domestic consumers an economic advantage over its competitors. There are some areas (e.g. environment and animal health) where exceptions are allowed.

The agreement on agriculture decided:

1. that domestic support to the agricultural sector as a whole should be reduced by 20% over the six year implementation period;
2. that export subsidies should be reduced to a level 36% below the 1986–1990 level and the quantity of subsidized exports reduced by 21%;
3. that levies should be replaced by tariffs which should be reduced by 36% over six years;
4. that minimum access tariff quotas be established where current imports are less than 3.5% over six years.

Special provisions were made for developing countries.

Davenport (1994) assessed typical reactions that the whole package would have 'very modest inflationary effects on world prices'. His own findings show:

reductions in self-sufficiency ratios for the EU, the USA and many other countries;
major effects in China, India and Indonesia;
production of dairy products boosted in Africa;
some damage to developing countries as a whole.

Examples are shown in the table below (after Davenport, 1994).

Estimated effects of Uruguay Round on self-sufficiency ratios [i.e. ratio of production to utilization (production + imports − exports)]

	Wheat	Dairy products
EU	−4.0	−6.4
USA	−14.3	−6.7
Developed countries	−6.6	−6.7
Developing countries	1.6	5.2
China	2.9	9.0
India	1.2	6.2
Japan	−3.4	−11.5

In 1995, the GATT was replaced by the new World Trade Organization, but in some areas (such as tourism, financial services and the protection of intellectual property) it will operate in partnership with national capitals.

Chapter three

Feeding the world

Whether elephants make
love or war, the grass suffers.
African proverb

The first thing to be clear about is that feeding the world is not the same as producing sufficient food: it is, in fact, a much more difficult problem. This is because people are hungry only because they are poor, and it is estimated (Tribe, 1994) that, today, over a billion people live in acute poverty. If they had money, the food they need would be produced for them and delivered to them. There are, of course, a few exceptions to this, mainly in the immediate aftermath of disaster, natural or human-made, when supplies may be disrupted. But, in general, the rich do not starve or even go hungry and this is true in the same countries where the poor and starving are found in large numbers. The reason is quite simple: if the poor have no money, they cannot buy food and no one can produce food without cost, so they have to sell it to stay in business.

Clearly, the only ways to feed the poor are: (1) to give the food away or (2) to give them the money to buy it. Both have been tried (see Chapter 4). Of course, it would be ideal if the poor could be employed at a wage that allowed them to buy food – but then they would no longer be 'poor'. If money is available, it is still necessary to ensure that enough food is produced, otherwise scarcity causes prices to rise too high. In this chapter, therefore, we will concentrate on whether sufficient food can be produced to meet the **needs** of the world population, recognizing that these are not the same as the economic **demand** (i.e. the quantity of food that people are able and prepared to pay for).

It is obvious that demand will be affected by food price and agricultural methods may influence this because they affect costs. Price and costs are not necessarily related, however, and the price is likely to rise if demand exceeds supply. Since a rise in price – if reflected in the rewards to the producer – will tend to result in increased production, the situation would be expected to reach a balance, given time and a reasonably free market. Since the price can never be zero, this does not mean that **needs** will be met, only effective demand.

Nevertheless, a sensible starting point is to consider the capacity that exists – or could exist – to produce food. Let us start at the most fundamental level, that of crop production by photosynthesis (see Chapter 5 for more detail).

It has been estimated (Duckham, Jones and Roberts, 1976) that only *c.* 2.8% of the possible world potential net photosynthate is actually formed in world farming, because less than half of the world's cultivable land is actually cultivated, only 70% of this bears a crop in any one year and only 70% of the growing season is actually used. About 22% of the potential net photosynthate is therefore used in the production of human food and the difference between this figure and the final 2.8% is reckoned to be due to losses, bad husbandry, weather uncertainty and inadequate crop canopies. There are further losses in harvesting, processing, storage, distribution and within the consuming household (see Table 3.1).

TABLE 3.1 Estimated losses in world agricultural production (after Duckham, Jones and Roberts, 1976)

	Energy	
	Per unit of cultivable land (GJ ha yr)	*Per head of total population (MJ day⁻¹)*
Photosynthate actually formed in food production and estimated as 'recoverable'	25.8	59.2
Photosynthate recovered as potential food products	9.1	20.9
Food products entering households	4.1	9.5
Food products actually eaten	3.7	8.6

There is every reason, therefore, to try and form a picture of the crop production process and to identify where losses and inefficiencies occur. The main process can be visualized as an energy flow diagram (Figure 3.1). It does not automatically follow that the efficiency of these processes can be increased (see Chapter 9 for a discussion of efficiency in agriculture) or that the losses can be avoided. But it is worth noting that some of the losses are enormous and so the question should at least be asked. Probably the greatest hopes lie in increasing the amount of crop grown (by using more fertilizer, better plant varieties) and avoiding the losses due to pests and diseases (see Box 3.1). All this only relates to crop production. Conversion to animal products (see Chapter 9) involves further inefficiencies and losses (see Box 3.2).

The importance of losses is that they reduce the amount grown, available as product and usable after storage, all from the same land area (and to a considerable extent the same labour and other inputs). So, if they could be greatly reduced, production would be increased without involving further land.

3.1 Land available

As shown in Chapter 2, the total land area (global) is *c.* 13 077m ha and, of course, will change little, though natural and human-made disasters could render some unusable. The recovery of land from the sea is only feasible where it will have exceptional value [to extend a town (e.g. Rio de Janeiro) or an airport, for example] or where the land area is very limited (e.g. the Netherlands). The recovery of usable land from deserts is a different proposition and could be a major world objective, given the will and ability to cooperate. Thus, land where

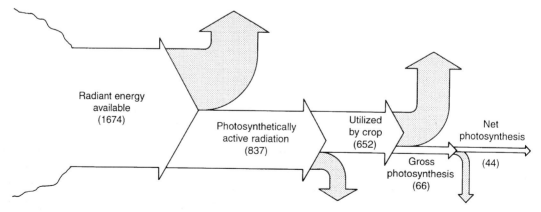

FIGURE 3.1 Energy flow in crop production. The numbers in parentheses are 10^8 J ha^{-1} day^{-1}. Losses are shown by the shaded areas. The proportion of the total incident radiation appearing as chemical energy in the crop = 2.7%.

Box 3.1 Losses in crop production
Walker (1983, 1987) discussed the enormous difficulties of assessing yield losses due to pests. He compiled a fairly comprehensive collection of data and published a selection of some 200 cases, where losses ranged from a few low figures (1–5%) to very large numbers from 10 to 30% and a significant number over 30%.

Just to take one example, Reddy and Walker (1990) studied the yield losses in graminaceous crops caused by the stem borer Chilo (27 spp.). Losses varied from 4 to 97% for *C. partellus* in maize, up to 88% in sorghum, up to 66% (for *C. suppressalis*) in rice and 16% (for *C. auricilia*) in sugar cane.

One of the few attempts to partition (%) global losses (pre-harvest) between weeds, insects and diseases is given below (after Cramer, 1967).

Cause of loss		*Crop*	
	Maize	*Wheat*	*Rice*
Weeds	37	40	23
Insects	36	21	58
Diseases	27	39	19
Total loss (*L*) pre-harvest (m tonnes)	121	86	207
Harvested crop (*H*) (m tonnes)	218	266	232
$\dfrac{L}{L+H} \times 100$	35.7	24.4	47.1

A more recent, and very comprehensive assessment is that of Oerke *et al.* (1994), who have published estimates of crop losses for all the major crops and all the major regions of the world. The following data are selected from their findings. Percentage crop losses in trials from agrochemical companies represent the difference between using crop protection chemicals and not doing so. The median losses in rice production for eight major rice-growing regions of the world were (for 1988–1990):

	From diseases (%)	*From insect pests* (%)	*From weeds* (%)
	16	14.6	30
For wheat (median for 15 countries)	15	6.4	12.8
World, overall			
Actual	12.4	9.3	12.3
Potential	16.7	11.3	23.9

Clearly, total losses are very high and would be much worse without the use of crop protection agents. It is estimated that the worldwide losses prevented by crop protection in 1988–1990 were 16.4% for weeds, 7.1% for pests and 4.2% for diseases.

the potential is limited by shortage of water is theoretically recoverable: land where use is limited by low temperature is not a feasible possibility, except on a very small scale, using glasshouses or other protection (see Box 3.3).

Arable land (1373m ha) and permanent crop-land (100m ha) could be made more productive by application of fertilizers and irrigation, as could the forests and woodland (4069m ha) but, in the latter case, not primarily for food.

Box 3.2 Losses in animal production

It is worth distinguishing between losses and wastage.

Losses can occur from systems of all kinds, but terms such as 'wastage' only make sense in relation to a purposeful system in which there is a desired output. It is not very helpful, however, to regard everything as lost or wasted that does not appear in the final product. The desired output, of meat or milk, for example, cannot conceivably contain all the energy used in its production. There are, in other words, production costs of many kinds, and at least a proportion of those incurred are inevitable and cannot sensibly be described as losses or wastage.

Wastage suggests losses that could have been avoided. These may be of two kinds: first, losses of product or intermediate product material that could have been harvested or converted; and secondly, production costs that could have been avoided. However, avoiding them might have cost more than they were worth, so we may be obliged to tolerate some losses for economic reasons. Avoidable production costs may have to be tolerated for similar reasons. For instance, very high growth rates in meat-producing animals lead to a higher proportion of the total food being converted into product. An unnecessarily high proportion of food being used for maintenance is thus avoided. But if the higher animal growth rates require more expensive foods, it may prove unprofitable to use them, even though they may be more efficiently used in terms of food conversion.

Loss and wastage, therefore, do not apply only to substances already formed, but include at least some aspects of reduced performance. This is important because, at first sight, it might seem sensible to exclude the latter. If we imagine one tree on an area of land that could support two, it does not seem very useful to describe the missing tree as a loss, unless it had been there once and had been removed. Incident light might nevertheless be considered wasted for the lack of photosynthetic material. Similarly, ungrazed vegetation, due to lack of stock, represents wasted vegetation rather than lost animal production. There are thus straightforward losses of what has already been produced and wasted resources that represent a loss of potential production.

In some circumstances these are described as 'direct' and 'indirect' losses. When a sheep is infected with internal parasites, for example, there may be small direct losses of substance (gut wall, blood, etc., in the case of stomach and intestinal nematodes), but, in addition, large losses of potential growth, due to reduced appetite and lowered digestive efficiency. As a consequence of a reduced growth rate, there is then a loss of nutrients because the proportion required for maintenance is increased.

The most satisfactory definition of 'losses' then appears to be 'diversion of a greater than optimum flow of matter (however described) from the main production process'. This means that it is necessary to state the product being considered and what sort of losses (of calcium, energy, dry matter) are being described; it is also essential to attach a meaning to the word 'optimum' in this context. For any of the material flows in a production system to have an optimum rate, it is necessary to describe the production required from the system, per unit of some or all of the resources.

None of this would matter greatly, if it were not for the fact that some deceptively obvious losses often turn out to be unavoidable or avoidable only with adverse consequences to the system as a whole. This will be clearer if the main sources of loss in production systems are examined. These may be identified from a diagrammatic representation of any particular kind of system such as that of goat production from browsing shown below. It will be seen that there is an opportunity for loss to occur throughout the production process. This is shown in the figure by broken lines. This process is continuous, although phases can be detected and points determined at which what has been happening may usefully be summarized. Losses may usefully be summarized at the same points, but this should not obscure the continuous nature of the wastage processes.

As is pointed out elsewhere (e.g. Chapter 9), conversion of feed by animals involves substantial losses, some unavoidable and some not. The efficiencies of energy conversion shown below illustrate that up to 90% of the energy in the feed may be lost.

Box 3.2 (cont.)

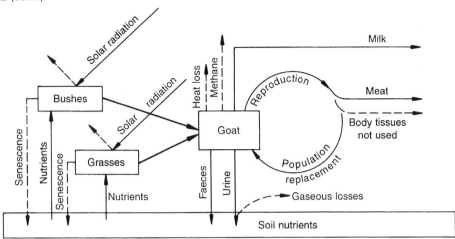

(a)

Efficiency values for the output of energy in the carcase of
independent meat-producing animals in relation to the
input of feed energy (from Spedding and Hoxey, 1975)

	E^a
Beef cattle	5.2–7.8
Sheep	11.0–14.6
Rabbit	12.5–17.5
Pigs	
pork	35
bacon	35
Hens	16
Geese	13.4

[a] E = (Total energy in the carcase/gross energy in the feed from
independence to slaughter) × 100.

Quite apart from these conversion losses, there are often enormous losses due to:

disease   causing mortality, causing reduced performance or production;
pests   resulting in reduced grazing time, for example;
parasites   internal and external causing loss of blood, reduced digestive efficiency and death.

In addition, over the whole food chain, there are losses in storage, processing, transport, preparation and
cooking.

Disease
This is an enormous subject and its importance is not really reflected adequately in this book. In
animals, the most important species have an infectious agent forming at least part of their cause (Biggs,
1985): most of the major plagues have been viral in origin because these are the most highly contagious
infections. The control of disease represents a major future challenge.

Box 3.3 Constraints on the agricultural use of land

Buringh (1985) calculated (from FAO data and other sources) that 78% of the total world land area (which he gave as 14 900m ha) was constrained by being covered in ice, or too cold, dry, steep, shallow, wet or poor. There is little that can be done about some of these constraints and, even where the problems can be resolved, it may prove too costly, both in monetary and energy terms.

Clearly, wet land can, in some cases, be drained, although in some situations this would be regarded as ecologically undesirable. Steep land can be terraced, and indeed this is carried out to a spectacular extent in many poor countries. There are other constraints too, such as the presence of boulders (solved impressively in Mauritius by stacking them in the centre of the field, where they conserve moisture).

Of the 3 278m ha remaining, 1 937m ha was regarded by Buringh (1985) as of low productivity.

The permanent cropland includes fruit orchards, oil palm trees, beverage trees, rubber trees and sugar cane, which cannot change rapidly.

The permanent pastures of the world (3214m ha) have some potential for increased production and, in some cases, could be converted to crop production.

There are increasing demands for land for urban development, recreation and amenity and this often erodes agriculturally useful land rather than areas that have little use. This can happen even in developing countries, in order to attract and provide for tourists.

3.2 Land use

Exploitation of unused agricultural potential is largely limited by economic factors but whether further capital investment can be justified depends on the economic demand for the product. If the food is not wanted, or no one is prepared to pay enough for it, it will not be produced.

A major problem in developing countries has been the need of many governments to provide cheap food to urban populations (poor and capable of revolt). The price received by the farmers is then so low that it provides no incentive to increase production. (When prices are high and guaranteed, production increases: in Europe, the CAP has illustrated this to the point of resulting in vast surpluses of the main commodities.) On the other hand, if prices are too high, those in need of food cannot afford to buy it. Furthermore, lack of continuing returns to reward production provides no incentive to the development of the essential infrastructure of communications (roads, transport, storage depots) and education that is needed for the creation of a modern food industry. Thus, such demand as exists for the products of a modern food industry has to be met by imports.

The problems of using land more productively are complex and it is unrealistic to consider the technical possibilities of increasing output without recognizing the costs associated with this and the probability of their being justified by the returns obtained.

Irrigation is one of the most effective technical interventions but it depends upon the availability and cost of the necessary water. One of its advantages is that it may make all the difference, in arid and semi-arid areas, between growing a crop and crop failure – all the work having been done and the other inputs paid for. Naturally, the rewards are greater for high-value crops, such as fruits and vegetables in Israel, or where money is plentiful, as in oil-rich Arabian countries. Irrigation is most widespread in Asia, which irrigates more than 60% of the world's total irrigated area (*c.* 227m ha).

Given all these possibilities, the question remains, could the world's needs for food be met from existing or potential resources? This depends upon the size of the population to be fed.

3.3 World population (Table 3.2 and Box 3.4)

It is frequently pointed out that world population doubled between 1950 (2.5bn) and 1988 (> 5bn),

TABLE 3.2 World populations (after Bunting, 1992)

| | Population (m) | Growth rate per year (% per year) | |
	1988	1950–1975	1975–1988
Developed nations	1235.303	1.13	0.73
Developing nations	3879.485	2.33	2.09
World total	5114.788	1.96	1.74

with that of developing countries (including China) increasing from 1650m to 3900m. However, as Bunting (1992) has pointed out, population growth rates (see Box 3.5) changed significantly during this period, decreasing everywhere except Africa (which increased from 2.54 to 3.03%) and the Near East (up from 2.48 to 2.82%).

At the same time, the proportion of economically-active people engaged in agriculture decreased from 58 to 47% in the world as a whole (developed countries: 28 to 9%; developing countries: 73 to 61%).

Populations, therefore, are still increasing, mainly because of a decline in death rates, especially of infants and children (Box 3.6). Birth rates appear to be generally decreasing. This still means that the number of females is increasing and there

---

Box 3.4 World population

Bunting (1992) calculated the increase in the populations of different regions of the world, between 1950 and 1988, as shown below.

| | Total population (m) | |
	1950	1988
World	2497.566	5114.788
Developed nations	847.674	1235.303
Developing nations	1649.877	3879.485
Developed market economies	574.167	833.447
North America	166.120	272.011
Western Europe	301.446	380.949
Oceania	10.082	19.693
Developing market economies	1064.214	2682.442
Africa[a]	176.650	496.554
Latin America and the Caribbean	162.110	430.171
Near East[b]	99.557	267.161
Far East	625.897	1482.375
Centrally-planned economies	859.170	1598.899
USSR and Eastern Europe	273.507	401.856
Asian cpes (mainly China)	585.663	1197.043

[a] Excludes Egypt, Libya, Sudan.
[b] Includes Egypt, Libya, Sudan.

It can be seen that the increases in the developing nations generally, and in the Far East particularly, were very large during this period. However, rates of increase have been declining and the projections recently made by Anderson (1994) suggest that the world population would increase from 5.8bn in 1995 to 8.5bn in 2025.

Again, it is projected that the 'less-developed countries' (Anderson terminology) would increase more, from 4.5bn in 1995 to 7.1bn in 2025, than the 'more developed countries' (1.2bn in 1995 to only 1.4bn in 2025).

is thus a lag effect, most of these additional females will also produce children for years to come, whatever the changes in future birth rates. This is part of the basis for the prediction of future population sizes.

The World Bank projections are that the total world population will increase to 6.2bn in the year 2000 and to 9.5bn by 2050. It is estimated that the figure may plateau at *c.* 11bn in about 100 years' time. (Of course, such predictions cannot take into account major wars and diseases such as AIDS.)

Box 3.5 Rates of population increase

Bunting (1992) calculated the following population growth rates for various regions of the world.

	Growth rate (% year)$^{-1}$	
	1950–1975	*1975–1988*
World	1.96	1.74
Developed nations	1.13	0.73
Developing nations	2.33	2.09
Developed market economies	1.13	0.70
North America	1.45	1.00
Western Europe	0.77	0.33
Oceania	2.02	1.26
Developing market economies	2.43	2.43
Africa[a]	2.54	3.03
Latin America and the Caribbean	2.74	2.21
Near East[b]	2.48	2.82
Far East	2.29	2.24
Centrally-planned economies	1.85	1.23
USSR and Eastern Europe	1.13	0.78
Asian cpes (mainly China)	2.14	1.39

[a] Excludes Egypt, Libya, Sudan.
[b] Includes Egypt, Libya, Sudan.

However, he also pointed out that, over the period 1960–1988, life expectancy had also increased significantly.

	Life expectancy at birth (years)	
	1960	*1988*
High-income nations	69	76
OECD	69	76
Developing	40	71
Middle-and low-income nations	49	62
Sub-Saharan Africa	39	51
East Asia	51	66
South Asia	44	57
Europe, Middle East, North Africa	54	64
Latin America and the Caribbean	57	67

Box 3.6 Mortality rates

Populations can increase by immigration from other countries and can decrease due to emigration but, apart from these, population increases are due to higher birth rates or lower death rates. Birth rates may be too high in many parts of the world but they are, on average, decreasing everywhere. This is indicated by the following changes (after Bunting, 1992).

	Birth rates per 1000 per year	
	1965	1987
High-income nations	19	14
Developing countries	35	29
Middle- and low-income nations	41	30
Sub-Saharan Africa	48	47
Latin America and the Caribbean	40	28
East Asia	39	23
South Asia	45	34

Population increases are, in general, due to lower death rates. Comparable figures for those given for birth rates are (per 1000 of the population per year).

	1965	1987
High-income nations	10	9
Developing countries	11	6
Middle- and low-income nations	15	10
Sub-Saharan Africa	22	16
Latin America and the Caribbean	12	7
East Asia	11	7
South Asia	20	12

Can these numbers be fed? is a question greatly affected by the arguments at the beginning of this chapter. We now have to ask: can enough food be produced?

3.4 Potential food production

Since there is no widespread famine except for disastrous situations brought about largely by human conflict, it could be said that the needs of the present world population are being met by current levels of production. But this is to ignore the variations hidden by average statistics.

In fact, hundreds of millions are hungry, at least seasonally, many of them chronically hungry. They include the poor, the old, ill and infirm, the disabled, the children and, often, the women.

Yet the output of cereals, for example has greatly increased (see Box 3.7), mainly due to greater yields per hectare brought about by better crop varieties, greater inputs of fertilizer and better methods of crop protection, as exemplified by the 'Green Revolution' (Box 3.8).

Fortunately, more productive methods tend to be more profitable. This is mainly because many of the costs (rent of land, cost of seed, basic labour) are incurred whatever the yields – indeed, even if the crop fails entirely. These overheads are spread over the entire output, thus, the higher this is, the lower the cost of each kilogram of product. Only the additional (so-called 'variable') costs rise as the

Box 3.7   Cereal yields (after Bunting, 1992)

	Yield (kg ha^{-1})	
	1948–1952	1986–1988
World	1134	2561
Developed countries	1292	3012
Developing countries	988	2257
Developing market economies	1625	3707
North America	1614	3849
Western Europe	1632	4377
Oceania	1095	1558
Developing market economies	911	1762
Africa[a]	640	1047
Latin America and the Caribbean	1113	2067
Near East[b]	1046	1627
Far East	928	1948
Centrally-planned economies	995	2916
USSR and Eastern Europe	876	2233
Asian cpes	1142	3822

[a]   Excludes Egypt, Libya, Sudan.
[b]   Includes Egypt, Libya, Sudan.

The remarkable increases that have occurred in recent years have to be seen against the very slow rate of yield increase from 1800 to well into the 1900s, illustrated for the UK, in figure (a).

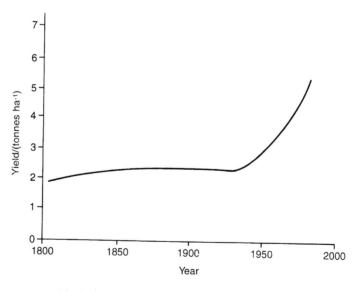

(a) The average yield of wheat grown in the UK in the period 1800–1983 (based on Buringh, 1985).

Box 3.8   The Green Revolution

In the 1950s, the International Agricultural Research Centres (see Chapter 16) embarked on a major programme of breeding for improved staple cereals that produced early maturing, day-length insensitive and high-yielding crop varieties (HYVs). These formed the basis of packages of improved seeds, fertilizers and pesticides, and the application of these constituted the 'Green Revolution'. Other inputs were often required as well – notably irrigation.

So, although these packages were widely available and began to affect developing countries in the 1960s, not every farmer had access to all the required inputs or could afford them. By and large, these tended to be the larger, wealthier farmers. The impact of the Green Revolution therefore varied considerably from one place to another.

The impact

In developing countries, the impact on wheat and rice production was tremendous. Between one-third and one-half of the rice areas in the developing world are planted with HYVs (Pinstrup-Andersen and Hazell, 1987) and in the eight Asian countries that produce 85% of Asia's rice, HYVs added 27m tonnes of rice annually, fertilizers added 29m tonnes and an added 34m tonnes due to irrigation. However, as Conway and Barbier (1990) have pointed out, there have also been 'significant equity, stability and sustainability problems'.

Also, where it has been most successful, it has tended to result in rice and wheat being grown instead of less productive crops, resulting in widespread monoculture and consequent pest problems.

A substantial literature (see Chambers, 1983) records those cases where some of the benefits have been exaggerated and where successes have benefited the 'haves' and exacerbated the gap between rich and poor. One has to remember that farmers – even neighbours – are often in competition with one another: improving the efficiency of farmer A may disadvantage the unimproved farmer B.

There has also been an ecological impact following the enormous increases in the use of pesticides and fertilizers (Conway and Pretty, 1991). Nevertheless, successes have been achieved. India is now self-sufficient in food and Harrison (1987) claimed that similar achievements could be obtained in Africa, based on improved varieties of maize, cassava and cowpea.

In the early 1970s the same grim predictions (about population growth, increasing food imports and impending famines) were made about India that are now made about Africa. The potential for accelerating food production in sub-Saharan Africa is discussed in detail by Mellor, Delgado and Blackie (1987).

yield increases (such as fertilizer to produce the increase and increased labour to harvest it).

Cereals represent the biggest food source for the world population: they are also a major feature of world trade, with surpluses in North America, Oceania and Western Europe being exported. Considerable stores are held: in developed countries, storage in 1990 amounted to 290m tonnes (eight weeks' supply); in developing countries unknown quantities are stored domestically. Table 3.3 shows the latest figures available.

Bunting (1992) has calculated that if the 1986–1988 average cereal yields of 353 kg head (692 kg per head for developed nations and 245 kg for developing ones) are projected to increase as they have done in the last 50 years, there would be an average

TABLE 3.3   World stocks of grain (1990/91) (after Sloyan, 1992; based on FAO, 1991)

	Grain (m tonnes)
Rice	61
Wheat	141
Coarse grain	132
Total (all cereals)	334
Stocks as % of annual consumption	19%
Tonnage in developing countries	144
Tonnage in developed countries	190

of 450 kg per head per year. Multiplying this (which appears to be sufficient for Western Europe today) by the 'plateau' population (11bn), the total figure would be 5bn tonnes to be obtained from 1bn ha.

The 1990 output was estimated at 1.9bn tonnes, so the average yield would have to increase to 5 tonnes ha^{-1} from its present level of just over half of this. A linear projection of current trends suggests that 5.3 tonnes ha^{-1} would be reached in about 100 years time.

This does not appear impossible, if you look at the averages already obtained in many parts of the world, shown in Box 3.7. Whether such increases can be produced within the political and economic framework that will exist is quite impossible to say, but it does not look to be physically impossible (see Box 3.9).

Indeed, other estimates have resulted in even higher figures for the total population that could be fed (FAO/AEZ, 1982). However, this does not mean that there would not be regional problems, since some areas have higher potentials than others, and no account was taken of possible major climatic change (see Chapter 10). There is also the problem of population distribution both within and between regions.

Bunting (1992) identified seven groups of factors that would work in favour of or against the necessary increases in food production.

They were:

1. effective demand (already discussed);
2. output delivery systems – roads and infrastructure (briefly mentioned earlier);
3. resources (see Chapters 5–7);
4. more productive technical methods (see later in this chapter);
5. policies and practices of governments (see Chapters 4 and 17);
6. foreign relations (see Chapter 4);
7. knowledge systems (see Chapter 15).

3.5 Resources needed

The extra output required to feed the projected population will need extra resources. It can easily be calculated that the extra cereal output will contain 45m tonnes of nitrogen more than is currently present. Some of this could be supplied by biological fixation (e.g. using legumes) (see Chapter 5)

Box 3.9 The prospects for producing the required output of cereals (from Bunting, 1992)

A yield of 5 tonnes ha^{-1} is well below many estimates of the potential yields of cereals. It may be compared with average yields of cereals in 1986–1988 of: 4 tonnes ha^{-1} in China and eastern Europe (not including the USSR), 5.7 tonnes ha^{-1} in Japan and 4.4 tonnes ha^{-1} in the USA and western Europe. Technically, it is possible to feel a measure of confidence about the possibility of producing 5 bn tonnes of cereals, especially as there is a century in which to find out how and where best to do it. It will require continuing advances in the techniques of production in suitable environments (considered below), and this in its turn will require larger supplies of plant nutrients, larger inputs of mechanical or muscular energy, and more effective means of protection of plants and animals against pests, diseases, competitors and predators.

Some may doubt whether the human species has the political, economic and managerial ability to achieve growth in output at this rate in the future. Competent government can powerfully aid agricultural progress, but in fact in many countries rural producers have been able to increase output to meet market need even where government has been less than fully effective. Perhaps they can be expected to show the same ingenuity and determination in the future.

It is not feasible to determine in detail where the additional output will be produced at different times in the future, or where it will be utilized. To do this would require predictions about the evolution of demand in time and space, and about the technical evolution of production systems in different regions and nations.

However, it seems likely that most of the extra output will be produced in the presently-developing nations, in tropical, subtropical and winter-rainfall environments. A substantial part of it will be produced in Africa. Within these regions, the prospects are most promising in those nations which seem most likely to have surplus productive capacity.

---

but it is unlikely that this could meet the need.

New technology may make it possible to breed better plants and to protect them better (see Chapter 17) but massive inputs of manufactured nitrog-enous fertilizer may still be necessary and all of these developments have implications for the energy supply (see Chapter 8).

3.6 Nutritional needs of people

This chapter has so far concentrated on the supply of cereals that would be needed to feed increased populations, partly because, quantitatively, they are the most important element and partly because the rest of the diet is made up of such diverse foods that a world picture cannot easily be constructed. Furthermore, in general, if the supply of energy is sufficient, the other elements of the diet are also usually adequate. However, most diets contain other foods and it is important to ensure that these, too, are available.

Meat is a major constituent of the diets of industrialized countries but much less so for developing countries (Table 3.4). Milk (Table 3.5) and milk products (Table 3.6) are also of considerable significance. World production of eggs was estimated in 1979 to be 26.6m tonnes, mostly in Asia, Europe and North America, but although amongst the most ubiquitous of foods – produced in some 180 countries – it can be imagined how accurate the statistics are for non-industrialized countries.

Of course, much of the grain included in the cereal data is actually used to feed poultry, pigs and cattle. This has often led people to believe that if grain was not so fed, it would be available to feed people. Of course, this is purely theoretical, in that the hungry could still not afford to buy it, quite apart from the fact that the cereals are in surplus chiefly in those countries that feed them to animals. In fact, if the feeding of grain to livestock was not allowed, livestock producers would not buy it, it would remain unsold and, in subsequent years, would not be produced. Sheep, goats and a great many cattle are primarily grazed on grassland (referred to in Chapter 2).

The major differences in diet between developed and developing countries are described in

TABLE 3.4 Production of meat (10^6 tons) contribution to total production (source: FAO, 1980a)

	World	Developed countries (%)	Developing countries (%)
Beef	47	24	11
Sheep and goat	7	2	3
Pig	51	25	14
Poultry	28	15	6
Total	133	66	34

TABLE 3.5 Milk production (1979) ('000 tonnes) (source: FAO, 1980b)

	Cows	Goats	Sheep	Buffalo
World	420	7.3	7.4	26

TABLE 3.6 Milk products (m tonnes) (source: FAO, 1980b)

	Cheese	Butter and ghee	Evaporated milk	Dried milk
World	11.01	6.92	4.57	5.72

Chapter 4 but this seems an appropriate place to describe the basic nutritional needs to be met, although the majority of people in developed countries are no longer much concerned about basic nutrition.

Standards of affluence in developed countries generally mean that few have to worry about getting enough to eat and people are more concerned about food safety and a healthy diet. That is not to say that they should not be concerned, since rich and poor alike may consume poorly balanced diets and the incidence of obesity (and overweight people) is rather high (the possible impact of agriculture on human health is discussed in Chapter 13). In developing countries, on the other hand, the problem is primarily one of getting enough to eat.

The intake of energy and protein needed to maintain good health in human beings varies with their size, age, sex, amount of physical activity and the climate. Average requirements are illustrated in Box 3.10 for energy and protein.

Box 3.10 Daily nutrient requirements of people in the UK (after CAS, 1979)

Age range (years)	Body weight (kg)	Requirements for:	
		Energy (MJ)	Protein (g)
Boys and girls			
0–2	7.3–11.4	3.3–5.0	20–30
2–5	13.5–16.5	5.9–6.7	35–40
5–9	20.5–25.1	7.5–8.8	45–53
Boys, 9–18	31.9–61.0	10.5–12.6	63–75
Girls, 9–18	33.0–56.1	9.6	58
Men, 18–35	65	11.3–15.1	68–69
		(depending upon activity)	
Women, 18–55	55	9.2–10.5	55–63
		(depending upon activity)	
Pregnant (late)		10.0	60
Lactating		11.3	68

In addition, there are daily requirements (again varying age and weight etc.) for, in this case, women 18–55 years, of:

Thiamine	0.9–1.0 mg
Riboflavin	1.3 mg
Nicotinic acid	15 mg
Ascorbic acid	30 mg
Vitamin A	750 µg
Vitamin B	2.5 µg
Iron	12 mg
Calcium	500 mg

The last five are increased greatly in pregnancy and at least doubled during lactation.

Since people in developing countries tend to be smaller and live in less cold climates, their needs may well be less, but their activities and the age structure of the population are also different. The ways in which these needs are satisfied affects the number of people that can be supported by 1 ha of land, because of the differences in resources (including land) required to produce different foods (see Table 3.7). This is one of the impacts of vegetarianism or, more generally, of the extent to which farmed animal products contribute to the diet (see Chapter 17).

The earlier discussion about whether these needs, qualitative and quantitative, can be met for future world populations suggested a fairly optimistic conclusion, from a technical point of view.

But, as has been repeatedly pointed out, producing enough food is not the same as feeding people and there are many constraints on achieving the latter.

However, we also have to recognize that there are severe constraints on whether the technical achievement can be brought about. Some of these constraints are also technical (see Box 3.11) and others are illustrated below.

3.7 Constraints on food production

3.7.1 Land use

The availability of land has been mentioned. One of the constraints on being able to use land is the

TABLE 3.7 Number of people who could be supported by the production from 1 ha of land (source: Spedding, Walsingham and Hoxey, 1981)

	Crude protein (kg ha^{-1})	Gross energy (MJ ha^{-1})	No. of people whose annual requirement could be met	
			Protein[a]	Energy[b]
Crop				
Cabbage	816	105 000	34	23
Field beans	613	43 466	26	9
Peas	566	40 805	24	9
Potatoes	522	102 080	22	22
Wheat	469	69 534	20	15
Sugar beet	416	152 469	17	33
Maize	392	75 905	16	17
Rice	375	87 768	16	19
Barley	350	59 274	15	13
Animal				
Beef	65	4 796	3	1
Lamb	65	7 486	3	2
Bacon	105	14 438	4	3
Rabbit	292	13 251	12	3
Chicken	135	7 056	6	2
Eggs	74	4 118	3	1
Milk	118	8 770	5	2

[a] Taking protein requirement as 65 g day^{-1}, i.e. 24 kg year^{-1} (assuming adequate protein quality).

[b] Taking energy requirement as 12.6 MJ day^{-1}, i.e. 4599 MJ year^{-1} (assuming all the energy is available, e.g. digestible).

possibility of extensive pollution – by heavy metals (e.g. cadmium, lead) from industry, by contamination from nuclear accidents or oil spills, by loss of soil due to erosion. All these are occurring at the present time and all are very difficult, costly and time-consuming to reverse: in the case of nuclear contamination, the time-scale is very long indeed.

Brown *et al.* (1990) have attempted to assess the loss of productive capacity due to environmental degradation. Their estimate of soil erosion is alarming: for the world as a whole, it is *c.* 24bn tons of topsoil annually (this equals 1 inch on 61m ha). Rising concentration of salt in irrigated areas and waterlogging are judged to result in an annual 1% decline in output on 24% of the 180m ha of irrigated grain.

Deforestation may alter the hydrological cycle locally and lead to severe flooding of agricultural (and other) land. Other causes of concern include ozone, sulphur dioxide, nitrous oxides and climatic change. Brown *et al.* (1990) suggest that the total of all

this environmental degradation is *c.* 14m tons of grain annually, roughly half of the gain from the use of more fertilizer and other inputs.

3.7.2 Political action

Entirely different constraints stem from human actions (or lack of them). Famine is now restricted to the African continent. Webb and von Braun (1994) argue that it need not, and should not, be occurring at all. They consider that the three pillars on which a food-secure future rests in Africa are:

1. **Good governance** Implying planned, efficient use of resources and their allocation in a transparently non-discriminatory fashion in regional adoption and ethnic terms.
2. **Sound growth policies** Mainly economic in order to bring about the adoption of improved technology.
3. **Active preparedness** Implying a coordinated

Box 3.11 Factors governing potential food production

Achieving such yields will depend upon many factors, the main ones will certainly include:

1. solar radiation (the quality of this may depend, for example, on changes in the ozone layer);
2. temperature (yield estimates have not generally taken account of possible 'global warming');
3. water supply (also subject to major climate change);
4. land area available (also affected by climate, erosion, overgrazing, 'Chernobyl'-type accidents);
5. soil type and fertility;
6. 'support' energy supply (mainly inputs from fossil fuels which may be constrained by pollution controls on the emission of 'greenhouse' gases);
7. genetic efficiency of plants and animals at capturing and converting solar radiation.

All these factors govern the potential production of biomass and it must be remembered that biomass is required for many more purposes than food production. Fuel and timber are two major needs but fibre for clothing and many others are important and may become more so.

intervention strategy aimed at protecting vulnerable households.

These go far beyond technical advances. However, economic approaches may also need to be rethought.

Brown *et al.* (1990), for example, argue that the currently-used economic indicators are fundamentally flawed. They take the gross national product (GNP) as an illustration of this. GNP simply totals the value of all goods and services produced and subtracts depreciation of capital assets, but it ignores the depreciation of natural capital, such as non-renewable resources. Thus, overexploitation of forests increases GNP and well-managed forests do less well. The same short-term view applies to oil extraction and the overuse of irrigation.

3.8 The global picture

Anderson (1994) attempted to put the global picture into perspective. He reviewed the 1480m ha of cropland used in 1990, estimated that an additional 1900m ha of land was not cropped in 1990 but was technically arable but concluded that only 150m ha of that would actually become used up to the year 2030.

Similarly, for irrigated land, the figures were 256m ha used in 1990, plus 134m ha potentially irrigable, but only 34m ha actually irrigated by 2030.

His overall conclusion was that 'we are going to make it' through socially acceptable conversion of land to cropping, some additional irrigation, sustained investment in new agricultural knowledge (see Chapter 16) and an open trading regime.

Questions

1. What is the optimum rate of population increase?
2. Why do cereals dominate world food supplies?

Helping developing countries

> Wisdom is rarer than
> emeralds, and yet it is found
> among the women who
> gather at the grindstones.
>
> *Book of Ptah-Hotep,* The Husia,
> Egypt

4.1 What is a developing country?

Some countries are more industrialized than others and are frequently wealthier with a better developed infrastructure. But, clearly, 'development' can refer to different aspects of a country, including social and cultural states, and one country may be more 'developed' than another in only some of these aspects. Terms such as 'less developed countries' (LDCs), 'undeveloped', 'underdeveloped' and 'Third World' were attempts to describe those countries, which, for one reason or another, lacked many of the characteristics of the wealthier, technically-advanced and industrialized countries (the so-called 'developed' nations).

Many of these terms were seen as offensive in one way or another and the current (imperfect) choice is 'developing'. There probably isn't a completely satisfactory term, if only because there are so many features at various stages of evolution. It may be that it would be preferable to be more specific about the feature being identified and to stop artificially lumping such different entities together. However, that makes it extremely difficult to construct any general picture, so the term is retained here, recognizing that all nations ought to be developing and, in some senses, the 'developed' countries are already 'overdeveloped'. Examples of the last point are in terms of fossil fuel use, with consequent pollution, 'developed' countries using far more energy per head of population than the 'developing' ones. The evolution of the terms used reflects much more important changes in the way 'development' is defined, its objectives stated and its progress monitored.

Many authors have reviewed all this and it is a very large subject. Clayton (1983) pointed out that the earlier concept of development being represented by increases in GDP had long been superseded and George (1985) has repeatedly stressed the damage that may be done by thinking only in monetary terms, arguing that the activities of the International Monetary Fund (IMF) (and the banks) have created a 'debt crisis' of enormous proportions.

One of the problems of simply aiming to increase GDP is that it ignores the needs and vulnerability of the poor. Between 1950 and 1975, the average increase in GDP for developing countries was 3.4% per annum, but more people appeared to fall below an absolute poverty line (Conway and Barbier, 1990) and the disparity between rich and poor increased. This led to a concept of 'growth with redistribution' and subsequently a shift in perspective to meeting 'basic needs'.

One way of defining a developing country is in terms of lacking a safety net for those in need – the hungry, the poor, the sick and infirm and the old. There would probably be some general agreement that one measure of a civilized society is how well it looks after its vulnerable members. Unfortunately, a great deal of help to nations has to go through their governments and may not reach the vulnerable at all. Indeed, some would go further and argue that aid may actually allow unpleasant and undesirable regimes to stay in power by: (1) preventing uprisings of the vulnerable and (2) by allowing governments to take credit for whatever they do distribute. A further dimension to defining developing countries is the north–south debate.

4.2 The north–south debate

The Brandt Report (1980), published under the title *North–South: A Programme for Survival*, was something of a milestone in dividing the world into two blocks (the north – relatively affluent – and the south – extremely poor) and emphasizing the importance of the relationship between the two and the need to improve this.

Three years later the Brandt Commission (1983) published an update (*Common Crisis*) and reported that the crises referred to in the earlier report had got worse. Although the first report had not been made 'obligatory reading for all citizens in the world', as was suggested by some, there was a massive public reaction and a general sympathy for the proposition that we are all in the same boat – there is little comfort in the fact that it is the south end that is sinking.

This north–south concept represented a different way of looking at the world (see Box 4.1), based on the major discrepancies in the average standard of life between people of the wealthy north and the impoverished south (see Box 4.2).

The Brandt Reports made clear recommendations as to what was needed as **emergency** measures. They were in four main areas:

1. A global food programme to stimulate world food production and to begin to abolish world hunger.
2. A global energy strategy to accommodate the need for security of both producers and consumers.
3. Additional financial flows to ensure the stability of national economies strained by precarious balance of payments and mounting debts.
4. Reforms to achieve broader participation in international financial institutions and more balanced conditions for world trade.

It is clear that food production is only one part of an interlocking package and it is important to be aware of all this even though, in this chapter, we will have to concentrate on the question of ensuring adequate food supplies for the hungry. Within the urgent, but broad, Brandt Recommendations, the following specific needs were identified that relate directly to this question.

1. 'Employment-creating agricultural development is the key to growth in low-income countries.'
2. The need to focus on helping small-scale farming, instead of concentrating on the support of large-scale farming.
3. The need for programme aid for agriculture to cover local-cost and recurrent expenditures.
4. The need for technical assistance, training and education.
5. The need to review the UN food and agricultural agencies.

In addition, whilst recognizing the dangers and problems, **food aid** remains an important and necessary tool.

Box 4.1 Ways of looking at the world

Figure (a)
This map shows the normal Mercator projection, which distorts the sizes of the land masses towards the poles. In this, Brazil appears to be about twice the size of Scandinavia.

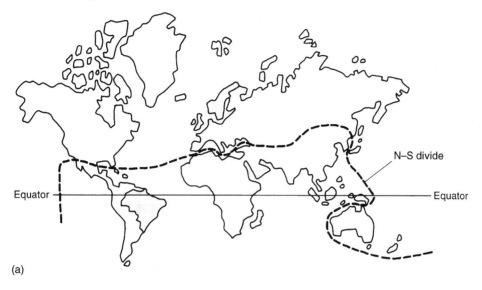

(a)

Figure (b)
This map is the Peters Projection, which reflects more accurately the relative sizes of the land masses. Brazil now appears closer to its real relative size, some 17 times that of Scandinavia.
The north–south divide is shown as a broken line, from which one can see that the line lies well north of the equator except for the Antipodes.

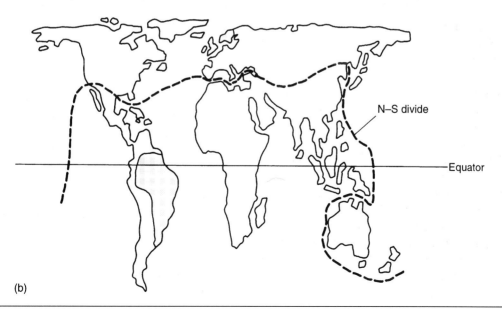

(b)

Box 4.2   North–south differences (from Hadjor, 1988)

North	South
Overproduction	Underproduction
High yields	Low yields
Overeating	Starvation
Obesity	Low body weight
Affluence	Poverty
Meat consumption	Crop product consumption
Oil powered	Animal powered
High percentage of old people	High percentage of young people
22% of population rural	79% of population rural
Good hygiene	Poor hygiene
Pure water	Contaminated water
Good control (over conditions)	Poor control
Most food processed by a food industry	Food processed in the home or on the farm
Sophisticated infrastructure	Poor infrastructure

All these issues have been debated at length and the discussions cannot even be summarized here, but food aid is such a specific issue that it cannot be ignored.

4.3 Food aid

The Brandt Reports called for **more** food aid (with proper safeguards) and recent experience in Africa has demonstrated: (1) that food aid is essential for the immediate relief of famine; (2) that this is occurring on a vast scale; and (3) that it is unlikely to prove a **short-term** crisis.

By contrast, Jackson (1982) has argued that food aid is generally counter-productive, due to distortions of trade, the corruption it causes, the fact that it may put the local producer out of business, may result in dependence, may itself be unsuitable and very often does not reach those in most need.

It is probably unwise to generalize about food aid, as if all foods, the people who are hungry and their circumstances, were all the same. It is also unwise to assume that because food-aid schemes have failed in the past there is no possibility of better, more effective ones being devised and implemented in the future.

It would be a terrible conclusion to have to reach, if we were forced to it, that there is no way in which the excess food capacity in the north could help those in need of food in the south. Whether the paradox of surplus and starvation is really any worse than all the other contrasts between rich and poor, oversupplied and needy that occur between so many countries and within most, is a difficult question. It is certainly one about which people feel very strongly but it is not really very logical. Once it is recognized that the problem of hunger is primarily one of poverty, some of the uniqueness disappears. The problem could be solved by monetary aid that could be spent on buying food, perhaps produced locally.

Furthermore, if those who overeat forego some of their consumption, it does not help the hungry and, if persisted in, by reducing effective demand, may simply result in a subsequent reduction in production. The same is true of grain fed to animals. If it was not so fed, it would feed no one and it would reduce the scale of agriculture. Whereas, if **expenditure** was foregone by the rich and the money transferred to the poor (which is at least possible), the poor could buy food and there would be a shift of resources towards food production.

But, when the complexity of the crises is revealed

and people begin to wonder whether their gifts have only prolonged the lives of some for a few weeks, then there is a powerful feeling: (1) that we need to think out what we are doing and (2) that we need to find ways of **preventing** crises rather than ameliorating them.

Of course, there are unhelpful governments, military actions and political manipulations that may defeat our best intentions. But this is the world in which we have to live and operate: our opportunities are rarely to **solve** problems but simply to help, to ease burdens and move events in a better direction. Hence, the general conclusion that we must aim to help people to help themselves.

Harrison (1987), like many others, recognized that, at that time, in a country like Ethiopia, food aid or cash aid (for peasants to buy food) was unavoidable: mass mortality from starvation was the only alternative. He thought that food-for-work was better than free relief food.

The problems with giving food aid are many, including the following:

1. It may put out of business those who are struggling to sell their own locally-grown produce. These are just the people one would wish to encourage, in order to produce more food where it is needed. However, if the hungry did not have the money to buy their foodstuffs, food aid to them can hardly have this effect.
2. If the food aid is channelled through government, or other agencies, little may actually reach the hungry and more food may then come on to the market and compete with that of local producers.
3. Any form of aid can confer power on the agency used to deliver it and this power may not be used wisely or in the interests of those in need.
4. Any form of aid offers opportunities for corruption and this has often been the case.
5. However, if aid is **not** channelled through the Government, it may be interpreted as interference in the affairs of another country.
6. Food aid may involve different kinds of food from those normally consumed: indeed, where

developed countries have surpluses, they are often in this category. This may bring about, or reinforce, a shift in consumer preference to a food not currently produced (e.g. wheat) in the recipient country, or costly to produce. (Examples of this are given by Lele, 1987). The food may also be nutritionally unsuitable: remember that many Asians are unable to digest cows' milk, for example. Unfortunately, the food most readily donated is that accumulated as surplus in a country that wants to get rid of it.

Several authors have concluded that food aid has done more harm than good and it certainly does not appear to be the ideal solution to hunger. On the other hand, there have also been successes and there is no other way of meeting the needs of victims of catastrophes. In the case of natural disasters there are no disadvantages either: for human-made catastrophes, however, such as war, it may remove part of the price of waging it, if someone else is going to pick up the pieces. This can become very complicated indeed, especially where those distributing the aid may become engulfed as hostages of one sort or another.

If food aid **is** given, it would be better and more efficient if it came from buffer stocks close to where the need is greatest. Not only is this practicable but improving storage facilities in vulnerable countries is a useful investment anyway.

One of the problems of food aid is the high cost of transporting and distributing it when needed and the same disaster may have damaged the infrastructure needed for distribution. The fact is, of course, that transporting food does have a high cost and takes time, whereas money can be moved by telephone. Money suffers from many of the same problems (e.g. corruption) but, if it reaches the right people (including the 'right' members of the family), it does create the 'effective demand' that the local economy actually needs.

There are no easy or clear-cut answers to the problem and the problem is not confined to food shortage.

Major characteristics of the countries of the

south are poverty, powerlessness, lack of fuel, lack of water (especially clean water), medical facilities and many other shortages (such as those illustrated in Box 4.2), and all these have to be tackled at the same time. The problems are dauntingly immense but, as Chambers (1983) pointed out, action is vital, and can and should be undertaken by vast numbers of people, all accepting that progress has to be made in many small steps. But there have to be some major **reversals** in attitude, he argued, 'putting the last first' (i.e. focusing on those people, needs and activities that are normally put last). This was one of the objectives of 'intermediate technology.'

4.4   Intermediate technology

One of the problems of trying to help a country (or a person) that is less advanced in some important ways is the temptation to show them how you do it, as if that is necessarily what they should do. This has often happened in the past. In education, students have been trained in developed countries to use techniques and facilities neither available nor appropriate in their own countries. Very often, they have not returned to their own countries anyway.

In research, reputations tend to be built on work at the scientific frontiers rather than tackling the immediate, very applied problems in developing countries.

In extension/advisory work, the problems are in some ways even greater. Education (Chapter 15) and research (Chapter 16) could be said to train the mind in ways that ought to be useful in whatever context and environment, but extension is concerned with the application of what is known to the adviser.

To take a simple agricultural example, consider an agricultural graduate going out to a developing country, recently qualified. The qualification will almost certainly be fairly specialized: if it is a PhD, for example, it would be in animal production (and probably in only one species) or in crop production (again, probably in only one or a limited number of species). Genuinely anxious to be helpful and

being in a strange country, modesty and caution combine to ensure that each sticks to what he/she knows about. So, an animal production graduate (A) will want to improve the livestock and the crop production graduate (B) will want to improve the crops.

However, a better animal to A is more productive, needs more food to be so and probably larger. A better crop plant to B will all be ready to harvest on the same day – because with the machinery of a developed country that is what is wanted – and most of the growth will be in that part of the plant that is the product. A does not yet know that his bigger, better animal will still have to live on the crop by-products (just reduced in quantity by B) and will be, if anything, less productive than the original. B does not yet realize that, without harvesting machinery (or storage capacity), the farmers would rather have something they can collect every day over quite an extended period: and that the by-products feed the animal that may also do the soil cultivations. This example is illustrated in Chapter 7.

Endless examples exist of attempts to help countries move forward by introducing technology that is too sophisticated. Sometimes, the problem is that there is no infrastructure to maintain or repair machines and the latter, including tractors, can be seen rusting in the fields.

Sometimes, a whole package of related inputs was required to make it all work and only some could be afforded. This was the case with the 'Green Revolution' (see Box 3.8), where improved seeds needed to be supplied with fertilizer, water and protection from pests and diseases.

The whole point of intermediate (or appropriate) technology was to avoid these traps and to ensure that not only was the technology appropriate to the needs, skills and resources of the people for whom it was intended, but that appropriate technology was devised where it was not already available (see Box 4.3).

There are difficulties, however, in meeting the real needs of people who have very little of anything, including land. There is a tendency for small prob-

Box 4.3   Intermediate technology

Technology is best considered as a combination of tools (including equipment), techniques (the knowledge, skills and facilities needed to use these tools) and organization (the processes for deploying techniques and tools). There is a tendency to associate sophisticated technology with developed countries and it is sometimes even regarded as a measure of development (or progress). But technology has to be **appropriate** to the needs of the user and this includes being affordable.

For example, cooking is one of the oldest and most widespread technologies but the needs of developed countries (met by gas- or electricity-powered appliances) are quite different from those of the developing countries (the 'south'), where cooking is predominantly done on open fires or in simple stoves using wood (or charcoal) as the fuel. There is no electricity or gas and nor could it be afforded.

This well illustrates the need for technology to be appropriate. It is not that current methods cannot be improved – many of them are very inefficient – but the improvements have to meet the real needs of those who will implement them.

The same is true for manufacturing in the south, which is characterized by small-scale, decentralized, low-cost, labour-intensive methods of production geared to local markets.

This philosophy has been embodied in the Intermediate Technology Group, a limited company based in Rugby (UK). It is an international development agency, founded in the 1960s by the late E.F. Schumacher (the author of *Small is Beautiful*) and works mainly in Africa, Asia and South America.

It 'puts people first, ensuring that communities design and control productive technologies for their own development' and recognizes that women's contributions have been greatly undervalued in the past. The main areas of technology focused on are:

energy;
transport;
manufacturing;
water and sanitation;
building materials and shelter;
mining;
food production;
agro-processing.

As an independent charity it is dependent upon donations from individuals, trusts, companies and other

lems to be ignored – until they become big ones. Yet big problems often simply result from the aggregation of small ones. Large-scale starvation can result from the multiplication of individuals who are hungry but it is hard to make an impact on those who can help until the problem becomes big. At this point, it seems to require a 'big' solution.

Thus, the water needs of an individual could perhaps be met by supplying a container to collect and store rainwater off the roof, but the water needs of a region (the total of all these individual needs) will justify (and appear to require) a new dam. This is costly, takes time, has all sorts of environmental consequences (and human consequences downstream) and may silt up after a while.

But the money required can be mobilized for such a large-scale problem. By comparison, who is going to fund and distribute the plastic containers to millions of individual households? The project does not have the same appeal – especially to large-scale engineering firms etc.

Yet the idea of saving rainfall off a corrugated iron roof is not new and plastic containers have been available for a long time. Furthermore, help of this kind immediately, and virtually permanently, raises the level of the household to a better state without making it any more dependent on regular inputs.

It will be clear that this is an elementary example but even this has potential. Water collected will immediately breed insects, so why not also create a fish pond (and plastic sheets are even easier to handle and transport) with fish feeding on the insects? And, as the Chinese have done for years, why not attract insects, by light or bait, so that they fall into the water.

However, imagine a world bank having to organize the distribution of large sums of money in this way. Of course, it could be done: for example, by using other agencies, including local universities and research institutes, to trickle down the funds, from sizeable grants, to the household level. But, it needs a change of attitude, to recognize, for example, that if electricity is in short supply, it may be because too many switches are turned on and it may be better to turn them off than to build another power station.

Many workers in the field have stressed the need to identify what poor people really need and the need to involve them in the processes of identification and implementation of solutions.

There is, incidentally, no problem about spreading successful solutions. It is said that one plant breeder simply grew better varieties of a crop in full view of, and with full access to, everyone. The successful varieties disappeared (mainly at night!) at such a rate that there was no separate problem of extension or technology transfer!

Very often, because of this tendency to focus on large-scale solutions, the main beneficiaries are large-scale farmers. Yet the developing countries are full of small-scale farmers (see Table 4.1). In fact, in many, probably most, developing countries, smallholders account for the bulk of agricultural production and make up the largest segment of the rural population (Ruthenberg and Jahnke, 1985).

4.5 Helping small-scale farmers

It should not be thought that, just because they are small in area (maybe only 1–2ha), that small farms are simple. Indeed, they may be extremely complex and very diverse, thus making it very difficult to generalize about them. Small-scale farmers usually have multiple objectives, including producing suf-

TABLE 4.1 Approximate numbers of small farmers

	No. (m)	% of total no. of farmers
Developed countries	21	75
Developing countries	112	91
MSA	83	97

ficient staple foods for their families in all years, producing cash surpluses, having adequate leisure, providing for the future and avoiding risk (Ruthenberg and Jahnke, 1985).

Generating technical innovations for such farmers is a major priority if developing countries are to improve their productivity. Ruthenberg and Jahnke's (1985) paradigm for agricultural development involved four basic steps:

1. understand the systems being operated (see Chapter 7);
2. design and conduct appropriate research (see Chapter 16);
3. assess the most promising innovations – including those most attractive to farmers;
4. consider the best way of introducing these innovations.

Chambers (1983) suggested six approaches:

1. sitting, asking and listening;
2. learning from the poorest;
3. learning indigenous technical knowledge;
4. joint R&D;
5. learning by working;
6. simulation games (see Chapter 7).

There is no reason why small farms should not produce at least as much per hectare as larger farms, provided that the necessary inputs can be afforded. But profitability is likely to be higher on larger farms (Pearse, 1980) and as soon as small-scale farmers produce a surplus they are usually at a disadvantage in terms of storage, transport and weakness in relation to purchasers generally and middle people in particular.

Some of these weaknesses can be overcome by cooperation (see Box 4.4) but this may be unfamiliar and easily dominated by one or two individuals.

To be effective a cooperative probably has to be large and able to deal with the purchase of inputs as well as the sale of outputs. In these circumstances, the greatest need for help may be in simple book-keeping, something that large companies could easily help with.

One difficulty, already mentioned, is that innovation and technology generally favour the large farmer and, where farmers compete, this may be to the disadvantage of the small farmer. What is needed is technology especially appropriate to the small-scale farmer.

4.6 Technology for small-scale farmers

It is quite hard to devise – or even imagine – technologies or systems that will benefit small-scale farms that will not benefit large-scale farms even more. But perhaps that is because we have got into the habit of thinking mainly in terms of big units.

One possibility worth exploring is to turn small-scale to advantage. For example, if we place a beehive on $1m^2$ the honey produced (collected, of course, from up to a 3 mile radius) will not be increased at all by **owning** any more of the surrounding land. In fact, we will have to pay for the extra land and its upkeep. Pigeon lofts operate in the same way.

These are two small examples but they illustrate a possibility that has been little considered.

But the need is enormous. What more could be done? There are two easily-identified areas in which we (in the developed countries) could help further: (1) relevant research and (2) the involvement of people.

Box 4.4 Cooperatives

The most successful cooperatives, in terms of guaranteeing a market at a fair price to small farmers with only sporadic supplies to sell, have been producer-owned and retained control of the entire process through to retailing. When they work, they tend to become centres of advice, and not only on technical matters: in fact, they may become social centres. They may also supply, at a fair price, the inputs that small farmers may begin to afford with their production-generated income. However, the farmers in most need may have no knowledge of how to set up a cooperative and a great deal of valuable help can be rendered in this phase. Training in simple techniques of accounting and management may also be needed.

Cooperatives can also lead to actual cooperative ownership and use of machinery or draught animals, of storage capacity and transport. Centres can represent points of contact for outside agencies offering help, including technical expertise. The Plunkett Foundation exists to help in the development of such cooperatives and there are agencies, from ODA to ITDG, that supply technical help, including some research back-up.

Most developing countries are poorly endowed with research facilities and those that survive from the past are usually concerned with plantation crops rather than local needs. The R&D capacity of the 'north' has been predominantly concerned with the needs of large-scale, high-input farming. One of the results of the International Agricultural Research (funded by the CGIAR system) was the Green Revolution. This was based on high-yielding varieties (HYV) of rice and wheat – varieties capable of very high yields per hectare, provided that they were adequately supplied with water, fertilizer and pesticides. Such inputs can only be afforded by the larger farmers and one effect of the Green Revolution was to increase the competitive advantage of the larger farmers over their smaller neighbours.

Today, attempts are being made to focus research on the needs of these smaller farmers, but it is not just technical help that they require. Technical advance, by itself, is rarely sufficient, even if it is relevant: in this case, relatively few relevant technical developments are available and they are insufficient by themselves. One reason is that farmers just beginning to emerge from subsistence farming need a guaranteed market for their produce, even though they cannot guarantee the supply, for they are still dependent on the weather, on pests and diseases, and must still first supply their own needs. Unless they can sell their produce at a reasonable price, there is little hope of significant progress. This can only be guaranteed by Government or by producer-cooperatives.

4.7 Relevant research and the involvement of people

There are three related problems with regard to research that is relevant to the needs of poor, small farmers. The first is how to identify it, the second is to carry it out and the third is to develop a relevant research capacity in the country that needs it. All three involve people but the first one is rather different in nature.

The identification of problems for research is not simply a matter of pinpointing the obvious bottleneck. Very often, this only leads to a sequence of such bottlenecks, each one emerging as the last is 'solved'. Furthermore, such a process can lead to dangerous consequences. An example of this can be found in the supply of wells to relieve an obvious shortage of water for livestock, resulting in overgrazing and destruction of the environment.

Somehow, it is necessary to identify what action or input will lead to real and continuing improvement in the agricultural system being considered. Furthermore, the action or input has to be affordable, preferably without creating dependence upon it, and must not result in other, different problems.

All this is extremely difficult and, as with so many research areas, it is necessary to take the phase of problem identification and definition more seriously. Generally, there is a rush to get on with solving a problem, often failing to recognize the importance of correctly identifying and stating exactly what the problem is.

To do this in developing countries, full of unknowns and imponderables, requires many people on the ground, with the humility to recognize that, whilst they have a contribution to make, they also need help from those who live and work there. Not only is this recognized by many aid agencies (e.g. VSO) but there are now several voluntary movements of farmers keen to help in these ways. In the UK, 'farmer-twinning' is only loosely organized; in France, the equivalent is much more formally organized. Both are based on volunteers who visit specific localities to try and help identify households, villages or communities.

When it comes to carrying out research, a major problem has always been that the research capacity is mainly in the north whilst the most urgent need is in the south. Partly because of this, the main emphasis in agricultural research has been on higher output per hectare by the use of higher inputs and higher output per person by reduction of the labour force. Neither of these emphases is appropriate for developing countries, where high inputs cannot be afforded and agriculture is the major source of employment. (Neither of them appears quite so relevant at the present time to developed countries either.) Furthermore, there has been an orientation towards the needs of large-scale farmers, whereas the farms of developing countries, as has already been pointed out, are very small.

It would be possible to reorient major parts of our research capacity towards the problems of developing countries, paid for as part of our overseas aid. This might be a very cost-effective form of aid, aimed at helping developing countries, in the longer term, not only to produce their own food but to develop their own research capability.

Fortunately, universities are increasingly developing schemes whereby overseas postgraduate students in agriculture carry out the majority of their research in their own country and on a relevant topic. By collaborative research and joint supervision of postgraduate students, the local research capacity is encouraged. But an imaginative use of our 'spare' research capacity could have greatly accelerated such progress (as well as preserving our own capacity). In this way, getting the relevant research done would also contribute to the development of the local capacity. The latter is not merely a question of equipment and laboratories, or even of trained staff, it is also a question of security and confidence.

A simple example will illustrate the point. In most hot countries, flying insects are major pests of crops, animals and humans. They cause losses, they cause and carry disease, and they exist in vast numbers. It would be possible to establish a useful research programme, based on this observation, in any country simply by posing the sequence of questions listed in Table 4.2. Much local knowledge and experience could be drawn upon at each stage, thus establishing a valuable contact and dialogue between researcher and farmer.

The whole sequence focuses attention on the essential research procedure, starting from observation and asking a sensible sequence of questions. The relevant experiments arise from a need to answer these questions for the pest that is important in the area. It combines the control of a pest with its use as a resource and would cost relatively little.

Even so, it is unlikely that overseas graduates, anxious to establish themselves as active scientists, would choose such topics or be funded to carry out the work by existing agencies (generally accustomed to handling much larger sums and thus focused on large, not to say grandiose, projects).

The best hope would be if collaboration with a developed country university or research institute could underpin both the confidence and the credibility of such relevant research. It may not seem a spectacular topic but it illustrates the need for research that may seem simple and almost trivial, until the potential is recognized for replication in village after village throughout the tropics.

The need in developing countries is for precisely these topics, small and cheap and capable of almost infinite replication if successful. Even so, all this has to take place against a larger picture and a recognition that agriculture is only part of it.

So, we are faced with the paradox that many

TABLE 4.2 Sample sequence of questions for a fly-trapping research programme

What is the major flying pest?
What attracts it? (to its mate or its food source)
What kind of trap works? (and what sort of attractant?)
Where are traps best placed?(and how far apart?)
How much (and what) do they catch? (including beneficial insects)
If pests are fed to poultry (or fish), what do they produce? (i.e. are they an adequate feed or feed supplement?)

developed countries, especially in North America and Europe, are worried about their food surpluses and their excess food production capacity, whilst malnutrition affects hundreds of millions in developing countries.

The problem is not unique to food: exactly the same is true, for example, for shoes – or any other commodity that poor people lack. People do not lack shoes because there is a shortage; even less because we do not know how to produce them. We may easily have a surplus of unsold shoes, but they are not available to those who cannot afford to buy them.

Furthermore, it is not really true that food is essential in ways that shoes and other commodities are not. Clean water, clothing and medicines may be just as essential: so may fuel to cook the food. The problem is to a large extent one of poverty, but it is not quite as simple as that because people with no money can often grow their own food or catch fish or graze livestock on natural vegetation.

So there is still a role for agriculturalists in countries where malnutrition is rife, provided that people have land. But the number of hungry people is expected to rise and the availability of land may be further reduced.

Furthermore, the technical capacity already exists to increase yields still further: there is no evidence that we have yet reached a plateau (Figure 4.1). Overproduction has been made possible by technical advance and the use of fertilizers, agrochemicals to control weeds, pests and diseases, and improved crop and animal varieties; but the **cause** has been the deliberate use of economic incentives. This itself carries an important message to those developing countries which reward the farmer with too little in order to keep down the cost of food to urban communities. This policy nearly always results in reduced food production and, when farmers are rewarded with higher prices, more food is generally produced, even though the farmers have only the same resources as before.

The next chapters examine the resources that all farmers have to operate with.

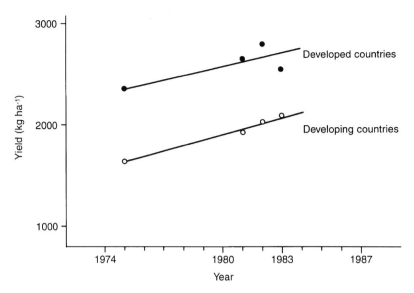

FIGURE 4.1 Yields of cereals.

Questions

1. If the productivity of a subsistence farm is increased, will the farmer produce more or work less hard?
2. If developing country farmers are in competition with each other, which one do you choose to help? Or can you help them all?
3. Is a shortage of food different from – or more serious than – a shortage of clean water, or medicines or fuel for cooking?
4. Would control of population growth make things better or worse for a developing country family or village?
5. Why are pestiferous insects **not** used for food or animal feed?

Chapter five

The role of plants

> When a neighbour is in your
> fruit-garden, inattention is the
> truest politeness.
>
> *Chinese proverb*

Green plants are the foundation upon which life on earth rests. They are the only significant means by which sunlight is used to create the organic compounds upon which we (animals and humans) depend for our food. There is, currently, no other way in which we could be fed.

The process by which green plants make use of solar radiation is photosynthesis (Box 5.1). Essentially, sunlight is absorbed by the green system of the plastids, primarily chlorophyll, and the energy thus liberated is used to produce mainly sugars and carbohydrates but also protein, fatty acids, fats and a variety of other compounds, including those needed for the regeneration of chloroplasts.

Because green plants are at the bottom of the 'food chain' they are sometimes portrayed as the base of a 'pyramid of numbers' (Figure 5.1), so-called because the number of animals that can be supported by the plant base represents less 'biomass' than the plants themselves (because there are losses in the conversion process – see Chapter 9). The biomass of predatory animals that can, in turn, be supported by the herbivores is much less again.

However, it is by no means as simple as this. The term 'food chain' can sensibly be used to describe the food supply to people (see Chapter 2) because it is a purposive use of resources to that one end. Ecologically, the term 'food web' better describes the complex network of interactions between plants and animals that occurs in natural systems. The 'pyramid of numbers' has to reflect this complexity (Figure 5.2) where, for example, many animals (including humans) are omnivorous

---

Box 5.1 Photosynthesis

Photosynthesis is the process by which green plants absorb sunlight (in the 0.4–0.7 μm wavelength region of the spectrum) and use the energy liberated to produce sugars and carbohydrates, proteins, fatty acids, fats and a variety of other compounds.

The mechanism for absorbing light is the green system of plastids. At low light intensities the photosynthetic function produces about 3.6 kg of carbohydrate $ha^{-1} h^{-1}$ for each 0.042 J $cm^{-2} min^{-1}$ of solar radiation. The maximum for individual leaves of agriculturally important species has been estimated at $c.$ 20 kg of carbohydrate $ha^{-1} h^{-1}$.

The light intensity on a clear day is often some four times what a leaf needs to operate at this maximum level. There is thus, at these times, more solar energy being received than large, horizontally displayed leaves can use. That is why many plants thrive with overlapping layers of leaves (as in trees) or with near-vertical leaves (as in grasses).

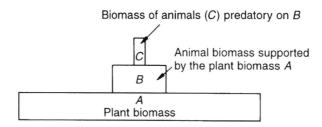

FIGURE 5.1 The simplest version of the pyramid of numbers. Due to inefficiencies of conversion, only *c.* 10% of food consumed forms the bodies of those feeding.

and consume a mixture of foods of plant and animal origin.

For people, it would seem very inefficient to feed on predatory animals since they require such large areas to support the animals on which they feed. But there are exceptions and we regularly consume predatory marine fish, mainly because (1) we do not bear the cost of the large areas involved (as we would for land) and (2) we could not practicably harvest ourselves the vast quantities of extremely small plants and animals on which the predators depend.

Not all plants are green, however, and many, such as the fungi, derive their nourishment from decaying plant and animal matter. Since all living things eventually die, one way or another their remains are reincorporated in natural cycles.

The plant kingdom is vast. Wilson (1992) has estimated that of the known total number of living species of all kinds (1 413 000 species) there are about 344 thousand species of plants (excluding bacteria). These can be subdivided into higher (see Table 5.1) and lower plants and so on. We have little idea of the numbers of species not yet discovered. For example, about 1.5 million kinds of fungi are believed to exist, of which only 5% have been given a scientific name. It is hardly surprising, therefore, that we have very little knowledge of the numbers of individuals, with the minor exception of large individuals where the numbers are relatively small.

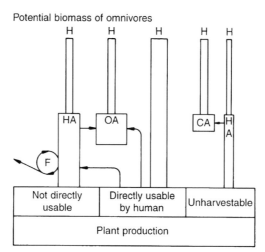

FIGURE 5.2 A pyramid of numbers that recognizes the actual biological complexity (the dimensions have no significance except that the base has to be relatively large). H, human; HA, herbivores; CA, carnivores; OA, omnivores (other than humans); F, faeces.

TABLE 5.1 Classification of higher plants and number of species

Plant group	Total no. of spp. of living higher plants currently known (after Wilson, 1992)
Dicotyledons (e.g. roses) [The flowering plants]	170 000
Monocotyledons (e.g. grasses)	50 000
Bryophytes (mosses)	16 000
Ferns	10 000
Minor groups	1 300
Gynnosperms (e.g. pine trees)	529

Individual plants vary greatly in size and structure, method of reproduction, longevity, growth rate and environmental requirements. An environment has to be very harsh indeed for it not to be able to support some form of plant life but, in general, adaptation to such environments usually involves small size (in cold places) and/or slow growth (as in deserts).

5.1 Plant needs

Solar radiation is not only needed for photosynthesis – to supply the energy needs of the plant – but also to evaporate water from the plant and to provide warmth. Without the warmth of the sun, present forms of growth would not exist and without the 'greenhouse effect' (Box 5.2) the earth's temperature would be some 30°C lower. One component of the greenhouse effect is moisture in the air (much of it as cloud) and this, too, is due to solar-powered evaporation, partly from water surfaces (notably the sea) and partly from vegetation (transpiration).

Transpiration (or more strictly, evapo-transpiration) is the process by which plants lose water, usually from their leaves, through apertures in the cuticle, i.e. the outer layer of cells. These apertures are called stomata (see Chapter 8, Box 8.4) and vary in shape, size and position between plant species. Carbon dioxide and other gaseous exchanges also take place through the stomata.

Box 5.2   The 'greenhouse effect'

The earth is surrounded by an atmosphere through which solar radiation is received. The atmosphere is not static but contains air, in constant motion, being heated, cooled and moved, with water being added and removed, smoke and dust. Only a tiny proportion of the sun's energy reaches earth and some of this is reflected back into space (from clouds etc.). When the radiant energy reaches the earth's surface most of it is absorbed, being used to heat the earth, evaporate water, power photosynthesis and so on.

The earth also radiates energy but, because it is less hot than the sun, this is of a longer wavelength and is absorbed by the atmosphere. The earth's atmosphere thus acts like the glass of a greenhouse – hence the 'greenhouse effect'.

The so-called greenhouse gases [CO_2, methane (CH_4), oxides of nitrogen (N_2O) and the chlorofluorocarbons (CFCs)], are those that absorb the earth's radiation and thus contribute to the greenhouse effect. But water vapour is also a major absorber of energy.

Where there is an increase in greenhouse gases (as with CO_2 due to the burning of fossil fuels), this results in an enhanced greenhouse effect – which is the cause of concern because it could lead to global warming. It is thought that agriculture contributes c. 9% of the effect due to greenhouse gases (CSTI, 1992).

One of the best known illustrations of the ways in which stomata can control water loss is provided by marram grass, which is often planted on sand dunes to prevent sand blowing inland. Here the stomata are placed on the inside of each leaf, which is curved in cross-section. The plant can conserve water by curling the leaf more tightly, thus reducing the exposure of the stomata to the air. Very often the 'guard' cells surrounding the aperture can alter the size of it, with the same effect.

Plants take in through the roots much more water than they require, in order to extract the necessary dissolved salts, especially nitrate for agricultural crops. Transpiration evaporates the surplus water but also has, at times, an important cooling effect.

The amounts of water evaporated are substantial and, when a full canopy of leaves is attached to a root system well supplied with water, the loss of water is about the same as that by evaporation from an open water surface.

'Potential' transpiration is that which is possible under these conditions, allowing for the usual closure of stomata at night. 'Actual' transpiration is what actually happens, due to restrictions on water supply or bare ground etc. If the soil water supply falls low enough, plants wilt.

Most crop plants require annual rainfalls of between 250 and 750 mm but the distribution of that rainfall over the year can be crucial. To produce 1 g of dry matter, pasture grasses require from c. 300 to c. 1000 g of water. In the case of grain, the requirements vary from c. 513 to 1026 g for wheat to c. 710–1420 g for rice. The literature contains a great variety of such figures, so these should be taken as illustrative only.

The amounts of water evaporated depend upon the energy available to bring it about and this is determined, directly or indirectly, by solar radiation.

The essential water supplies for plants thus depend upon the sun, and the water cycle (Figure 5.3) is a major feature of the global environment. In addition, plants need carbon dioxide (CO_2) and oxygen (O_2), which they derive from the atmosphere and nutrients (see Box 5.3), normally taken

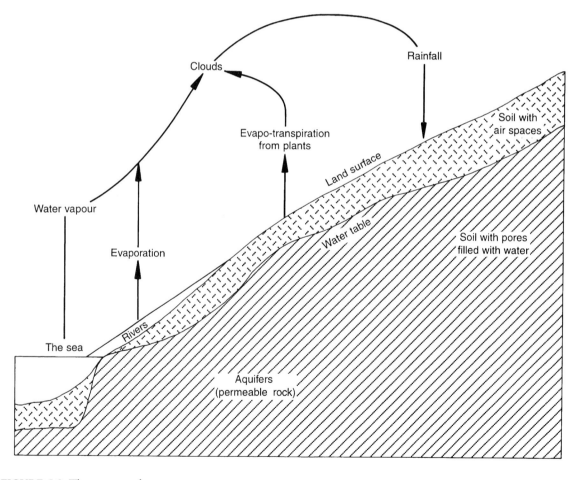

FIGURE 5.3 The water cycle.

in by the roots in the form of salts dissolved in water. Nitrogen is slightly unusual in that some plants (notably the legumes, such as peas, beans and clover, in agriculture) can 'fix' the gas directly from the atmosphere. This they are able to do because of the presence of nodules on the roots, which contain nitrogen-fixing bacteria (Box 5.4). The nitrogen cycle (Figure 5.4) is one of the most important features of the environment for plant growth but, of course, all the nutrients contained in plants are eventually released by digestion or decay and are therefore mostly recycled. Some, however, may become fixed in enormous deposits (as of calcium, for example, in limestone, by the skeletons of small organisms falling to the bottom of oceans).

A very topical case is that of carbon. Carbon is an essential part of the structure of most, and especially, woody plants, being a component of cellulose compounds (found in plant cell walls). Carbon is thus a major part of, for example, tree trunks and, in this form, may be locked up for hundreds of years, partly in the living tree, partly in products made from it and, massively, in the fossil fuels (coal, oil, gas) derived from past vegetation. It is the release of carbon dioxide from burning these fossil fuels that may cause an 'enhanced' greenhouse effect and result in global warming (see Chapters 8 and 10).

So, most plants consist mainly of water, carbon and minerals but, agriculturally, the most important elements are water, protein and energy (Table 5.2). In a great many cases plants occur not as isolated individuals but as communities (this is illustrated for Europe in Box 5.5), which harbour their own characteristic animal populations and may contain several species of plants occupying different layers of the vegetation (such as mosses

---

Box 5.3 Plant nutrients

These vary with species and stage of growth but the range and magnitude of the nutrients required can be illustrated by their presence in the plant. The following table shows the mineral composition of two major herbage species, perennial ryegrass and white clover (after Spedding, 1971).

Mineral	Ryegrass (% in DM or ppm where stated)	White clover (% in DM or ppm where stated)
Phosphorus	0.26–0.42	0.25–0.40
Potassium	1.98–2.50	2.09–3.11
Calcium	0.4–1.0	1.36–2.10
Magnesium	0.09–0.25	0.18–0.24
Sulphur	0.13–0.75	0.24–0.36
Sodium	0.10–0.57	0.12–0.41
Chlorine	0.39–1.30	0.62–0.91
Iron	50–200 ppm	117–291 ppm
Manganese	22–200 ppm	51–87 ppm
Zinc	15–60 ppm	25–29 ppm
Copper	5.4–8.5 ppm	7.3–8.7 ppm
Cobalt	0.15–0.16 ppm	0.13–0.24 ppm
Iodine	0.22–1.45 ppm	0.14–0.44 ppm
Selenium	0.1–1.0 ppm	0.005–153 ppm
Lead[a]	0.3–3.5 ppm	0.3–3.5 ppm
Silica	0.6–1.2 ppm	0.03–0.12 ppm

[a]This illustrates the fact that composition may not always reflect need but sometimes external contamination.

Box 5.4   Nitrogen fixation

Nitrogen (N) is an abundant (c. 80% of the atmosphere) but nearly-inert gas, consisting of molecules with two atoms each (hence N_2). In industrial fixation (e.g. by the Haber–Bosch process) the energy needed to break the chemical bond between the atoms is provided by fossil fuels. The process is carried out at high temperature and combines the nitrogen with hydrogen (H) to form ammonia (NH_3).

In biological fixation, ammonia is also the initial product. This is only done by bacteria possessing the enzyme nitrogenase, but they are found in a variety of sites. Some of the best known, and agriculturally important, are bacteria of the genus *Rhizobium*, which includes those found in the legumes (such as soyabeans, peanuts, peas, beans, clover), where they inhabit nodules on the roots. These nodules are packed with nitrogen-fixing bacteria (the species varies with the species of plant) and the nitrogen is released directly to the plant, where most of it is formed into proteins, or indirectly to the soil, following death of the root.

Non-leguminous plants (such as the Alder tree and the aquatic fern *Azolla*) also contain nitrogen-fixing bacteria and there are many independent, free-living bacteria that can fix nitrogen. The latter include the genus *Azotobacter* (which is aerobic) and *Clostridium* (anaerobic). The blue-green algae (cyanobacteria) are of particular importance in rice paddies and applications of these algae (containing nitrogen-fixing bacteria) can result in the fixation of c. 25–30 kg of $N.ha^{-1}$ per cropping season, increasing yield by 10–15% (Agarwal, 1979). Alternatively, they can be used to reduce the input of chemical (nitrogenous) fertilizers.

and flowers at or near ground level, shrubs and bushes as an 'understorey' and trees with most of their leaves placed near the top of the canopy).

Individual plants are structured in order to satisfy all their needs, typically with a more or less extensive root system to collect water and the mineral salts dissolved in it, a stem to hold the leaves where they are exposed to sunlight (though not necessarily direct sunlight) or to creep along the ground or climb up other plants.

Reproductive organs, such as flowers, are also designed to allow or encourage pollination (by wind or insects, for example) and to spread the seeds or spores from which new plants can grow.

Reproduction is important both biologically and agriculturally. Quite a number of agricultural crops are reproduced vegetatively, i.e. from leaves, stems, roots or tubers. These include:

	Organ involved
Potato	tuber
Cassava	shoot
Yam	tuber
Banana	shoot (sucker)
Strawberry	rooted runner
Sweet potato	shoot
Sugar cane	shoot
Pineapple	shoot or crown of leaves on fruit

For many plants, however, the starting point for a new individual is the seed: the variety of seeds is enormous (some designed for dispersal, some for survival), in size, number, shape, structure, etc.

Where the seed forms the agricultural product, it has been selected over a long period, mainly for size, but it is also remarkable how even tiny seeds may form a staple food (the grass 'tef' in Ethiopia is an example).

The number of seeds is of obvious importance since the greater the number, the smaller the proportion that has to be retained to produce the next crop, although more small seeds may have to be

TABLE 5.2   Composition of agricultural plant products (examples from the literature)

	Moisture (%)	Crude protein (% of DM)	Energy (MJ kg^{-1} DM)
Wheat	16–24	8–20	18.4
Maize	15–35	8–15	19.0
Sorghum	15–20	8–20	18.8
Soya-bean	10	25–52	14.0
Beans (broad)	11.9	24–35	13.3
Chick-pea	10.7	17–26	13.0
Potatoes	80	7.7–8.8	1.4
Cassava	62	2.4	7.4
Lucerne	57–85	13–27	8.5
Lettuce (wet)	95	1.2	0.46
Apple	84	1.5	11.9
Brasil nuts	4.6–8.5	14	30

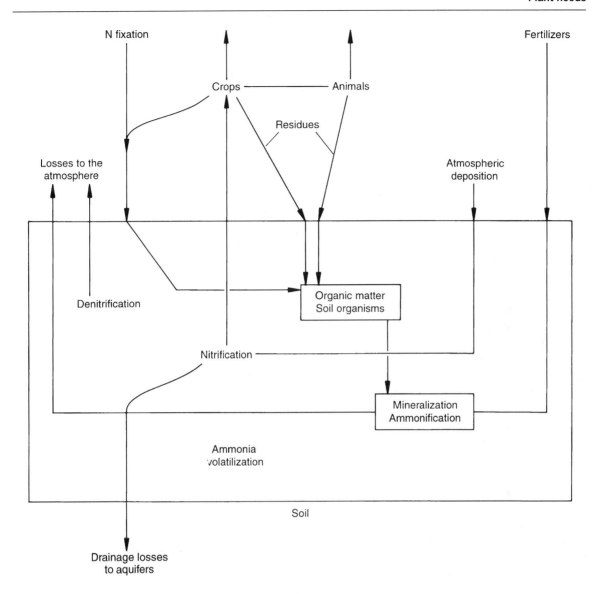

FIGURE 5.4 The nitrogen cycle (see Goulding and Poulton, 1992).

sown in order to withstand the ravages of pests. Large seeds contain more nutrients to aid rapid growth of the seedling but they may also be more attractive to pests.

Reproduction may be spread over a long period of the year and crop breeding has tended to concentrate the period in order to make it easier to harvest crops efficiently by machine. This is some-times an example of **inappropriate** technology (see Chapter 4) as far as developing countries are concerned. A subsistence farmer with no machines, no great storage capacity, substantial losses from any harvested product, shortage of labour at times of peak demand and a need for food every day for as long as possible, would do better with crops that continue producing their products over a very

Box 5.5 The major plant communities of Europe (Polunin and Walters, 1985)

1. Arctic vegetation
   Composed of dwarf evergreen shrubs, sedges, rushes, mosses and lichens, with some grasses, adapted to the short growing season and permafrost. The main types are:

   Tundra   treeless plains
             moss tundra
             lichen tundra
             dwarf heath tundra

2. Boreal vegetation
   Evergreen coniferous forests (called the 'taiga'): where drainage is impeded peat mires occur. Montane and alpine heaths also occur but the grasslands are due to agricultural activity.
3. Atlantic vegetation
   Originally extensive deciduous forests, now mostly semi-natural woodlands. Some of the latter have been cleared for agricultural grassland.
4. Central European vegetation
   Originally forest, most have been cleared for agriculture, mainly grassland.
5. Mediterranean vegetation
   This is quite distinct, dominated by evergreen trees and shrubs. Herbaceous plants die down in the long, hot, dry summers. Grasslands are maintained by grazing.
6. Pannonic vegetation
   Takes its name from the Roman province of Pannonia and stretches from southern Romania and northern Bulgaria down on either side of the Danube. Steppe-grasslands are characteristic but there are also steppe-woodlands.
7. Alpine plant communities
   These occur above the tree-line (e.g. 1000 m, but this varies greatly) and consist mainly of low shrubs and tall herbs.
8. Freshwater wetland communities
   Similar throughout Europe, mainly continuously flooded with some mires and rivers, lakes and ponds. Aquatic plants, rushes, willow and alder, depending on water levels.
9. Coastal plant communities
   Plants tolerant of salinity, often forming saltmarshes and mudflats. Sand dunes and shingles have their own characteristic flora.
   It is worth noting, at this point, that some plants are more tolerant of salt than others. Amongst agricultural crops, for example, corn (maize) and rice are described as 'moderately sensitive' to salt, wheat as 'moderately tolerant' and barley as 'tolerant'.
   Where irrigation results in salinization (see Chapter 8), this can be of great importance but it is also possible to develop unfamiliar, salt-tolerant species as crop plants (e.g. *Leptochloa fusca* and *Atriplex* spp. for fodder, *Salicornia* spp. for oil and *Distichlis* spp. for grain production).

extended period. However, reproductive processes are often governed by day length and other factors beyond the control of the farmer.

The vegetative parts of plants (roots, stems, leaves) tend to grow whenever conditions allow and, equally, to die, senesce and fall to the ground fairly continuously.

The growth of plants can often be spectacular. Very often, as in a bud, for example, cells already formed are packed together and, when temperature allows, very rapid growth occurs by the uptake of water and the expansion of these cells to form leaves and shoots. In very simple plants, and especially single-celled organisms, growth may occur by simple division into two. If anything divides continually in this way, the increase in numbers can be dramatic (e.g. one becomes two, two become four, and the sequence proceeds to eight, 16, 32, 64, 128, each successive step taking only the same time as the first). This is called **exponential growth** and can be seen in the way duckweed (*Lemna* spp.) covers the surface of a pond (see Box 7.1). Very often, temperature is a major controlling factor.

Grass, for example, grows whenever the temperature is above about 4°C and when water and nutrient supplies are adequate but, even if it was not cut or grazed, leaves and roots would be continually dying and being replaced by new ones. This death and decay of vegetation, then processed by 'detrivores' (e.g. fungi, bacteria, earthworms) plays a considerable part in the formation of soil. The roots of plants also have a significant role in holding soil together, thus preventing erosion, by wind and water.

In total, plants create the environments in which animals live and hide, and which form our landscapes. They also provide food for people and feed for their agricultural animals.

5.2 Crop plants

Agricultural crops are grown for a number of different purposes, the main ones being:

food;
feed for animals;
fibre;
fuel;
amenity (including gardening and decoration);
recreation (including golf courses, playing fields, etc.).

5.2.1 Food crops

Many different crops are grown across the world (Table 5.3) but, as pointed out in Chapters 2 and 3, a very high proportion of the world's food comes from remarkably few – some 60% from only nine species. The areas sown to the main crops (Table 5.4) are in a slightly different order due to the fact that some areas (e.g. paddy rice) may produce more than one crop per year. (At this point, it is perhaps necessary to restate the reservation made in Chapter 2, that statistics on matters of this kind are very hard to obtain, often out of date and may well be inaccurate.)

In general, crops are grown on land that can be cultivated, although there are notable exceptions, such as maize grown amongst boulders in parts of China.

5.2.2 Feed for animals

Feed for animals comes from a great variety of sources, including crop by-products and roadside verges, but, over much of the world, where it is deliberately grown, it tends to be on poorer land that is difficult (or expensive) to cultivate. However, in developed countries, cereals (maize, barley, wheat) are intensively produced for processing by a well-organized feed industry. (Utilization by animals is dealt with in Chapters 6 and 7.)

TABLE 5.3   The main crop plants

Crop category	Examples
Cereals	Maize, rice, wheat, oats, barley, millet
Pulses	Bean, pea, peanut, soya-bean, cowpea
Forage crops	Grass, clover, lucerne (alfalfa)
Roots and tubers	Potato, cassava, sweet potato, turnips
Leafy crops	Cabbage, spinach
Fruits	Orange, lemon, lime, olive, apple, strawberry
Oil crops	Palm, peanut, olive, cotton-seed, linseed, sunflower
Nuts	Almond, filbert, pecans
Sugar crops	Sugar cane, sugar beet
Beverages, spices, etc.	Coffee, tea, cocoa, grape, perfumes, peppers
Fibre crops	Flax, jute, hemp, sisal, cotton
Fuel crops	Hardwoods, softwoods

TABLE 5.4 The global harvested areas ('000 ha) for the main cereal crops (see Table 2.4 for more details)

Type of cereal	Area harvested ('000 ha)	
	Average 1979–81	Average 1987
All cereals	719 450	694 000
Wheat	235 000	221 600
Rice paddy	143 750	141 500
Maize	126 100	126 000
Barley	81 000	78 100
Sorghum	44 100	42 700
Millet	40 100	36 300
Oats	25 700	23 500
Rye	15 100	16 300

5.2.3 Fibre crops

The main fibre crops are given in Table 5.5. Many of them have local importance but, in terms of world trade, the dominant crop is cotton.

5.2.4 Fuel crops

Wood from trees still supplies a very high proportion of the fuel in places like Africa (Box 5.6) but few of these trees are deliberately produced for this purpose and fuel collection may take a great deal of time and effort (see Chapter 8). As an illustration, it has recently been reported (Seager, 1995) that in Nepal it takes 200–300 person days to forage for a typical household's annual fuel supply.

TABLE 5.5 Fibre crops and the part of the plant used

Crop product	Part of the plant
Cotton[a], kapok, flosses	Seed or fruit
Flax[a,b], hemp[b,c], jute[a], ramie[a]	Inner bast tissue or bark
Abaca[b] or Manila hemp, sisal[b]	Leaves
Trees used for paper making	Woody fibres of the stem
Coir	Husk (mesocarp)
Mexican whisk	Roots

[a] Cotton, ramie, jute and flax are mainly **textile** fibres (for clothing and bags).
[b] Sisal, abaca, hemp and flax are mainly **cordage** fibres (for ropes).
[c] Hemp is also usable for newspaper (it is lighter than wood pulp) and has even been used for tea-bags and cigarette paper.

Box 5.6 Fuel wood in Africa

Whether there is a major fuel crisis in Africa is debated. Recently, for example, Pearce (1994) reported the findings of several studies which revised upwards the amount of wood in Africa to about double the previous estimates. Such a store of 70bn tonnes of carbon stored in trees is relevant to calculations about the CO_2 balance in the atmosphere and may help to explain the 'missing' 1bn tonnes of carbon unaccounted for each year. Some of the errors may have come from foresters' calculations of tree mass that focused on the trunks and ignored the branches: others, it is suggested, derive from roadside estimates that only looked at the somewhat depleted, accessible verges and not the denser growth further from the road. There are also apparently more trees on farms than was thought.

All this illustrates yet again, the difficulty of compiling statistics on anything like a global scale. Certainly, there are many families who have to spend time and effort collecting wood over increasing distances. Masai women have been reported (Harrison, 1987) as spending up to 4 h day^{-1} collecting fuel wood. But shortages of wood are acute in the drier parts and in the humid parts of Africa wood may be very plentiful.

It has been estimated that 16–18 countries (out of 45) have serious deficits or acute scarcity of fuel wood and that more than half the regional population of c. 320m suffer from this.

Harrison (1987) reported that in the Majjia valley of Niger, women commonly spent half a day scouring the hillsides for kindling: yet, in two villages, living hedges of the thorny leguminous tree *Prosopis juliflora* had run wild and could hardly be cut fast enough (for fuel but also for poles and fodder from the pods). As in many places, in Niger all the trees are government owned and this can affect the willingness of people to plant trees.

Wood is said to be the main energy source for nine out of every 10 Africans and accounts for more than 58% of the total energy consumption on the continent. In general, however, fuel wood is not used very efficiently in cooking (only 5–10% of the energy content of the wood going to heat the pot). There has therefore been much work devoted to more efficient, but still cheap, stoves.

A great deal of research effort has gone into the development of other fuel crops but even in developed countries they tend to be of marginal economic efficiency, largely because oil is still remarkably cheap. This is mainly because it does not generally have to be produced, only harvested or collected. South Africa has been an exception, producing oil from coal.

However, a low oil price may not last indefinitely and, if fossil fuels should become very expensive, it may be necessary to use cultivated fuel crops and there are good arguments for learning how to do so before the need arises. Box 5.7 illustrates the crops currently being considered in Europe.

5.2.5 Amenity crops

In some countries, especially in Europe, ornamental plants (pot plants, cut flowers, garden flowers, shrubs and trees) are of great commercial importance. In fact, horticultural plants, including cuttings, are even transported by air over substantial distances.

5.2.6 Recreational crops

Grass is by far the most important. Because its growing point is so close to the ground (see Figure 5.5) it can be cut or grazed very short without destroying it and thus killing the plant. In fact, it responds by vigorous branching (called 'tillering') and the formation of a short, dense turf, highly suited to many recreational purposes. Ploughed land is not recreationally attractive, except to look at, and even long grass is not nearly as useful for recreation as short grass (although it is possible to imagine – or remember – occasions where longer grass carried some advantages!).

5.2.7 Other crops

There are many, relatively minor, crops that may not be of world significance but which are important locally. Many of them have a very high unit

---

Box 5.7 Fuel crops

Many trees have historically been used for fuel and most wood can be burned as logs, chips or charcoal (which involves partial burning in the near absence of oxygen and involves the loss of some 55% of the energy in the wood).

But many other biological sources of fuel exist, including: (1) the fossil fuels (coal, oil, gas) derived from long past and highly compressed vegetation; (2) dry waste products, such as cereal straw and dried dung – all of which can be burned in an open fire or in a stove.

Vegetable oils can also be used for fuel and wet material, such as leaves or fresh manure, can undergo bacterial digestion (anaerobic fermentation) to generate methane (a burnable gas). In some cases, wet material such as sugar cane or potatoes can be fermented to produce ethanol but this requires a great deal of support energy to distil the alcohol from a watery solution.

Indeed, in all cases it is necessary to calculate the energy costs of actually producing energy. The following table illustrates the support energy needed to operate a fuel wood forest (see Spedding, Walsingham and Hoxey, 1981).

	Support energy usage $(MJ\ yr^{-1})$
Sowing and planting trees	Negligible
Machinery	
Manufacture	4.15×10^6
Fuel used	1.88×10^7
Transport in logging trucks	8.46×10^6
Chipping	
Machinery manufacture	1.84×10^5
Fuel used	1.84×10^6
Fire protection	
Chemical production	2.90×10^6
Application	1.90×10^6
Total input	36.52×10^6
Yield of wood	$1.84 \times 10^9\ MJ\ yr^{-1}$
Energy input per MJ wood produced	0.02 MJ

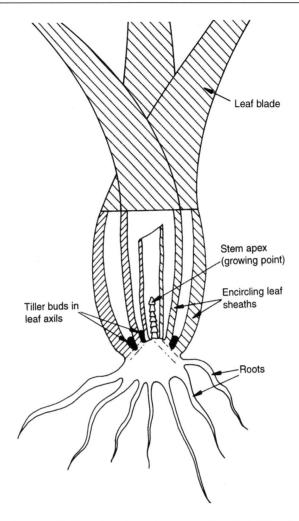

FIGURE 5.5 The structure of grasses (half-section through a vegetative tiller) (after Spedding and Diekmahns, 1972).

value, especially those which produce perfumes, for example. Some of these products are included in Table 5.6. But such slight mention as some of these receive does little justice to their importance or potential. A good example of this is in the medicinal plants (see Phillips, 1992).

5.3 Crop production

The plants themselves are only part of a crop production system and production implies supplying them with appropriate soil conditions, plentiful nutrients and protection from their many diseases, pests and parasites.

Management of the system created requires labour, machinery, transport and capital: the complexity of the resulting system is difficult to comprehend without forming some kind of a picture of how it all works (see Chapter 7).

5.4 Crop products

The products are rarely derived from the whole plant and plant breeding has concentrated on chan-

TABLE 5.6 Crop products

Crop products for human consumption	Examples
Cereals	Wheat, rice, maize
Starchy roots	Potatoes, sweet potatoes, yams, cassava
Sugar	Cane, beet, palm, maple
Pulses	Beans, peas
Nuts	Groundnuts, coconuts, walnut
Oils	Olive, coconut, sesame seed, sunflower seed
Vegetables	
Roots, bulbs, tubers	Beets, carrots, onions, potatoes
Leaves	Cabbage, celery, spinach, lettuce
Other	Cauliflower, beans, peas, tomatoes
Fruits	
Starchy	Bananas, plantains
Citrus	Lemon, orange, grapefruit
Fat rich	Avocado, olive
Other	Apples, grapes, mangoes, melons, plums
Dried	Dates, figs, prunes, raisins
Beverages	
Non-alcoholic	Tea, coffee, cocoa, maté, soft drinks, cola
Alcoholic (beer, spirits, wine)	Barley, hops, potatoes, grapes
Flavourings	
Spices	Peppers, mustard, cloves, vanilla, nutmeg
Herbs	Parsley, rosemary, mint, thyme
Tuberous roots	Ginger, horseradish, liquorice, turmeric
Medicines	Quinine, opium, cocaine
Fumitories	Tobacco
Masticatories	Betel nuts
Feed for livestock	
Fresh green feed	
Grasses	Ryegrasses, guinea, pangola, wheatgrass
Forage legumes	Clovers, lucerne, stylo
Cereals and grain legumes, used green	Oats, maize, peas, beans
Brassicas and other leafy crops	Cabbage, kale, rape
Leaves of root crops and waste from vegetable food crops	Sugar beet tops, peas and bean haulm
Leaves and shoots of trees and shrubs	Acacia (Mulga, boree), saltbush, heather
Bulky conserved feeds	
Hay	Grass, legume, cereal
Dried crops	Grass, legume
Silage	Grass, legume, cereal, crop by-products
Roots	Potatoes, swedes, turnips
Bulky by-products	Sugar cane bagasse, beet pulp, straw, chaff, bran, citrus pulp and skins, olive residue
Concentrate feeds	
Cereal	Barley, maize, millet
Pulses	Beans, peas, lupins
Oil-seeds, cakes and meals	Linseed, soyabean, groundnut meal
Miscellaneous nuts and seeds	Acorns, chestnuts, locust beans
Materials	
Construction	Timber
Paper	Newspapers, books, packaging
Textiles	Cotton, hemp
Rubber	Gloves, boots, insulation, seals
Household goods	Cork, woven utensils

TABLE 5.6 (Cont.)

Crop products for human consumption	Examples
Fertilizers	Green crops, crop wastes
Fuel	Wood, charcoal, alcohol, methane
Industrial oils	Linseed, cotton-seed, corn oil
Essential oils	Perfumes, camphor
Gums	Gum arabic, gum tragacanth
Resins	Lacquer, turpentine
Dyes	Logwood, indigo, woad
Tannins	Hemlock, oak, mangrove, wattle
Insecticides	Roterone, pyrethrum

nelling a high proportion of crop growth into the part of the plant desired, whether seed, root, stem, leaf, etc. Table 5.7 illustrates the main uses of plant parts, most of which involve processing before consumption or use.

In the very early days of agriculture a very high proportion of our raw materials was derived from crops and animals. With the development of the chemical industry, many of these raw materials have been displaced and it is relatively recently that there has been some move back towards naturally-occurring materials. One of the safest (or most easily acceptable) applications of genetic modification of plants falls into this category. For example, it has recently been found possible, by genetic engineering, to produce crop plants that contain a plastic usable by industry (Kleiner, 1995). The plastic [polyhydroxybutyrate (PHB)] is a natural, biodegradable material.

Looking ahead, there is an increasing number of exciting possibilities for altering the genetic make-up of existing crop plants to produce (or produce more of) compounds of high unit value. For example, some strains of oilseed rape produce erucic acid, used as a substitute for whale oil. It may be possible (Stevenson, 1995) to replace one of the enzymes, an acytransferase, by a more active version of the same enzyme from meadowfoam (*Limnanthes*), thus greatly increasing the erucic acid content. Crops like potatoes could be altered to produce specific starches for specialized industrial use.

It might be thought that medicinal products fell into this category, and indeed this will undoubtedly be so but, in fact, the pharmaceutical industry currently still derives many of its drugs from plants.

None the less, one of the main arguments used for the conservation of our heritage of plant species is that we should screen them for useful substances before they die out. The debates about biodiversity (see Chapter 10) also apply to animals (see Chapter 6).

TABLE 5.7 Plant parts used for food

Plant part	Example
Reproductive organs	
Tubers	Potatoes, artichokes
Nuts	Coconuts, walnuts
Seeds	Cereals, beans
Fruits	Apples, dates
Flowers	Globe artichokes, cauliflowers
Bulbs and corms	Shallots, garlic
Fruiting bodies of fungi	Mushrooms
Vegetative parts	
Leaves	Cabbage, lettuce
Stems	Sugar cane, asparagus
Roots	Carrots, cassava

Questions

1. Why don't we crop successful weeds, such as nettles?
2. Why don't we crop plants from the sea?
3. Can plants feel and, if so, can they suffer?

Chapter six

The role of animals

You will walk for a long time
behind a wild duck before
you pick up an ostrich feather.

Danish proverb

Of the total number of living species of all kinds that are currently known (about 1 413 000), the majority are animals (1 032 000). Since all animals are, directly or indirectly, supported by plants, it may seem surprising that there are some four times as many species of animal as there are of plants. However, every plant has many different parts (roots, stems, leaves, seeds) and there are often different species of animals living on them. Then, of course, there are animals that live on other animals, as predators, pests or parasites. In fact, the number of **large** animal species is relatively small – only 4000 mammals, for example (see Figure 6.1). We know little about the actual numbers of individuals, except for relatively large species and especially if they are rare, but they are unimaginably large for most species.

Even for the numbers of **species** on earth, referred to above, Wilson (1992) suggests that we do not know for certain what the order of magnitude is: it could be 10 million or 100 million! As with plants, animals have been classified on the basis of common features and evolutionary links (see Box 6.1). But belonging to the same group does not necessarily mean that they live in the same way or in the same environment. One only has to think of birds, with many recognizable features in common, there are species that are now flightless whilst

others may spend most of their lives in flight, there are birds that can swim under water and cannot walk too well, while others run at tremendous speeds and cannot swim at all.

It has been said that, so widespread are the parasitic nematodes (roundworms, such as occur, for example, in puppies), that if the rest of the living world was eliminated, the ghostly shape of it would remain sketched out by their bodies.

6.1 What do animals do?

Animals spend substantial parts of their lives feeding or seeking food. Herbivores, from caterpillars to cattle, spend very long periods in eating plants, whereas predators may feed infrequently. Almost every part of every kind of plant provides food for some animal and virtually every animal is eaten by another.

Indeed, if the layperson does not know where to start in trying to comprehend the animal world encountered by any observer of their own environment, not even knowing what questions to ask, this is a good starting point. On encountering any animal, first ask 'what does it eat, and what eats it?'. This immediately leads on to a series of questions, such as 'how does it find (collect, digest) the

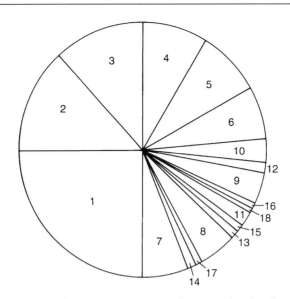

FIGURE 6.1 Numbers of known animal species (see Box 12.1 for more detail) (after Wilson, 1992). Major group: 1, Coleoptera; 2, Lepidoptera; 3, Hymenoptera; 4, Diptera; 5, Homoptera and Hemiptera; 6, Arachnida; 7, smaller insect order; 8, other Arthropods; 9, Mollusca; 10, other Chordates; 11, Platyhelminthes; 12, Nematoda; 13, Annelida; 14, minor Phyla; 15, Cnidaria and Ctenophora; 16, Echinodermata; 17, Porifera; 18, Mammalia.

first and how does it avoid the second?' and thus to a reconstruction of its world.

Wild animals have to survive in a fiercely competitive world and endeavour to grow and reproduce. Many of their innate behavioural patterns are designed to these ends. When such animals are domesticated, even for many generations, they may retain the urge to behave in the same ways, even when the need has gone. So, hens and pigs will make nests if they are able to do so and depriving them of the opportunity may cause stress. If animals, such as pets, are kept as individuals, clearly there may be no opportunity to hunt, mate or even collect food. These issues are discussed in more detail in Chapter 12 (Animal welfare).

The processes of growth have been exploited for meat production, reproductive processes have yielded eggs and milk, and the strength of animals has been used as a source of power. Most of this has been through agriculture but prior to that (and still continuing in many parts of the world) by simply harvesting animals from the wild: marine fishing being the largest-scale example of this currently.

As with plants, agricultural animals have been bred to accentuate the most valuable parts or processes. Many wild animals also live in families, groups, flocks, herds and populations. Behaviour in such groups may be very hierarchical and facilitated by threatening, but not damaging, attitudes. When domesticated and crowded together, at higher densities and with unfamiliar companions, the innate behavioural patterns no longer serve and this can give rise to fighting (in dogs and pigs, for example), feather pecking and cannibalism (in poultry) and bullying (in cattle).

Understanding wild animals may help in the management of animals brought under human control. One of the characteristics of domesticated and farmed animals is that **reproduction** is controlled. In wild animals, although there is no human control on reproduction, there are many ways in which some regulation of reproduction occurs.

Since all kinds of hazards exist for the progeny of all animals, the numbers produced are often greatly in excess of what is needed to replace losses and maintain, or even increase, population size.

Box 6.1 The classification of animals

There are many ways of classifying things (books, for example, could be classified by colour, weight, height, subject matter, authors, etc., and it is possible to think of a reason for doing it in each of these ways). Animals are classified in great detail by zoologists for their purposes and such classification schemes can be found in relevant textbooks. Here, the purpose is to illustrate what kinds of animals there are and thus to see which are used in agriculture.

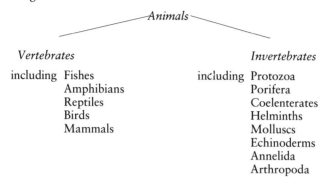

Animals

Vertebrates

including Fishes
Amphibians
Reptiles
Birds
Mammals

Invertebrates

including Protozoa
Porifera
Coelenterates
Helminths
Molluscs
Echinoderms
Annelida
Arthropoda

Each of these classes is further subdivided. For example, the Arthropoda contains the insects, crustaceans and spiders. The relative numbers of species within these groups are shown in Figure 6.1 and details are given in Box 12.1.

Agriculture focuses mainly on mammals and birds but many other organisms are important components of agricultural systems.

Each species thus provides food for many others: if it were not so, the world would become grossly overpopulated, resulting in death by disease or starvation.

In many wild species, there are feedback effects (see Chapter 7), so that, for example, if there are too many progeny they may eat each other. In addition, the food supply may influence how many eggs are laid or progeny produced. Quite commonly, such mechanisms fail and overpopulation occurs, resulting in population peaks followed by crashes. Thus, the 'balance of nature' may not be a continuous condition but may only work over long periods of time: in any event, there is nothing idyllic about it.

In the animal kingdom, there are many methods of reproduction but, for the majority of farm livestock, it is restricted to egg-laying (for poultry) and live birth (for all the mammals).

For both mammals and birds, the reproductive rate has been naturally limited to the numbers of offspring that the parents can look after. This imposes limits because of the number that can be carried during gestation and then suckled (in mammals) or because of the number of eggs that can be incubated (in birds mainly).

The result is that livestock husbandry has developed practices to remove these limitations – for example, by artificial rearing away from the mother (common with calves from the dairy herd) or by artificial incubation in controlled temperature containers (as with poultry eggs).

As it happens, the mammals used in farming are mainly large and able to defend themselves, so that they do not need a high reproductive rate. Furthermore, the period of greatest vulnerability is at or just after birth and there are great advantages in well-developed offspring able to run very shortly after birth. This favours small numbers of progeny: where this argument is less strong (as in pigs), a powerful parent makes a nest and hides it well, something that is not possible for grazing animals.

TABLE 6.1 Number of progeny per parental unit

Animals	Male (a): female (b) ratio	No. of progeny[a] per parental unit per year (c)	c/(1 + a/b)
Cattle (using AI)	1:2000	1.0	0.9995
Sheep	1:40	1.5	1.46
Rabbits	1:20	40.0	38.1
Hen	1:15	240.0	225.0
Goose	1:4	40.0	32.0

[a] These vary, but typical values have been selected for the purposes of this table.

Now, because animals are protected and cared for within agriculture, it is possible to cope with a higher than natural reproductive rate. More than that, it is actually desirable as it increases the overall efficiency of food conversion (see Chapter 9). This, as will be explained later, is simply because food is the major cost (50–85%) in animal production and much of it is used simply to maintain the parents. This is a kind of 'overhead' charge which has to be spread over the output and, for meat animals, this is dominated by the number of offspring.

Agricultural animals have therefore been bred for increased reproductive rates and, to reduce the parental cost, the number of males needed has been reduced by the use of artificial insemination (Table 6.1). Indeed, most aspects of reproduction in farm animals are gradually being brought under control. Some of these may be manipulated by modern forms of biotechnology (see Chapter 17). Others are already controlled to some degree but this varies with the species: a good example relates to the breeding season.

6.2 The breeding season

Obviously, most animals breed at particular times of the year, usually related to food supply. Where this is vegetation, reproduction tends to occur in the spring, to coincide with fresh and plentiful supplies of plant material. This is so for large grazing animals, such as sheep and cattle, but also for caterpillars feeding on oak leaves, for example. In this latter case, the reproduction of blue tits, which feed on the caterpillars, is also timetabled appropriately. However, giving birth in the spring does not necessarily

mean mating at the same time, since gestation periods vary with species (see Table 6.2).

Furthermore, not all animals are able to mate at any time of year: where the breeding season is limited, it is often determined by day length. Examples of these variations are given for farmed animals in Table 6.3. Where breeding is controlled by day length, as in sheep and poultry for egg-laying, it is possible to change the date of onset by controlling day length artificially by altering lighting patterns, though this is only practicable for housed animals. Even then, the exclusion of even small amounts of light during the 'dark' periods may be crucial.

Since the relevant physiological responses in animals involve hormones, this provides an alternative method of control. Ewes, for example, can be treated with hormones to bring them on heat at a particular time, out of the normal breeding season: this has the effect of synchronizing lambing and thus reducing the length of this very busy period.

TABLE 6.2 Gestation or incubation periods in agricultural animals

	Gestation/incubation period (days)
Buffalo	314
Camel	396
Cow	280
Goat	150
Horse	338
Pig	114
Rabbit	31
Red deer	236
Sheep	147
Duck	28
Goose	30–33
Hen	21

TABLE 6.3 Breeding of female animals

	Duration of oestrus	Oestrus cycle (days)	Season[a]	Breeding Life (years)
Cattle	18 h	21	All year	8–14
Sheep	26 h	16.5	Mainly autumn to spring	6
Goats	28 h	19	Mainly autumn to winter	6–10
Rabbits	Inducible	–	All year	2
Horses	6 days	21	Spring and summer	16–22
Pigs	2 days	21	All year	3
Buffalo	12 h	21	All year	Up to 20 +
Hens	–	–	All year	1.5
Turkeys	–	–	Spring	1
Geese	–	–	Spring	4–5
Guinea fowl	–	–	Spring	?

a Breeding seasons tend to be less marked in the tropics.

In cattle, this is not necessary, since they are able to reproduce at any time of year. Although it would still seem more natural to calve in the spring, this would affect the supply of milk for human consumption. Since the demand for milk is not as seasonal as that, cows are in fact calved all the year round in order to even out the supply. This also reflects the fact that milk is not produced at a constant rate but follows a characteristic lactation curve (see Figure 6.2).

In order to function in all these ways animals have to satisfy their needs.

6.3 What do animals need?

Like plants, animals have nutritional needs (see Box 6.2), including water – though their needs for this vary greatly with the species – and they are adapted to a wide range of climatic conditions, such as temperature, humidity and oxygen supply.

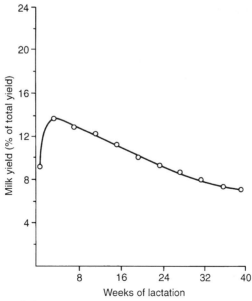

FIGURE 6.2 The lactation curve of the cow.

Box 6.2 Nutritional needs of farm animals

Farm animals require water, energy, carbohydrates, fats, protein, minerals and vitamins. Energy needs are related to the 'maintenance' requirements and to that needed for production. Maintenance (just to maintain body weight and function) is related to size: this is not quite the same as weight but needs can be viewed in this way (see below).

Type of animal	Energy needed for maintenance $(MJ\ kg^{-1}\ L/wt)$
Warm-blooded	
Beef cattle	0.20
Fattening sheep	0.26
Pigs	0.19–0.22
Cold-blooded	
Trout	0.078

Proteins are essential components of the animal body and they are needed in quantities related to performance. For example, a dairy cow giving no milk would need 430 g day^{-1}: giving 30 kg milk day^{-1}, it would need 2180 g day^{-1}. The concentration of protein in the feed would need to be 86 g kg^{-1} DM for the dry cow but 117 g kg^{-1} DM for the high yielder.

Vitamins Few of these organic compounds can be synthesized by animals but several can be converted from precursors present in plant foods. They are required only in very small amounts: for example, the pig's daily need for cobalamin is $c.$ 20 μg (or about one part in 100m of the diet). The most important vitamins in animal nutrition are:

	A	formed from carotene
	D	formed from sterols
	E	tocopherol
All these can be	B$_1$	thiamine
synthesized by	B$_2$	riboflavin
bacteria in the	B$_6$	nicotinamide
rumen of ruminants	B$_{12}$	cobalamin

Minerals These are divided into major and minor (trace) elements, as follows:

Major elements	Calcium	Ca
	Phosphorous	P
	Magnesium	Mg
	Sodium	Na

Trace elements	Iron	Fe
	Copper	Cu
	Cobalt	Co
	Zinc	Zn
	Iodine	I
	Manganese	Mn
	Selenium	Se

Any of these may be deficient in the diet.
    In addition, there are major requirements for

	Potassium	K
	Chlorine	Ch
	Sulphur	S
	Molybdenum	Mo
	Chromium	Cr

but these are rarely deficient in the diet.

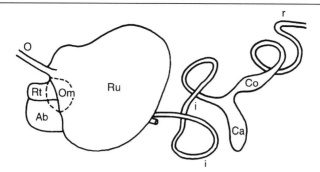

FIGURE 6.3 Digestive system of the cow. Ab, abomasum; Ca, caecum; Co, colon; i, ileum; j, jejunum; O, oesophagus; Om, omasum; r, rectum; Rt, reticulum; Ru, rumen.

Virtually all environments support animals of one kind or another, adapted also to the food sources available. They are themselves fed on by others, including pests and parasites, and subject to many diseases, including those caused by fungi, bacteria and viruses: they vary in their resistance to all of these.

Adaptation to the food source involves the behaviour and structure (legs, necks, jaws, teeth, etc.) of animals and, of special importance, the structure of their alimentary tracts and digestive systems (Figures 6.3–6.6).

The relevance of these features to agriculture is immediately apparent when it is recognized that some animals, such as pigs and poultry, may consume the same kind of staple foods as us, and are therefore in competition with us, whilst others, such as cattle, sheep and horses, utilize plants, such as grass, that we are unable to utilize to any worthwhile extent.

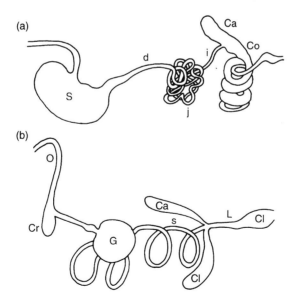

FIGURE 6.4 Diagrammatic representation of the digestive tracts of: (a), the pig; (b), the fowl. Ca, caecum; Cl, cloaca; Co, colon; Cr, crop; d, duodenum; G, gizzard; i, ileum; j, jejunum; L, large intestine; O, oesophagus; s, small intestine; S, stomach.

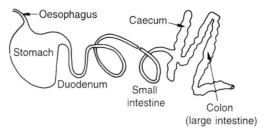

FIGURE 6.5 The digestive tract of a horse.

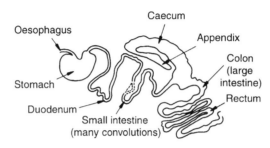

FIGURE 6.6 The digestive tract of a rabbit.

6.4 Agricultural animals

Only a very small proportion of the animal species in the world are used for agricultural purposes, almost entirely from:

mammals;
birds;
fish;
insects (of which there are effectively only two, the honeybee and the silkworm).

The groups used are illustrated in Figure 6.7 and the main species and their main uses are listed in Tables 6.4–6.6.

There are other, relatively minor uses and a few important, but secondary, ones (Table 6.7). Amongst the former is the increasing use of insects and some other groups for the biological control of pests (see Chapter 17) and for pollination. These operations may involve large numbers of small animals, not kept for the products they produce, but in some circumstances of great importance.

Secondary purposes include all the by-products which, although the animals are not kept primarily for them, are often of considerable economic im-

portance. Dairy cows, primarily kept for milk, may depend for their profitability on the production of calves, which enter a meat production enterprise (see Box 6.3). Few animals are kept primarily for their skins, so leather is usually a by-product, but fur-bearing animals (especially mink, fox, chinchilla, sable and rabbit) are an exception. This is not to say that the rest of the animal is not used, but the high value of the fur makes this the only product of significance.

As shown in Table 6.7 there are a great many different by-products, even from one kind of animal (see Box 6.4). Even faeces (for fuel or fertilizer) may be an important secondary product (see Chapter 8).

A wider variety of species are eaten by people but they are usually hunted, fished, trapped or simply collected. Not all the animals that are farmed are necessarily domesticated, e.g. many fish species (see Box 6.5), snails, foxes, and this is also true for some pets (e.g. tropical fish). It is clearly not the case for animals kept in zoos.

It must not be thought that the only animals of importance in agriculture are the ones that are farmed. Not only do they suffer from a wide

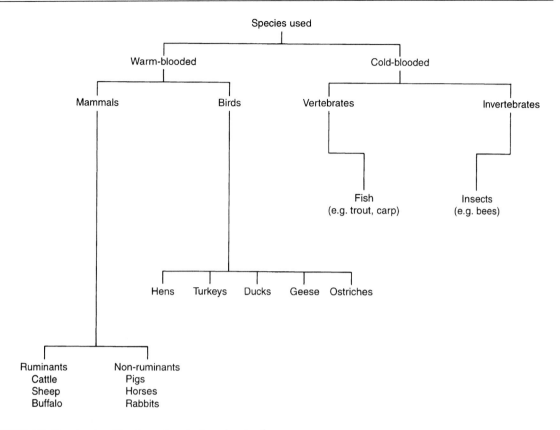

FIGURE 6.7 Simple classification of agricultural animals.

range of pests and parasites (as well as diseases caused by viruses, bacteria, protozoons and fungi) but the agricultural system depends upon other animals to maintain nutrient cycles, soil condition and fertility.

The least obvious of all these other animals are those that inhabit the soil. They are usually small but present in enormous numbers. They are responsible for incorporating organic matter into the soil, they play an important role in the decay and decomposition of dead plant parts, burying large animal excreta (cattle dung, sheep droppings), and they may also feed on plant roots. Their importance is illustrated in Box 6.6.

It is clear that the role of agricultural animals has changed greatly with time. For example, the use of animals in battle (e.g. horses and elephants directly but also mules for carrying supplies) is now outdated but the role of the horse in past military conquests and the extension of empires was of enormous importance.

Many animals are still used for transport, especially over difficult terrain, and they still provide the power for cultivation in many countries (see Box 6.7).

One consequence of this is that a significant amount of land may be cultivated, not to feed people, but to feed such animals. As has been pointed out, agricultural animals provide a great variety of products, but the majority are kept for food. Two important distinctions need to be made about these: (1) some animals have to be killed to provide a product, whilst in others the product comes from the live animal; and (2) some animals are adapted to live on herbage that we cannot utilize.

TABLE 6.4 Agricultural animals and their uses (based on Spedding, 1975)

	Milk	Meat	Eggs	Leather	Fibre	Power (traction and transport)
Ass	x			x		x
Alpaca		x			x	
Buffalo	x	x		x		x
Bees (honey)						
Camel	x	x		x	x	x
Cattle	x	x		x	x	x
Deer		x				
Dog		x				
Eland	x	x		x		
Fish		x	x			
Goat	x	x		x	x	
Guinea-pig		x			x	
Horse	x			x	x	x
Llama		x			x	x
Mule						x
Musk oxen		x			x	
Ostrich		x	x	x		
Pig		x		x		
Poultry		x	x			
Rabbit		x			x	
Reindeer	x	x		x	x	x
Sheep	x	x		x	x	
Silkworm					x	
Yak	x	x		x	x	

6.4.1 Food from live animals

The disadvantages of having to kill the animal in order to obtain the product (meat, skins, etc.) are that it then has to be replaced (at a frequency which varies with how long it takes to reach a marketable weight) and, whilst it is growing, there is no usable product but there is a continual cost (of food, labour, housing, medicines, etc.).

Food from live animals, chiefly milk and eggs, but blood has also been important in the past (e.g. to the Masai in Kenya), overcomes these disadvantages.

As already mentioned, where egg production is high, only a tiny proportion have to be used to provide replacements. The production of milk requires reproduction so, here also, the replacements arise as part of the whole process. In both cases, the main snag is that roughly equal numbers of each sex are produced and only females need to be replaced in numbers. This is because so few males

are needed for reproduction (see Box 6.8). There is therefore a waste or, as in the case of cattle, the surplus calves (see Box 6.3) are used in meat production (some of them for veal).

It also happens that the output per unit of feed is higher for production from live animals than is the case for slaughter animals, but this is dealt with in detail in Chapter 9.

6.4.2 Adaptation to a herbage diet

As was pointed out earlier in this chapter, some animals, chiefly the ruminants (e.g. cattle and sheep), have digestive systems designed to utilize fibrous herbage on which we cannot live. Some non-ruminants (e.g. geese, rabbits, horses) are also adapted but in rather different ways. Contrasted with all of these are the so-called simple-stomached animals (e.g. ourselves, pigs and poultry) that cannot cope with such diets.

TABLE 6.5 Products from agricultural animals

Product	Used for
Milk	Liquid consumption
	Manufacture of butter, yoghurt, cheese
Meat	Fresh or frozen for:
Beef	Direct consumption in the home, restaurants, institutions
Pork	
Lamb	
Poultry	Manufacture of pies, sausages, pet food, other processed meats, e.g. pastes and minces
Rabbit	
Goat	
Horse	
Eggs	
Hen	Shell eggs for direct consumption
Duck	Manufacture of processed products, e.g. dried egg
Goose	
Fish	Fresh and frozen
Fish eggs	E.g. caviare

TABLE 6.6 Animal products (sources: Masefield *et al.*, 1969; Rao, 1976)

Animal products	Examples
Meat	Beef, lamb, pork, camel, buffalo, goat, horse, rabbit
Poultry	Chickens, turkeys, geese, ducks
Game birds	Pheasant, partridge, grouse, guinea-fowl
Offal	Liver, heart, kidney, brain
Blood	Cattle blood, sausages
Eggs	Hen, duck, goose, turtle
Fish	Trout, carp
Shellfish	Oysters, mussels, shrimps, prawns
Other aquatic products	Frogs, turtles
Milk and milk products	Whole milk (cow, goat, sheep, buffalo, camel), cheese, yoghurt
Fats	Butter, ghee, lard, suet, tallow
Sugar	Honey
Miscellaneous	Snails

TABLE 6.7 Animal by-products and their uses

Animal by-products	Examples
Meat by-products	Meat, bone, blood and feather meals
	Offal
	Tallow and other fats
Milk by-products	Skim milk
	Butter milk ⎫ Liquid or dried
	Whey ⎭
Faeces	Processed or fresh poultry, cattle or pig faeces
Worms and other invertebrates	

Box 6.3 Calves from the dairy herd

In order to stimulate a cow to produce milk, it has to produce a calf and, in general, this has to happen as close to once a year as possible, in order to maintain the cow in milk for a high proportion of its time (for economic reasons). As pointed out in Box 6.4, almost half the calves born to dairy cows in the UK are surplus to requirements. Some of these surplus calves are grown and fattened for beef but not all are of suitable breeds and crosses. The male calves are not wanted at all for replacements of dairy cows but they tend to be more suitable for beef production, if the breed is appropriate.

About 25% of UK calves born to dairy cows are currently only suitable for **veal** production. Unfortunately, veal means different things in different countries and is produced in many different ways. The welfare implications are discussed in Chapter 12. In the rest of Europe, white or 'clear' veal is produced, mainly from calves reared in narrow crates and fed a milk-only diet, deficient in fibre and iron (this gives the pale colour). In the UK such systems were effectively banned in 1990 and British veal is produced in loose-housed straw yards with access to fibre: the meat is therefore pink. But UK consumption of veal is very low [only *c.* 55 g (2 oz) per person per year] and we therefore have a surplus of *c.* 500 000 calves (only 4000 are needed to supply the veal consumed here) per year. As a consequence, by 1995, most of these calves were being exported (live) to be reared on the continent, mainly for veal production. This gave rise to demonstrations calling for the banning of live exports and a massive rethinking within the UK industry, aimed at finding ways of solving the problem.

Box 6.4 The importance of by-products

The animal populations used for production vary greatly but they do embody a common structural pattern, (illustrated in figure (a) below. As pointed out in Table 6.1 and Box 6.8, the need for replacements varies with species but is predominantly for females. This is particularly so for mammals (such as the cow) where AI can be used and very few males are required. This means that almost half the calves born in the dairy herd are surplus to requirements: they may be used for rearing for veal or beef, depending on the suitability of the breed and the demand for those products.

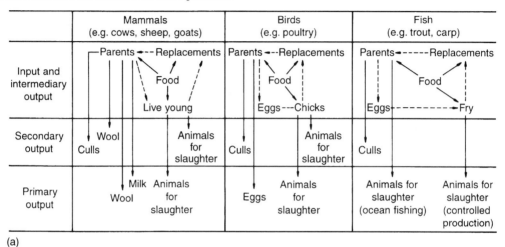

(a)

In the case of poultry, the situation is even more extreme and, again, about half the newly hatched chicks in breeding flocks are unwanted, except for immediate slaughter and use as food for zoo animals or the manufacture of pet food. In the case of laying hens, the eggs are unfertilized and no male is required at all (although there is a separate breeding system for replacing the laying hens themselves).

In figure (a), continuous and broken lines show both primary and secondary production processes: the broken lines indicate the replacement process. The 'culls' are those animals that are removed for slaughter (or 'culled') at the end of their useful lives, due to age, disease, injury, etc. These also may be of economic importance.

There are, in addition, other losses (not shown) due to disease and mortality, where no product results: there is thus no return and there may well be a cost of disposal.

In addition to secondary products from the production system, there are many by-products from an individual animal at slaughter. This is illustrated on the next page for a beef animal.

It is sometimes said that these latter animals compete with people for food, since they live on the same kinds of food. They do not, of course, demand the same quality and standards of hygiene and, indeed, were kept originally as scavengers precisely because they could utilize the leftovers from the same diet.

As agriculture has developed, they no longer live on our waste products but they do not necessarily compete either. However, they cannot derive their main diet from, say, grassland and, as was seen in Chapters 2 and 3, there is a great deal of land which comes into this category.

Adaptation to a diet of grass and other grassland plants, including clover, extends to many parts of the animal but the most important features are the jaws and the alimentary tract. In cattle and sheep the jaws and teeth are specially designed for harvesting grass: cattle get their tongues round longer grass and pull it; sheep are equipped for shorter

Box 6.4 (cont.)

Useful by-products of the beef animal[a]

Product	Parts of the animal
Edible offal	Liver, heart, tongue, kidney, brains, cheek and head trimmings, tail, stomach
Leather	Hide
Insulin and pharmaceuticals	Pancreas, pituitary, thyroid and thymus glands, genitals, spinal cord, gall and gallstones
Edible fats, tallow and for processing and manufacture	Internal fat
Glue, gelatin, tallow, meal, pet food, buttons and handles	Bones
Pet food, animal feed, fertilizer, glues and fire-suppressing foam	Blood
Cow heel jelly, neatsfoot oil, etc.	Feet
Meal	Horns
Surgical ligatures and pharmaceuticals	Small intestines

grass, which they nip off between lower incisors and a horny pad in the upper jaw. Obviously geese behave differently and are able to crop the grass very short – but so can horses.

But ruminants **eructate**, i.e. they bring back up into the mouth a bolus of partly digested grass, grind it up again with their powerful molars and swallow it. This process is repeated and is the familiar 'chewing the cud'.

The digestive tract for cattle is illustrated in Figure 6.3, those for pigs and fowl in Figure 6.4 and those for the horse and the rabbit in Figures 6.5 and 6.6. The essential feature in the ruminant is the division of the stomach into four parts, by far the biggest being the enormous rumen. This is like a large vat full of bacteria and other micro-organisms which actually digest the fibrous plant material. In cattle, the rumen can hold up to 200 litres (l) of contents. The horse has no rumen, has a true stomach of similar capacity to that of cattle, but has a greatly enlarged colon (over 30 l), some three times the size of the colon in cattle.

So ruminants digest their fibrous food at the beginning of the process, horses do it towards the latter part of the gut – as do rabbits, with a greatly enlarged caecum. The disadvantage of later digestion is that there is less opportunity for absorption, through the gut wall, of the proceeds of digestion. The rabbit solves this by engaging in eating some of its faeces in order to put the material through the whole sequence a second time.

Geese and Chinese grass carp have no specialized equipment and simply eat a lot and live mainly on the cell contents rather than the fibrous cell walls.

One measure of the success of all this is the digestibility coefficient, representing the percentage of the food eaten that is not passed out in the faeces (all expressed as dry matter) (Box 6.9). The percentage retained varies with the animal and the feed but will commonly be 60–70%.

Of course, nutrition is more complex than this simple expression and it is possible to measure separately, for example, the percentage of energy

Box 6.5 Fish farming

The farming of fish has a long history. In the UK, trout and carp have been the predominant species and, in tropical countries, species such as Tilapia have been used. But the list of species farmed is very long (see Spedding, Walsingham and Hoxey, 1981, where 57 species are given) and includes herbivorous, carnivorous and omnivorous fish. Fish culture is especially common in the Far East.

Historically, fish farming has been predominantly in fresh water and usually requires a plentiful supply of neutral or slightly alkaline water. Marine fish farming is generally carried out in flexible mesh cages suspended in coastal seawater. More recently, indoor fish farms have been developed in Europe, especially in Denmark, to mass-produce both seawater and freshwater species, such as cod, eel, turbot, sea bass, salmon, catfish and lobster.

So-called 'fish factories' (Lloyd, 1995) are housed in large hangars and employ computer-controlled feeding, heating and water circulation. Filtration means that water consumption is low, since it is all recirculated (four times an hour). It is claimed that all-the-year-round production is possible without the depletion of stocks, high cost and danger of fishing from trawlers.

Arguments continue about the problems of environmental pollution, welfare and eating quality but, by analogy with poultry, such intensive methods are likely to prosper since they will produce cheaper products and the alternative of deep sea fishing is hard to defend on environmental, efficiency or welfare grounds: taste may be another matter.

Marine farming is not confined to fish and a variety of crustaceans and shellfish are also farmed, often producing high yields. Seawater ponds are used in Japan for cultivating the prawn (Penaeus japonica) and the cultivation of oysters (Ostrea edulis) is widespread. Filter-feeds, such as oysters and mussels, are, however, very sensitive to pollution and may accumulate toxic materials.

Box 6.6 Examples of soil animals

	No. ha^{-1}	Wt ha^{-1}(kg)
Collembola (springtails)	144m	750
Earthworms (0.1–3.0 g each)	7m	c. 14 000
Moles (85–110 g each)	10	1

As Darwin pointed out in 1881, earthworms may move a vast amount of soil in a year. In England, this has been estimated at c. 27 000 kg ha^{-1} (11 tons acre^{-1}), or 2 cm in depth every 10 years (Lewis and Taylor, 1967).

The numbers of small, living organisms are normally very great. For example, bacteria are commonly in the range of 10^7–10^9 g^{-1} of soil, equivalent to c. 3 tonnes ha^{-1}, but less than 1% of the soil organic matter.

Fungi are also numerous, and somewhat between them and the bacteria are the actinomycetes, present in the order of 10^5–10^8 g^{-1}.

Algae are usually only a small part of the soil biomass but may represent 10–300 g ha^{-1}.

All these feed on decaying roots and other vegetable matter, on animal faeces and on each other.

One of the consequences of being warm-blooded is that food intake has to be sufficient to maintain the characteristic (and remarkably constant) body temperature at all times. Unfortunately, at the coldest times, the need is greatest and the supply least. Farmers have a major problem in supplying their animals' food requirements in winter and in drought because they keep more animals than the natural environment would normally sustain.

Wild animals, of the kind from which farm animals have been derived, lay down fat reserves to tide them over such lean periods (and many simply do not survive) but, agriculturally, this is a rather inefficient process.

Incidentally, the fermentation that goes on in the rumen generates a great deal of heat so that ruminants are not greatly troubled by even quite low temperatures (e.g. below freezing).

or protein digested. Food supply has both quantitative and qualitative aspects and these vary with the season.

Box 6.7 Animals for power

The main uses of animal power to perform useful agricultural work are:

Traction, cultivation and clearing of land	by horses, oxen, buffalo, elephant
Transport	by horses, donkeys, mules, oxen, buffalo, llamas
Herding	by dogs
Biological control of insect pests and weeds	by insects, arachnids, ducks, geese, fish, amphibians
Pollination of flowers for fruit and seed	by bees

In the first two cases, the animals are quite large and have to be fed, so there may be a substantial cost. This is either in the form of labour and time, where animals are herded or grazed on common land or roadside verges (at no monetary cost), or purchased forage (e.g. hay), or land set aside for annual feed. In some cases, a substantial proportion of the available land may be used to feed non-food animals (e.g. 10–20%).

Both biological control (Debach and Rosen, 1991) and pollination are now very highly developed, with insects and other animals being bred and moved about to ensure that they are present where they are needed, in adequate numbers and at the right time.

Where animals work, there is usually a saving in 'support' energy (see Chapter 8): biological control and the use of bees for pollination save agrochemicals and labour, respectively.

Transport by animals is found in a vast variety of forms, some requiring speed, some strength and endurance. The attributes of such animals are:

Buffalo	docility, sure-footedness, strength
Camel	feet adapted to walk on sand, ability to withstand heat and lack of water
Dog (husky)	thick coat, ability to work as a team
Donkey	strength and endurance for small size
Horse	strength and speed
Mule	endurance
Reindeer	speed and adaptation to cold
Yak	adaptation to high altitude and difficult terrain
Llama	endurance at high altitudes

For traction, different animals are similarly adapted to different conditions, e.g.:

buffaloes for cultivation in wet rice fields;
elephants can handle timber even in deep water.

Speed of work varies from 0.7 m s^{-1} for cows and donkeys to 0.8–1.0 m s^{-1} for mules, buffaloes and horses. The power developed by a light horse is c. 735 W.

All these features emphasize that the main role of agricultural animals is to help humans use the available land (and sometimes water) to supply their needs, mainly for food.

Where land can grow crops for direct consumption, animals may convert to meat, milk or eggs, the crop products and by-products that are not wanted or surplus to human needs.

The animals may be ecologically necessary to maintain soil fertility or to facilitate rotations, but in general they are used to produce products that are preferred, where these can be afforded.

Where land is too barren, stony, steep, etc., to be economic for crop production, ruminant animals can be used to provide human food where otherwise none would be available. Other resources, such as forests and roadside verges, can be similarly utilized.

Ruminants can even provide for nomadic people, not having to be carried, sometimes serving as a living store of food (until required). They and their owners may then move from one source of feed to another, according to season. This arrangement may apply to mountains and lowland with the people

Box 6.8 The need for males

Males are obviously needed in agricultural animals:

1. to perpetuate the species (as in breeding replacements);
2. to produce progeny for slaughter at an appropriate weight (as in fat lamb and beef production);
3. to initiate milk production (as with dairy cows).

They are not needed in substantial numbers for (1) (see Table 6.1).

For meat production they are often the preferred sex, since males commonly grow faster than females and thus reach slaughter weight earlier. Uncastrated males also tend to be leaner, producing more muscle and less fat: but in some species (e.g. bulls) they are more difficult to handle.

For use in milk production, very few males are needed where AI is employed, because semen can be diluted to cover many more females. For example, a single ejaculate from a bull can be used to inseminate up to 200 cows. AI in sheep and pigs is currently limited because their semen cannot be so extensively diluted and their spermatozoa are damaged by current freezing techniques for storage.

Cows could go on lactating for more than one year (though this might reduce overall yield) and goats commonly go for longer. This further reduces the number of males needed.

	Males (%)
Sheep	49–50
Cattle	52
Pigs	50
Goats	57
Hens	48

The problem is that the number of males born is often more than half of the total, as shown above: In some circumstances, these males may not be wanted and disposal of them creates a problem.

It is possible to use dual-purpose breeds, in which the male calf is suitable for rearing for beef, but the value of the calf may not compensate for the reduced output of milk. Most of such problems can be solved but they take time.

This is one example where biotechnology might make it possible: (1) to determine the sex of the calf born or (2) to initiate milk production without mating. Whether such techniques would be acceptable to the public is another matter.

living for a time in each location (known as **transhumance**). This was the pattern, for example, in the Swiss Alps, utilizing the alpine pastures in the summer and storing the output as portable cheeses.

In fact, animal products, being a more concentrated form of nutrients than are plants, can be stored and transported more efficiently than bulky crops. Grain, however, is relatively easy to store and transport, being relatively concentrated and dry.

Finally, animals have cultural significance: they may represent wealth and be necessary for celebrations and ceremonials.

Attitudes to animals thus vary greatly across the world and this, of course, affects the way they are treated.

6.5 The status of animals

As interest in – and concern for – animal welfare increases (see Chapter 12), so the precise status conferred on animals has become of vital importance.

Currently, the Treaty of Rome classifies animals as 'goods' or 'agricultural products' and those most concerned with animal welfare find this wholly inappropriate.

To an agriculturalist, it seems perfectly natural to describe the animals that they have produced for sale as breeding stock as 'products'. This does not imply to them any diminution in their status as living and feeling creatures. However, this may not be the case with those who then handle, market

Box 6.9 Digestibility

The importance of digestibility is easily illustrated: suppose all our energy needs could be met by straw, what would be the relevance of this, given that we cannot digest it? Digestibility is usually defined in one of the following ways:

1. 'true' digestibility = the feed that is digested and absorbed (%);
2. 'apparent' digestibility = (amount eaten − quantity of faeces)/ amount eaten × 100.

Both are expressed in the same terms (e.g. dry matter, organic matter, energy or protein) and measured for the same period.

   Digestibility is thus an efficiency ratio (see Chapter 9): the difference between the two expressions largely reflects the fact that faeces contain dead bacteria and sloughed-off gut wall tissue, in addition to the undigested parts of the feed. The faecal output therefore overestimates the amount that has not been digested; nevertheless, the apparent digestibility may be used as an approximation to digestive efficiency and, as it is more easily measured, most data are available in this form.

   Digestibility may vary with the age of the animal, the level of feed intake and even the level of internal parasitism. There is not just one figure that will be valid for all animals, even of animals that are of the same species and breed and are similar in physiological state, eating the same diet.

   The following examples, taken from the literature, illustrate the range of values that may be obtained:

1. for the digestibility of the crude fibre part of the diet:

Feed	Rabbit	Pig	Sheep	Cattle
Maize	−11.8	41.1	−48.1	3.6
Soyabeans	50.6	55.3	84.7	55.7

2. for the digestibility of crude protein ($CP = N \times 6.25$):

Feed	Rabbit	Pig	Sheep	Cattle
Maize	24.8	70.0	48.7	52.0
Soyabeans	74.2	79.5	90.9	91.6

Negative values are obtained when the amount in the faeces (including that from bacteria and gut wall cells) exceeds the low amount taken in. They therefore tend to occur in low-fibre feeds. Alfalfa (Lucerne) is a relatively high-fibre feed:

	Sheep		Horse		Rabbit		Guinea-pig	
	CP	F[a]	CP	F	CP	F	CP	F
Alfalfa (pellets)	69.7	50.5	74.0	34.7	73.7	16.2	69.0	38.2

[a] Fibre – in these cases, measured by the method known as 'acid-detergent fibre' determination.

There are many factors that affect the value obtained, including the species and age of animal, the form of the feed, the amount eaten and the internal parasite burden of the animals. The figures quoted should not be taken as representing anything other than the orders of magnitude likely to be found.

and transport them and their categorization as 'goods' may encourage the view that they can be treated in the same way as non-living goods.

There are therefore pressures for livestock to be given a new status in the Treaty of Rome as 'sentient animals', to which different rules and regulations would apply. When the Maastricht Treaty was agreed, in 1991, a 'Declaration on the Protection of Animals' was added to enshrine in it 'full regard to the welfare requirements of animals'. This is not just a matter of trying to change attitudes but also of altering the rules that apply.

For example, the 'free trade' that is the essence of the GATT, sometimes works to the disadvantage of animal welfare and it is held by welfarists that the requirements of free trade should not override the necessity of protecting living animals. Of course, to ensure this worldwide is a major challenge but it is part of the wider view of agricultural systems that is now needed. Another dimension of this wider view relates to biodiversity (see Chapter 10).

6.6 Biodiversity in agricultural species

When most agricultural animals were kept extensively, a great variety of different breeds were found, mainly because each was adapted to the particular conditions and climate of the locality in which they were kept. This was probably accentuated by localized breeding and preserved by fierce rivalries between breed societies.

However, as animal production became more intensive, more animals were housed, where the 'climate' could be controlled and feeding standardized. There was thus less need for local adaptation and greater advantages in the standardization of products to meet the needs of supermarkets. The great variety of breeds tended to disappear.

Thus, in the UK, there are still about 50 breeds and crosses of sheep but the majority of the dairy cattle are black and white Friesian/Holsteins. The number of beef breeds, by contrast, has been ex-

tended by the importation of continental breeds, such as the Charollais, Limousin and Simmental. Pigs and poultry, however, are hardly ever referred to by breed names and these industries use a few standardized hybrids known by numbers.

Even within a breed, there has – particularly in dairy cattle – been a concentration on a few genetic lines (made possible by the use of artificial insemination (AI) which enables one bull to cover many cows and produce many thousands of daughters). This has happened to the point that there are worries about the narrowness of the genetic base and the loss of genetic diversity. Interestingly, the Banner Committee (HMSO, 1995) considered that the application of new technologies, such as AI and embryo transfer, offers ways of introducing genetic variations.

One aspect of diversity not discussed so far is that of size.

6.7 The size of animals

It is obvious that different species of animals have quite different characteristic sizes, from microscopic to huge (e.g. elephants and whales). This reflects the results of evolution and the adaptation of species to particular environments. There are many answers, therefore, to the question of why size variation exists: more interesting is to examine some of the consequences of size.

For example, a large body has to have some mechanism for getting the necessary oxygen to all the active tissues. Lungs and the circulation of blood achieve this for warm-blooded animals, such as mammals and birds, and for cold-blooded vertebrates, such as reptiles. Fish are also cold-blooded vertebrates but have to obtain their oxygen from water, so have gills instead of lungs.

Insects, on the other hand, have no equivalent systems and rely on the diffusion of oxygen through spaces in the body. This is a relatively slow process and effectively limits the size that insects can achieve.

Just as there is therefore an upper size for insects, there is a lower size for warm-blooded animals.

This is because the smaller the body, the greater its surface area relative to its weight or volume.

Moore (1995) gives the following illustration. A cube that is 1 cm high has a surface area of 6 cm^2 (= six faces of 1 cm^2). Glue 27 such cubes together to form a new, larger cube and it will have a height of 3 cm, a surface area of 54 cm^2 (= six faces of 9 cm^2) and a volume of 27 cm^3 (= 3 × 3 × 3 cm). Thus, a 27-fold increase in volume results in only a ninefold increase in surface area.

So, as an animal grows it has less surface area per unit of volume or weight. (It is worth noting that skin and wool production, for example, are related to surface area.)

The importance of this is easily illustrated, in two different ways:

1. Surface area affects air resistance and slows the fall of small animals. A mouse can fall quite safely from a considerable height: we cannot (unless we increase our air resistance, e.g. with a parachute).
2. Surface area affects heat loss, so a small animal has to eat incessantly to generate all the additional heat required to keep it warm. So there is a lower size limit for warm-blooded animals, close to that of a harvest mouse (5–6 g) – a house mouse weighs 35–40 g – or a humming bird.

Another interesting facet of surface area is that a coating of water is relatively much heavier for a small creature, which is why insects get trapped in water or when wet.

Conversely, a bird above *c.* 12 kg in weight has great difficulty in (flapping) flight, though it may be able to glide.

There are other consequences of great size and weight. For example, the bones have to be disproportionately bigger to support the weight because bone strength is proportional only to their cross-sectional area and not to their volume.

Muscle power is proportional to mass, so that animals built to the same design, whatever their size, can jump to about the same height. Thus, elephants can jump no higher than, for example

mice (they are not of the same design, of course, so compare, say, mice and rats).

One of the most important consequences of these relationships is that the basal metabolic rate (BMR) turns out to be proportional to three-quarter power of body mass. This exact relationship is not fully understood but it is not too far from the two-third power of body mass to which surface area is related. The metabolic rate affects rates of growth and of intake, both of great importance in agricultural animals.

But other, quite practical, criteria must have operated in the choice of animals to domesticate for agricultural purposes. Big animals produce more but need more food, smaller animals are a problem to protect and to confine. Fierce animals are better avoided, animals that tolerate others are to be preferred. Grazing animals also satisfy the need to supply food, since herbage grows naturally and the animals harvest it for themselves.

By contrast, domestication for companionship (or even protection) would use other criteria, including ability to live on the same food as human beings (in the form of household waste).

Questions

1. Why have we domesticated so few animal species?
2. Why have we not exploited highly successful animals, such as mice, rats, houseflies, greenfly?
3. Are all domesticated animals tame and are all tame animals domesticated?
4. Why aren't draught animals replaced by tractors – thus releasing all the land used to feed them?
5. Is a companion animal just a form of entertainment?
6. Why are horses and dogs eaten in some countries but not in others?
7. Are pests and parasites 'natural' and should they be accepted? On/in pets as well?
8. If a harvest mouse is about as small an animal as can exist, how do its offspring survive?

Chapter seven

Agriculture as a system

The camel driver has his plans – and the camel has his!

Arabic proverb

When *An Introduction to Agricultural Systems* was first published (Spedding, 1979), the idea of a systems approach to agriculture was relatively novel and confined to a small number of enthusiasts (who came, incidentally, from a range of different disciplines). The second edition (1988) was primarily aimed at updating the first but retained the same main objectives and the same introductory character. Since that time, there has been widespread recognition of the need for a systems approach and a broad acceptance of its value. There is, therefore, nothing like the same need for an introductory text.

However, other relevant changes have also occurred, particularly the increased attention paid to the interactions between agricultural systems, the environment and society in general. Indeed, it is increasingly inadequate to look at agricultural systems in isolation, quite apart from the massive development of the food industry in developed countries.

The operational units of agriculture may certainly be described as agricultural systems, including all the variations in size and complexity of unit that are called enterprises, farms, plantations, regional and national agricultures. It is useful to have a term of this kind that does not imply a particular size or level of organization. After all, the essential

activity [of agriculture] does not imply anything of this kind and it is desirable to be able to discuss operational units in general. However, there is much more to the deliberate use of the word 'systems' in this context, because the 'systems approach' is now recognized as a distinctive way of looking at things. This is based firmly on the concept of a system and it is important to be clear about its meaning.

7.1 Definition of a system

So many different things can legitimately be regarded as systems (a bicycle, a car, a cow, one's own body, a farm, a sewage works) that it is tempting to conclude that anything can be a system. This, however, is not so and if it were so the concept would be useless. If it were not possible to distinguish 'systems' from 'non-systems' the concept could not be used and if the distinction did not involve important properties it would not be worth making. It is the properties of systems that chiefly matter and they may be summarized in the phrase 'behaviour as a whole in response to stimuli to any part'. Thus, a collection of unrelated items does not constitute a system. A bag of marbles is not a system: if a marble is added or subtracted, a bag of marbles remains and may be almost completely unaffected by the change. The marbles only behave as a whole if the whole bag is influenced, for example by dropping it, but if it bursts the constituent parts will go their own ways.

Of course, any collection can be transformed into a system by building into it such relationships between the components as are required to give it characteristic system properties. A simple, if rather trivial, example would be nailing together separate pieces of wood: once joined they would behave as a whole because of their physical connections.

However, not all sets of components that are joined together constitute systems. They may be joined up to other units and be incapable of behaving independently and thus be incapable of responding to stimuli at all. So a bicycle is a system but, if it was joined up to another one or to a sidecar, it would then be necessary to regard the new combination as a system and the original bicycle either as a subsystem (but this ought then to mean something special) or simply as part of a system.

An animal is a good example of a living system. It has an obvious structure, it behaves as a whole in response to major stimuli and it is relatively easy to see where it begins and ends (i.e. to distinguish the boundaries between the system and its non-system surroundings). This last point is very important. It is of only limited value to be able to identify a system if it is not possible to say where it ends. In the case of an animal, such as a hen, the obvious boundary would be just outside the external layer of feathers, including the thin layer of air that forms a kind of 'private' microclimate around the hen. Figure 7.1 illustrates this in an outline diagram, showing the major inputs (food and water) and the major outputs (excreta and heat). Now, in most situations these outputs do not immediately, directly or automatically influence the hen producing them. The atmosphere is so enormous that the heat output of the hen does not appreciably influence it. But suppose that the hen was confined in a small, sealed box. This immediately reminds us of the limitations of our model: for example, oxygen was not mentioned and will clearly run out, with disastrous consequences. However, let us confine ourselves to the heat output in order to illustrate the main point. The atmosphere in the box is soon heated up by the hen and immediately and directly affects the hen and the rate at which it produces heat. This is called a 'feedback' mechanism or loop (see Box 7.1) and, if we ignore it, we are going to be misled about how our system (the hen) will respond to any stimulus.

Atmosphere

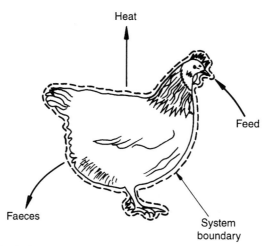

FIGURE 7.1 The hen as a biological system.

Box 7.1 Feedback

Feedback is simply the mechanism whereby action of one kind is modified by the results of that action. If the effect is to increase the action, it is called positive feedback: if it is to decrease the action, it is called negative feedback. This is often shown diagrammatically, in the following way:

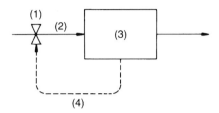

where (1) represents a valve-like device for controlling the flow (2) of something into a collecting reservoir (3). The feedback loop (4) shows that the amount in the reservoir (3) is affecting the rate (1) of flow (2).

A ballcock is about the simplest example.

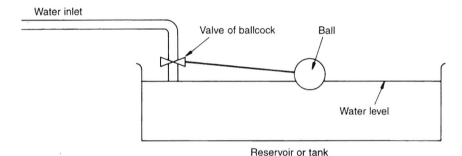

As the water level rises, so does the ballcock, shutting off the valve and reducing or stopping the flow of water from the inlet. This is a negative feedback loop.

An agricultural example would be the deposition of cattle excreta on the grass – as a result of eating herbage – which soils it and reduces the amount eaten subsequently.

Positive feedback is not so common but can be illustrated – at least for a time – by population growth. As the population increases, so the number multiplying also increases. Thus, the rate of reproduction is related to the number of adults and results in the latter increasing. This 'exponential' growth is readily seen in duckweed or cell growth.

If each individual divides in unit time, then at each successive division there are twice as many doing it. Thus:

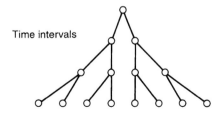

Time intervals

Box 7.1 (cont.)

Plotted on a graph, this becomes:

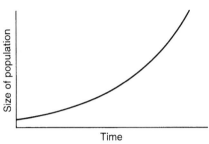

Norbert Wiener (1961) illustrated both processes, as used in picking up a pencil. We do this not by working out which muscles to use and trying to control each one, but by getting feedback from our senses (mostly sight) to which we respond either negatively or positively, as appropriate.

The hen in the box is no longer a sufficiently independent organism to be regarded as a system: it is necessary to consider the 'hen plus box' and place the system boundary around the outside of the box (see Figure 7.2). This correct positioning of the boundary is a way of defining the content of a system and, if it is not done, much of the value of a systems approach is lost.

Imagine that we studied different parts of the body (arms, legs, etc.) separately and in isolation. We would clearly never understand how the whole body worked, but we would not even understand how each part functioned when joined up to the body: and it is precisely in these latter circumstances that we particularly wish to understand the function of parts. In fact, they cannot function

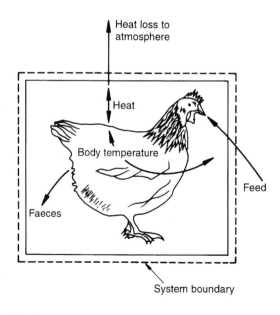

FIGURE 7.2 The 'hen-in-a-box' system.

separately but, in any of the senses in which they might, it would be quite irrelevant to their role as parts of the body system.

Thus, it is important to know when we are dealing with a whole system and when we are dealing with a part. Since it is not possible to study whole systems all the time, however, it is useful to be able to identify 'subsystems', i.e. parts of systems that are worth considering separately because their separate function **is** relevant to their role in the whole system. (We will return later to this question of subsystems.)

It is this importance of being able to tell when we are dealing with a system that makes the definition vital. There have been a great many attempts to define a system satisfactorily; the following contains the main elements discussed above:

> A system is a group of interacting components, operating together for a common purpose, capable of reacting as a whole to external stimuli: it is unaffected directly by its own outputs and has a specified boundary based on the inclusion of all significant feedbacks. (See Box 7.2.)

7.2 A systems approach (see also Box 7.3)

This way of looking at the world and of tackling problems is founded on the idea that it is necessary to identify and describe the system that one wishes to understand, whether in order to improve, repair or copy it, or to compare it with others in order to choose one.

Its main proposition is obvious and is common sense but it is not always clear how to operate on this basis in practice. Certainly, the approach does not come naturally or automatically and has to be cultivated. Fortunately, it is now backed up by techniques and methods that are very powerful, but this also means that they can be misused disastrously. So, as with all powerful tools, it is necessary to accept a learning period and to be very clear at the outset what it is we are trying to do and how we propose to do it.

A learner pilot does not jump into an aircraft and expect to fly immediately: even less do they dissipate their energy trying to flap the wings up

---

Box 7.2 Definition of a system

It may be helpful for those to whom this terminology is both unfamiliar and alarming to explain the concept very simply. The fact is that the concept is quite simple but allows us to deal with great complexity: examples therefore quickly become complicated and, for this reason, forbidding.

One of the examples briefly mentioned in the text is the human body. This is a good example because you can consider your own body and easily recognize that it is a system which conforms to the following definition:

> A system has a boundary (e.g. the skin) enclosing a number of components (e.g. heart, lungs) that interact (the heart pumps blood to the lungs) for a common purpose (to maintain and operate the living body).

It is quite obvious that there is a sense in which your body is an independent system – in ways that are not true of the circulatory system (heart, arteries, veins, etc.) or the skeleton: these may all be thought of as subsystems (see Box 7.5). Furthermore, there is no way in which you could think of an arm or even your head as separate systems, much less eyebrows or fingernails. The latter are clearly components and could not exist on their own. The same is true for animal bodies and, indeed, for plants.

Take a tree: although one might speak of its root system (really a subsystem), no one would do so in the technical sense being defined here, because it does not and cannot exist on its own. What happens within a system will be further considered in Box 7.5. What happens outside systems is also of interest, since systems can also become part of other systems.

For example, any one human being (already accepted as a system) can join with others to form a team (for games), a society, a church or a club. These groupings then each behave as a system, with all the components (systems) interacting and operating together for a common purpose. The connections do not have to be physical joins and the interactions can be largely of information flow, but the principles are the same.

---

and down because they have observed birds doing this. It is, in fact, quite an art to be able to use tools skilfully, for it is necessary to accept the limitations

---

Box 7.3 A systems approach

This is described in the text but deals more with the technical use of the methodology available, such as modelling. The latter also alarms some people, sounding rather theoretical. However, just consider a model aeroplane: very useful for showing someone how to identify enemy or friendly aircraft or for flying for recreational purposes. These are scale models that look rather like the real thing but are smaller. But models can be much more detailed and can even show the internal structure. These images frighten no one, so the idea of a model is quite appropriate.

Now, suppose you wish to understand how an animal body works, what kind of a model would help. Pictures and solid models are helpful but the problem is to show all the complex and dynamic interactions that go on. Somehow, a picture has to be built up of how it all works together. Fortunately, film-making has made possible very intimate views of the insides of living organisms and time-lapse and speeded-up photography can make movement very clear.

What a systems approach encourages is a recognition that all these components and activities are linked, they affect each other. It is not sensible to look at one component by itself without recognizing that what it does and what happens to it will affect other parts of the system.

Consider what happens when you stub your toe: the whole body may react and different parts respond differently. Eyes may water, the voice may make appropriate sounds, the pulse rate may increase and hands may try to rub the damaged toe. It would be very rash to alter any component of a system without regard to the consequences and reactions elsewhere.

You cannot, for example, improve a car (system) by doing research on one wheel and then making it rather bigger than the rest. Or increase the power and size of the engine without regard to the ability of the chassis to support it.

These things are common sense in such familiar contexts – they also apply to biological and agricultural systems.

---

of the tool and to use it when and where it is appropriate. The danger is then that one may only do what the tool allows, even when new tasks become more important. The tool-user has to be aware of the developing needs for tools and the development of new tools by those who concentrate on such innovation. A systems approach embraces the disciplined use of existing tools, the development of new ones (which involves studies of methodology) and the definition of tasks for which tools are required.

Since a systems approach can be applied to many subjects, and not just to agriculture, its characteristics tend to include methods and techniques. But many of these also apply to other approaches, so the essence is, in fact, more of a philosophy (it is well named an 'approach'), or a way of doing things and thinking about things. Its main propositions are listed in Box 7.4.

The central idea that one must understand a system before one can influence it in a predictable manner, has embedded in it the importance of recognizing a system, what is in it and what is not, and where its boundary lies. This is all included in the identification and description of a system and this represents a kind of model building.

7.3 Models

There are all kinds of models, but they are all representations of the real thing, simplified for some purpose: they include those features that are essential for the purpose and they leave out those that are inessential. Without a clear purpose, there are no criteria for deciding what is and is not essential.

So, if you wish to show someone what a tractor looks like, approximately, on the outside, you can show them a scale model. If you want to show them how it moves, you may push the model about. If you also wish to talk about it more generally, you can agree on the word 'tractor' to mean that sort of thing and you can vary the picture in your minds by changing instructions. If you wish to think about a red one, when your scale model was grey, you

Box 7.4 The main systems concepts

1. Systems are identifiable entities with important properties and attributes that are quite distinct from those of non-systems.
2. Systems can be of any size or complexity, from a molecule to an elephant or a universe, and any system can be either a component or a subsystem of another system.
3. When a system is a component of another system it has specific inputs and outputs resulting from its interactions with the rest of the system: an independent system is not bound by such interactions.
4. In practice, no system is completely closed or completely independent but the differences between an object as a component and as a system are usually very large.
5. All systems can be modelled but it is not always practicable to construct a model at any given level of detail.
6. The level of detail and the structure of the model should be related to the purpose of construction and should be as simple as will serve the purpose.
7. Models should always be capable of validation and, where possible, should be validated.
8. The required precision and accuracy of a model should be specified in advance: otherwise validation is not possible.
9. Where systems are too large or too complex to be studied in their entirety, subsystems may be identified that can usefully be studied separately.
10. Subsystems have a degree of integrity and independence of the whole system, such that they can be studied separately and the results of such studies incorporated into a model of the whole.
11. The independence of subsystems depends on the existence of only a few main interactions with other subsystems or components of the whole system.
12. Subsystems will usually have the same output as the main system but relate to only some of the components and thus to only some of the inputs.
13. Improvement (or indeed any response) of a system, due to changes in one or more components, cannot be predicted without the use of a model of some kind representing that system.

only have to say so. Thus, a 'mental model' is much more flexible: it can, in fact, do things that no real tractor can do (such as travel at 100 mph). Mental models can be experimented on, therefore, but not systematically or in a disciplined fashion, because you cannot attend to all the detail, or the consequences (of your changed speed, for example), and you cannot remember all that was done.

To be really useful, therefore, except in the most imaginative of contexts, a model is better stated. A number of advantages immediately follow. Everything is explicitly stated, so there is never any doubt about the structure and content of the model. Once written (or drawn) it does not have to be remembered, because it can always be referred to, and it is clear to anyone who looks at it: it cannot be different things in the minds of different people, and, of course, diagrams can and should be labelled and some indication given of what they are intended to represent. Nevertheless, it is extremely helpful if the viewer can immediately identify the meaning of any symbol used in a diagram. Lines to represent connections and arrows to represent directions are the most obvious and elementary conventions that we all accept and it is usual to extend these to include a whole range of components and processes. This is so, for instance, for representing electrical circuits, engineering diagrams and architects' plans, but it is also true in biology and agriculture. Flow diagrams, for example, are most simply drawn in the 'boxes and arrows' form shown in Figure 7.3. These imply that material (or information) flows along the lines, in the direction of the arrows, from one box to another. Since the amount leaving one box and arriving at the other depends upon the **rate** of the flow, this is commonly indicated as well. This has the advantage that the main factors affecting the rate can also be shown.

The drawing of such diagrams is therefore a matter of considerable importance and involves substantial skills. Final versions can be polished and made very attractive and this may best be done by someone good at such work, but the initial stages have to be done by the person whose thinking is being represented.

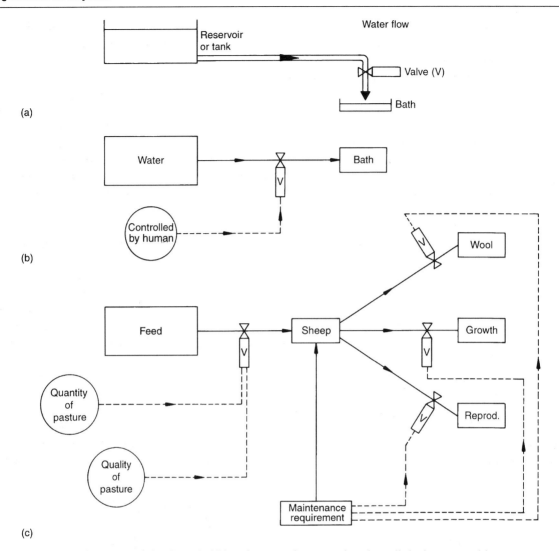

FIGURE 7.3 Flow diagrams. The flow could be of energy, for example, when all the boxes would represent quantities of energy in different forms. It is always worth considering whether such a diagram could be improved: this immediately demonstrates how diagrams help in thinking about how the system really works.

The best common example of how useful a diagram can be is the London Underground map. Imagine trying to express all that information – about how to get from any one station to any other – in any other way. Just consider how to write it all down in words!

It is a salutary challenge to try and represent in a diagram one's knowledge about something for a particular purpose. The last part is important be- cause it turns out to be virtually impossible to produce a picture of anything that includes all one's knowledge about it. If we know next to nothing about a gecko, for example, we cannot produce a picture at all. But if we take a familiar animal, such as a cow, superficially there appears to be no problem in producing a picture. Figure 7.4 is an illustration of the most common kind of British cow and it is clearly recognizable because we already know about such cows and the picture contains enough information to identify the animal. A moment's thought, however, reveals that it is a picture of one side of a stationary cow with its musculature exposed. Yet we know much more than this. We know it has other sides and that a great deal goes on inside; we know that it eats and breathes and moves about and we know that it reproduces and gives milk. None of these features appears in the picture and it is quite impossible to envisage a picture that did contain all that we know.

For any particular purpose, however, it is much easier to imagine a picture that contains all we **need** to know and it is better done as a diagram because it is clearer and less ambiguous. There are, nevertheless, many systems or organisms about which we know (and need to know) much more, even for a restricted purpose, than can be shown in a diagram. Furthermore, we may wish to represent processes quantitatively and dynamically over time.

It is this kind of complexity that is best dealt with by mathematical modelling in which the models are entirely abstract and expressed as equations of one sort of another. Amongst the advantages of this kind of modelling is that the calculating capacity and speed of computers can be harnessed to

operate the model and to try out all kinds of changes on it. This experimentation on the model is very useful but its value depends, of course, on the validity of the model itself.

There are two ways of assessing the value and usefulness of a model but it is important to make this assessment in relation to the purpose for which it was constructed. Obviously, if the purpose of building a model was to determine gaps in the information available, then it is no use complaining about the model's predictive powers. The two kinds of assessment both relate to the purpose, therefore; first, related to whether or how well the model achieves its purpose and, secondly, to the 'correctness' or 'accuracy' of the way in which it does so. If a model is being used to simulate a real situation and it always behaves in the same way as the real world, in the sense that it gives the same answer or responds in the same way to changes in inputs or conditions, then the model is clearly useful for this purpose. However, it may do this for the wrong reasons, because things are correlated or linked in some way.

For example, a model could be constructed that associated crop growth with the application of fertilizer nitrogen when, in fact, the real mechanism might involve a deficiency in some minor element

FIGURE 7.4 A cow (see text).

that was supplied as an impurity in the fertilizer. As long as the same conditions were obtained the same results would follow from applying fertilizer and the model would give correct predictions. So the model would be 'valid' for this purpose but it could not be 'verified' because the mechanisms assumed to operate would be found to be irrelevant and the assumptions false.

Models, of course, can relate to whole systems, or to parts of systems that are then regarded as systems in their own right, or as subsystems.

7.4   Subsystems (see Box 7.5)

It is highly desirable to use 'subsystems' to mean something distinguishable (1) from systems and (2) from components of systems. As with the word 'systems' itself, we should be able to use the definition of 'subsystem' to distinguish the things that **are** subsystems from the things that are not.

The main feature turns out to be the degree of independence. For example, if we look at a dairy farm as a system, we have no difficulty in recognizing that each cow is a component and could be taken out and viewed as a separate system. Thus, every system could, theoretically, be a component of another, larger system. It is also clear, however, that many studies could be made of one cow without necessarily learning anything about the dairy farm system. Studies could, for example, be concerned with the cow's reactions to climatic conditions or to feeds that it would never encounter in the dairy farm system considered. Any dynamic, practical view of the cow would need to take a food supply into account, however, and would have to recognize certain outputs (see Figure 7.5). Why not all inputs and outputs? Because they are only required if our purpose demands them and the practical view of a cow inevitably involves feeding it. But if it is a grazing cow, we cannot ignore the fact that its faeces may influence how much grass grows and how much of this grass the cow will eat. So we cannot relate our individual cow to the dairy farm system without enlarging our view at least to include these effects (see Figure 7.6). The question is,

Box 7.5   Subsystems

Box 7.2 referred to subsystems of the human (or animal) body to illustrate that they have many of the properties of systems but are not independent – they cannot function in isolation. They can exist, of course, as a skeleton may but it cannot even stand without support. This is curious, in the sense that the skeleton is the support for the body, yet it needs the musculature to keep it all in place. Similarly, the muscles would collapse in a heap without the skeleton.

It is these numerous and intimate connections that characterize subsystems. After all, it could be argued that the human body cannot function in isolation – it requires food and oxygen, at least. But not a precise piece of food or molecule of oxygen: so the body can be transferred to other climates and food sources – as it does on holiday. The subsystems of a human body illustrate all these features extremely well.

The example of a tree is instructive. Looked at physiologically, the root system is hardly a subsystem: it might be better to describe the whole water transport system (roots + xylem + stomata, etc.). However, if we are looking at a tree from a structural point of view, the roots can represent a support subsystem, designed to resist tensions and pressures (often of great magnitude) from all directions.

So, the way we describe systems and subsystems has to be related to the purposes for which we are doing it – an argument strongly made for modelling in the text.

how far does one have to go in this expansion and elaboration?

It is clearly extremely difficult to answer such a question without a picture of the whole system and it seems likely that subsystems have to be extracted from systems; they cannot be built up independently. We then have two closely linked problems: how to describe systems and how to extract subsystems from them (see Spedding, 1988).

7.5   The testing of models

The usefulness of models depends upon their being relevant to the needs of the user and this can only be

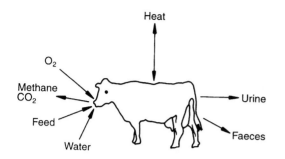

FIGURE 7.5 Minimum inputs and outputs in cow nutrition.

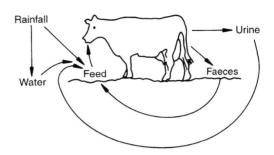

FIGURE 7.6 Minimum inputs and outputs for a grazing cow.

so if they are 'valid'. **Validation** is simply concerned with establishing whether a model behaves sufficiently similarly to the real system being modelled.

The model represents a hypothesis that, for certain purposes, the real system can be represented in that way and validation is a test of that hypothesis. It is only necessary, therefore, to demonstrate that the model gives the same output (or whatever is being assessed) as the real system, over the same range of variables, controllable and uncontrollable.

However, two conditions need to be satisfied: (1) that the model is tested against data not used in its construction; and (2) that what is meant by 'the same' output is specified in advance. In other words, the precision and accuracy required of the model must be specified in advance, recognizing that the **real** system performance may vary considerably.

Until a model has been validated, it cannot be used with any great confidence, although it can be used to generate other hypotheses based on the assumption that it **is** valid.

If the validation procedure shows the model to be unsatisfactory, it is necessary to have measured many aspects of its performance in order to pinpoint where it is wrong. This means assessing separate processes within the model and is similar to verification.

Verification is the process of establishing the truth or accuracy of the processes and interactions on which the model is based and which are represented by equations in the model. Such testing

usually requires the same kind of experimental procedures as are used in science.

7.6 The practical value of a systems approach

So far, the value of a systems approach has been implied on largely theoretical grounds and the case may sound logically convincing (as I believe it is, but this should not be accepted uncritically, of course). However, the approach has to be justified on grounds of practical relevance as well as on the basis that it helps us to think clearly about the subject.

Let us, then, consider how agriculture may be improved. First of all, someone has to have an idea that they think or believe would result in practical improvement. If they have no evidence to support the idea, it is still possible to explore its likely effect, provided that we have a good enough picture of the agricultural system or component that we wish to change. Nevertheless, this is still a theoretical exercise.

Suppose, then, that this idea is backed up by observations from practice or experiment. The difference is that experiments are carried out under more controlled conditions and their results are somewhat easier to interpret. In practice, it is difficult to connect an outcome with any particular thing that was done because a lot of other things probably changed as well.

Thus, an observation that the number of lambs

born per ewe was higher after feeding extra minerals to the ram probably does not indicate a causal relationship. It might also have been a very wet summer, or less fertilizer might have been applied, or the stocking rate might have been less, or more rams might have been used: previous knowledge would suggest that the ewes were probably in better condition when they were mated and it is usually only in experiments that such a factor can be isolated.

However, since the experiment will look different from the farm situation, one is left wondering whether the results will apply to the latter. In other words, someone has to say what agricultural situations (or systems) it is thought that the results **will** apply to and to argue that the experimental situation represented a sufficiently similar system to make this likely.

The issue becomes even clearer if we consider how we would copy an improved system. If we are shown a better (e.g. more profitable) system, whether in practice or at a research centre, and wish to adopt it, we have to know which bits to copy. We need to know whether we have to have the same cattle that were used (or just the same breed, age, weight and so on), whether the particular stocking rate, pasture species, fertilizer input, design of gates, method of ear-tagging, were all essential, whether the slope of the ground or the tree in the corner of the field were vital or trivial, whether we need to have the very person who operated the system successfully, and answers to a vast number of other such questions.

Clearly, we hope that the operator (or designer) will be able to tell us what the essential features are, so that we may copy only those. They may not be able to be sure about this, of course, but may say that they have repeated their findings in several years and on several different fields, that they have used different cowpersons and different cows. All this adds confidence, chiefly because they have developed a clear picture of the essentials (this is essentially the same thing as the models mentioned earlier).

But can we be sure that they are right? One other feature that should add greatly to our confidence is if they can also say: I have put all these essentials together in a mathematical model, given them the relevant values (for stocking rate, fertilizer application, milk yield, etc.) and calculated the results for different weather conditions, and the answers come out near enough to those I actually get when I try it in practice. This tells us that the results of combining what they have identified as the essentials do appear to agree with results actually achieved in practice. Such calculations about complicated systems would not be possible without modelling techniques and the use of computers, which is one of the reasons why it is now possible to use a different approach. Two things of great importance to practical improvement flow from a systems approach, therefore. The first is the recognition that it is necessary to identify and describe the system studied and those to which the improvement will apply. The second is that it is possible to make calculations about the outcome as a basis for confidence in the result of the proposed improvement, and not only in one year but over a series of different years with varying weather and changing costs and prices.

Now, of course, this kind of thinking could be attempted by a specialist in any one discipline, but it would not relate to the whole system and it is this that characterizes the systems approach. Consider the following example, taken from an attempt to improve livestock production – in this case, to produce a 'better' cow.

7.7 A 'better' cow

An animal production specialist may look at a rather primitive milk production system in Africa and, viewing the single small Zebu cow producing a very low milk yield, may understandably consider that surely a 'better' cow could be provided.

This quite natural thought has often been the starting point for a selection programme, a crossbreeding programme or the importation of exotic breeds, usually resulting in a cow that was both genetically capable of a higher milk yield and also

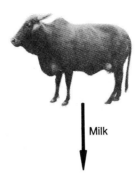

Milk

FIGURE 7.7 An unimproved zebu cow producing a poor milk yield.

larger in size. This immediately suggests that any increase in milk output is likely to involve an increase in the feed supply – yet the latter is not necessarily considered in detail by the animal scientist, particularly if the feed comes from other parts of the farm or from by-products of crop production.

Thus, an animal scientist may try to improve an agricultural system by 'improving' the animal component, without realizing that the role of the animal may not just be, say, milk production. The animal may also carry out the cultivations for crop production and produce manure for fertilizer or fuel. In addition, the animal has to live and produce or perform in a climatic environment to which it is closely adapted – in ways that an 'improved' animal may not be. It may also be adapted to the local pests, parasites and water supply and, above all, be able to live on the available feed supply, much of which may be derived from crop by-products.

Similarly, a crop specialist might be tempted to 'improve' the crop component in directions that maximize the proportion of human food and reduce the proportion of by-products on which the animals are fed (or which are used for fuel, construction purposes, etc.).

All this is illustrated in the sequence shown in Figures 7.7–7.10, for the substitution of a 'better' cow in a small-scale, milk production system. The 'better' cow may be genetically capable of greater milk production but, under the conditions of feed supply and environment in which it has to perform it may actually produce less milk. This is quite possible if, as is likely, the animal is larger, with a greater maintenance requirement, and has to live on the same quantity and quality of feed. If the feed supply also has to be improved, it is clear that animal science is not the only discipline involved.

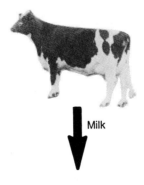

Milk

FIGURE 7.8 A 'better' cow (Friesian), genetically capable of a much higher yield.

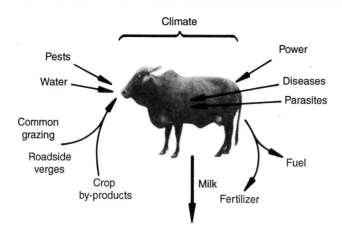

FIGURE 7.9 An indication of the environment within which the zebu cow actually produced its yield.

Equally, if the main effort is directed to improvement of the feed supply, it cannot be achieved solely by crop scientists, if only because the cultivations may depend upon available animal power, which is, in turn, influenced by the number, size and strength of the livestock.

It should also be clear that crop production has to achieve a balance between feeding livestock, feeding the farmer and his family, and selling surpluses to generate income. With no income, there can be no inputs and this imposes severe limitations on the viability of a farm or a family. Yet it must be obvious that the balance to be achieved will depend upon the size and structure of the family, of the animals kept and on the markets for the sale of produce.

Such complex interactions are characteristic of agriculture and are most marked in the agricultural systems of developing countries. Indeed, the interactions frequently go well beyond the agricultural system and involve water and fuel collection, competing for time and labour. This serves to emphas-

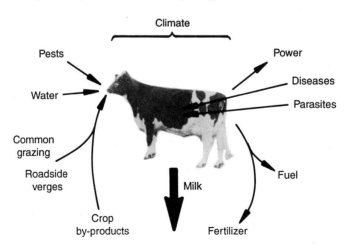

FIGURE 7.10 The 'better' cow in the same environment as that of Figure 7.9 with an indication of the milk yield it may be capable of if it survives.

ize the need for a multi-disciplinary approach and this can rarely be adequately embodied in any one individual.

So far we have been dealing with the systems approach in principle. In practice, it has been applied in several different forms by different groups of people. Since this has created some confusion, the essence of the main forms is given here.

7.8 Farming systems research (FSR)

This has been variously considered as being the same as 'a systems approach' or in opposition to it! It has further been divided into at least three subcategories. The following descriptions are taken from a review by Simmonds (1985).

7.8.1 FSR *sensu stricto*

FSR *sensu stricto* is the study of farming systems *per se*, as they exist; typically, the analysis goes deep (technically and socio-economically) and the object is academic or scholarly rather than practical; the view taken is nominally 'holistic' and numerical system modelling is a fairly natural outcome if a holistic approach is claimed.

7.8.2 New farming systems development (NFSD)

NFSD takes as its starting point the view that many tropical farming systems are already so stressed that radical restructuring rather than stepwise change is necessary; the invention, testing and exploitation of new systems is therefore the object.

7.8.3 On-farm research with farming systems perspective (OFR/FSP)

OFR/FSP is a practical adjunct to agricultural research which starts from the precept that only 'farmer experience' can reveal to the researcher what farmers really need; typically, the OFR/FSP process isolates a subsystem of the whole farm, studies it in just sufficient depth (no more) to gain

the necessary FSP and proceeds as quickly as possible to experiments on-farm, with farmers' collaboration; there is an implicit assumption that stepwise change in an economically favourable direction is possible and worth seeking.

7.9 Other views of a systems approach

There are, of course, many different views as to the way in which systems thinking should be applied. One important difference, derived from the work of Peter Checkland (1981) and Checkland and Scholes (1990), is that between 'hard' and 'soft' systems.

7.9.1 'Hard' systems

'Hard' systems are those involving industrial plants characterized by easy to define objectives, clearly defined decision-taking procedures and quantitative measures of performance.

Such systems tend to the mechanical, although biological systems are often of this kind (e.g. those represented by a single individual). Highly developed agricultural systems (e.g. battery hens) are also at this end of the range.

The more intimately people are involved as part of a system, however, the less appropriate this view becomes.

7.9.2 'Soft' systems

'Soft' systems are, by contrast, those in which objectives are hard to define, decision-taking is uncertain, measures of performance are at best qualitative and human behaviour is irrational.

In 'hard' systems, it is possible to focus on problems and endeavour to find solutions. In 'soft' systems, matters are rarely, if ever, quite so straightforward. In both kinds of system, however, it is still the case that 'improvement' is sought.

The distinction can readily be appreciated by focusing on extreme examples. A bicycle is a mechanical system and, if it suffers from a puncture, it is quite possible to seek a 'solution' by repairing the puncture. The problems of starvation in Ethiopia, on the other hand, had many dimensions (biological, social, military, climatic, etc.) and it is naïve to seek a 'solution'. At the same time, there are obligations to help and this will turn out to mean trying to make changes that will move matters in a better direction, such that the result can be regarded as an improvement.

Of course, it is important not to oversimplify the question of improvement: what is an improvement for one person, may not be so for their neighbour. That is why two questions have to be posed at the beginning of any attempt to apply a systems approach. These are:

1. what is the system to be improved?
2. what constitutes an improvement?

Improvement cannot be sought by any method until these two questions have been answered and neither is simple. Indeed, both pose great difficulty and, in some circumstances, they cannot be answered to everybody's satisfaction.

The first requires a description of the system to be improved, in terms of its essential components, interactions and processes, boundary, inputs and outputs and, preferably, in such a way that possible improvements can be examined theoretically. In other words, a model is required, upon which experiments can be carried out. This model should be as simple as will serve the purpose and the latter is to help in determining potential improvements.

Until the second question has also been answered, therefore, it is very difficult to build the model. The only criteria available for deciding what must be included in a model and what should be left out are the objectives and purpose of the model and of the activity or system being modelled.

If, for example, increased profit is the objective of improvement, then the model must be built in economic terms and contain all the factors relevant to the formation of profit.

It is often assumed that profit is the most likely objective and, indeed, that the aim is usually to maximize it. But in many agricultural systems, especially in developing countries, profit is not the most important objective. Stability of output or profit over a long period may be more important; or the objective may not even be expressed in monetary terms at all.

Very rarely, in fact, does anyone wish to maximize profit, except within a great many constraints, even where an increase in profit is sought. Peace of mind, security, reduction in drudgery and the satisfaction of current need may all count for a great deal. In general, an improvement in efficiency is sought, but this can be expressed in many different ways (see Chapter 9).

In many circumstances, as with most other human aims, the objectives may be neither simple nor single and they may not remain constant over time. Difficult as it may be to agree on objectives (which may differ markedly between producer, employee, consumer and the nation), it is essential to examine the issue carefully before embarking on an improvement programme.

In general, improvement is achieved by one of three main methods:

1. advice on component changes;
2. adoption of innovation; or
3. copying.

(a) Advice

Many producers rely upon advice as a basis for action designed to improve their crop or animal production systems. The sources of advice are numerous, some associated with government services, some with commercial concerns and some paid for from independent consultants.

It must be clear from what was said earlier that successful advisers must be able to relate their advice, which, of necessity, will usually concern only part of a system, to the functioning of the whole system. Otherwise, producers have to depend upon their own judgement and knowledge of the system they are operating in order to decide whether the advice is sound for their system, or which parts of it are relevant. In either case, a picture of the whole system and the way it operates has to exist, if only in the mind of the farmer or of the adviser.

Advice need not be concerned with new knowledge or practices and, in most circumstances, is based on ideas that have already worked in practice in other systems. Part of the judgement then required is whether the system now considered is sufficiently similar to those in which the idea has previously been applied successfully.

(b) Innovation

Improvement can also come from innovation, in the sense that new ideas are involved that have not been tested in practice before, although they may have emerged from research and development studies.

The pioneer farmers usually engaged in this kind of experiment tend to be interested in trying out new ideas – sometimes in originating them – and can often afford to take the risk of an innovation not working. Indeed, such farmers may be able to benefit by being in the van of progress when innovation is successful and thus able to survive the occasional setback.

In extreme cases, such trial and error can be carried out without any kind of systems approach, although it is hard to imagine successful farmers who do not have some picture of the system they are operating.

It is certainly important to realize that the picture or model required may be quite different (in content, detail, level of sophistication) according to whether the purpose is to operate, repair or invent new systems.

(c) Copying

Very often, practical improvement is achieved by copying what is done by others, whether they are other farmers or research workers.

The main problems of copying are of three kinds. First, there is a matter of confidence that the system to be copied is better than that already being operated. This has to be judged by the results that matter to the prospective copier and by their confidence that they have available all the information needed (and that it is reliable). Judgement may also require evidence over a period of years, involving a range of weather conditions (as discussed earlier).

Second, there is a matter of relevance: there has to be good reason to suppose that the system to be copied is relevant to the climate, aspect, topography, soil type and so on, of the copier's farm.

Third, there is the question of exactly what to copy. Part of the difficulty is that some things cannot be copied. It is not possible to have the identical animals, the identical pasture or the identical staff; and all of these may be important and any one of them vital to the success of the system to be copied. Another part of the difficulty is to determine which of the things that could be copied are important to the success of the system and which are not.

Related to this is the question of how similar the things that have to be copied have to be in order to ensure success. In animal production the similarity may relate to breed or strain of animal, level of feeding or fertilizer use, type of fencing used, availability of water, timing of mating, health care and veterinary treatment. Any one of these may matter greatly and success may depend upon getting them all right.

In general, one has to rely on the operator (of the system to be copied) to distinguish between what matters and what does not and to answer all the detailed queries. But, unfortunately, the operator of a successful system may not actually know these answers, even though they may think they do. Nor are the answers easily determined, even at a research institute.

One way of gaining confidence that the vital elements have been identified is to construct a model, as mentioned earlier in this chapter.

7.10 A model of agriculture

There is no way of producing an absolutely complete picture of agriculture or even a complete model. This is partly because the flows of material involved cannot be represented in all terms simultaneously. We have to decide whether it is the flow of money, or energy, or nitrogen or whatever, that we are most interested in. This illustrates that a model is always constructed for a purpose and that it is that purpose which tells you what you must include in a model, what you can leave out and what terms to express flows in.

For example, a very simple model (Figure 7.11) can be constructed to show that an agricultural system has a boundary within which a biological subsystem transforms resources into products. But the real system is vastly more complicated and any useful model would have to show flows of material (expressed in appropriate terms) and the factors that affect these flows. Figure 7.3 illustrated one conventional way of using symbols to do this. The complexity that accompanies any attempt to include flows of different kinds is indicated in Figure 7.12.

However, none of this illustrates the point that agriculture can no longer be visualized in isolation from its interactions with the environment and society within which it operates. Figure 7.13 simply foreshadows what would be needed. Within this kind of framework agricultural systems take an enormous variety of forms; the main types are briefly described below.

7.11 Crop production systems

These are based on photosynthesis (see Chapter 5) using solar radiation (or, more accurately, the c. 45% of it in the appropriate wavelength), to produce crop products.

The world's most important crops (wheat, rice, maize, pulses, roots and tubers, sugar and cotton) are grown in relatively simple systems, mainly planted and harvested annually.

However, there are important tree crops and other perennials, and there are grassland systems (included under 'animal production systems' below) and fuel crops (dealt with in Chapter 8). The main UK crop production systems are listed in Table 7.1.

In addition, many farmers operate 'mixed' farming systems, in which crop production and animal production are not only both present but may be closely integrated.

7.12 Animal production systems

Typical examples for the UK are given in Table 7.2. Descriptions of any of these systems would need to

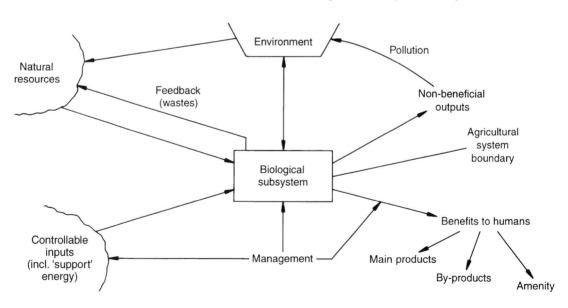

FIGURE 7.11 The place of a biological subsystem within agriculture.

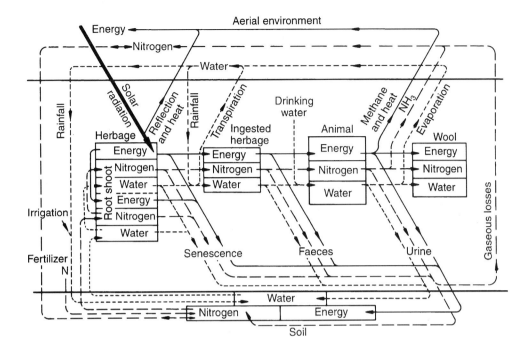

FIGURE 7.12 The flow of energy, nitrogen and water in wool production (Spedding, 1973).

TABLE 7.1 UK crop production systems

Cereals (wheat, barley, oats)
Pulses (beans, peas)
Oil-seed crops (rape, linseed)
Field vegetables (cabbage, onions, lettuce)
Fruit (top fruit, e.g. apples and pears; soft fruit, e.g.
   strawberries)
Roots and tubers (sugar beet, potatoes)
Hops (for beer making)
Protected crops (tomatoes, cucumbers, peppers)
Flowers (bulbs, pot plants, cut flowers)
Nursery stock (roses, trees, shrubs)
Forage crops (kale, rape, turnips)
Grassland
Fuel crops

detail the breeds used, how the animals are managed, the timing of events during the year, the equipment used, the disease prevention and control programme, methods of mating, feeding, rearing of young, stocking rates (if grazing) or space allocation (if housed), and a great many other attributes, all of which could differ from one kind of animal

to another but also within an enterprise. For example, there are a number of quite different dairy systems or beef systems.

It is obvious that one section of one chapter cannot mention even the main animal production systems of the world, never mind describe them. Let us take just as one example to illustrate the complexity, a sheep grazing system, since grazing systems are amongst the most complex, depending, as they do, on a close integration of both pasture and animal growth and reproduction.

7.12.1 A sheep grazing system

Sheep are kept mainly for meat and wool and breeds may be highly specialized for the production of one or the other.

If one wished to describe only the flows of energy, nitrogen and water in a wool production system, a diagram such as that shown in Figure 7.12 would suffice. These flows fit in to the whole wool production system as shown in Figure 7.13.

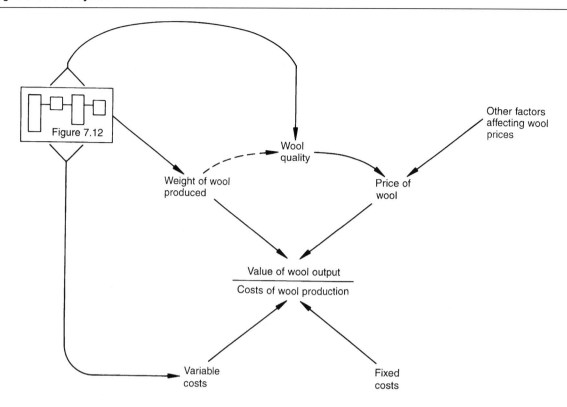

FIGURE 7.13 The relationship between Figure 7.12 and the agricultural system of wool production (Spedding, 1973).

TABLE 7.2   UK animal production systems

Dairying (milk and milk products)
Beef (suckler cows, calves from dairy herds)
Sheep (lamb, wool)
Pigs (meat: outdoor, indoor)
Poultry (eggs, meat: hens, ducks, turkeys, geese)
Deer, rabbits, fur animals
Horses (breeding)
Fish (freshwater, coastal)
Goats (dairy, meat)
Game (pheasant, partridge)

But these diagrams reveal little about one of the main management problems in grazing systems: how to equate the needs of the animals with the supply of herbage. Neither are simple – they often cannot even be described by straight lines.

Figure 7.14 shows the changing requirement of cattle (as they grow bigger) and ewes and lambs (following birth of the lambs in about March and their slaughter in late summer), all to be fed on the herbage which, in the UK, grows most rapidly in the spring and very little in the winter.

Clearly, grazing alone cannot solve this equation: one common solution is to make hay or silage in the spring when herbage is surplus to requirements and to feed it back in the winter. In fact, the presence of the animals recycles nitrogen in faeces and urine and may affect grass growth by defoliation and trampling.

Just as importantly, grazing animals pick up parasitic larvae with the grass they eat, which develop into adulthood within the animal, causing serious losses. The complexity for just one sheep parasite, and there are many, is shown in Figure 7.15.

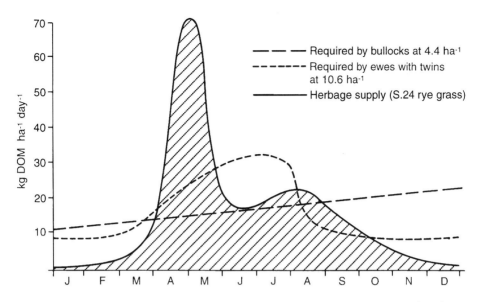

FIGURE 7.14 Balanced total supply and demand (5800 kg of digestible organic matter ha^{-1} yr^{-1}) (from Spedding, 1971).

It is clear that many pictures are always needed to describe a system. This is, in fact, everyday experience. Consider a horse. Can you imagine a picture of a whole horse? One side of the outside of a horse, certainly: but think of the number of possible outside views (from underneath?) and then contemplate describing the inside (alimentary tract, respiratory system, muscles, fat deposits, skeleton, etc.).

Even a sheep grazing system will be totally different in, for example, the tropics.

7.13 Tropical farm systems

Grigg (1974) included them in a world overview, Norman (1979) described their annual cropping systems and Ruthenberg (1976) covered the whole range of farming systems in the tropics.

There is general agreement on the main systems of the world and those found in the tropics (marked with an asterisk):

* shifting agriculture;
* wet-rice cultivation in Asia;
* pastoral nomadism;
  Mediterranean agriculture;
* plantations;
* ranching;
* dairying;
  mixed farming in Western Europe and North America;
  large-scale grain production;
* rain-fed annual cropping;
* irrigated annual cropping;
* fallow;
* ley systems;
* permanent upland cultivation;
* perennial cropping;
* mixed annual and perennial crops.

It can be seen that there are overlaps and mixtures and several systems occur at different intensities. For example, a category of semi-nomadism can be inserted between full nomadism and pastoralism. It is also clear that these 'names' of systems contain

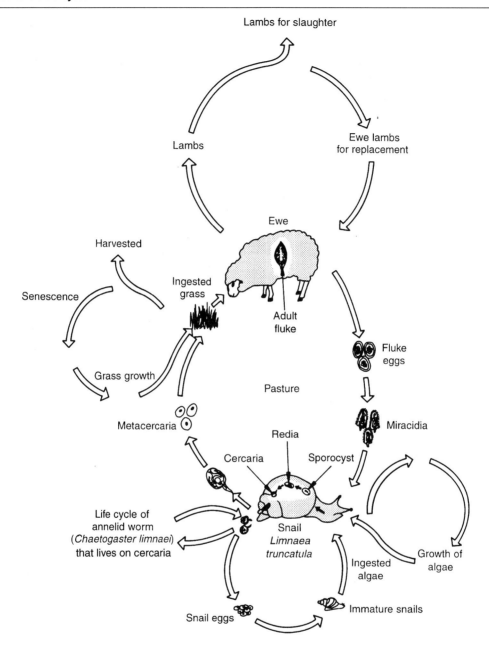

FIGURE 7.15 Life cycle of the liver fluke. Note problems of conveying: (1) scale (sheep are *c*. 1 m × the size of snails, which are about the same size as the adult fluke); (2) time (the rates of different processes differ widely and they vary with temperature, moisture, plane of nutrition and other factors: even seasons are not indicated: (3) number (the number of eggs, snails and sheep differ greatly and it is not clear whether the diagram is on a per sheep basis or per hectare per year: losses at all stages are unclear); (4) space (the areas or volumes occupied cannot be conveyed); (5) detail (there is a limit to the detail that can be included, either of important factors or physical feature: the latter means that the precise point of entry of miracidium to snail cannot be given – yet it matters a great deal).

little **description** but could be arranged in a hierarchical **classification** (see Box 7.6).

To the extent that some of these systems are virtually self-contained, they may be regarded as land-use systems. But the overwhelming number of people in the world are supported by agricultural systems that interact with the environment and society within which they operate.

The role of energy and water as the driving forces of all such systems is discussed in the next chapter.

Box 7.6 The classification of agricultural systems

There are many different ways of classifying things. Books, for example, may be classified by colour (which seems rather pointless), by size (not so silly if shelves are of different heights), by subject, author and so on. What governs the choice is the purpose for which it is being done: in the case of books, it is commonly to ease retrieval. Animals and plants tend to be classified on the basis of relationships based on evolution, but there are other useful ways.

An example of a world farming systems classification is shown in figure (a), aimed primarily at helping one to visualize the world picture.

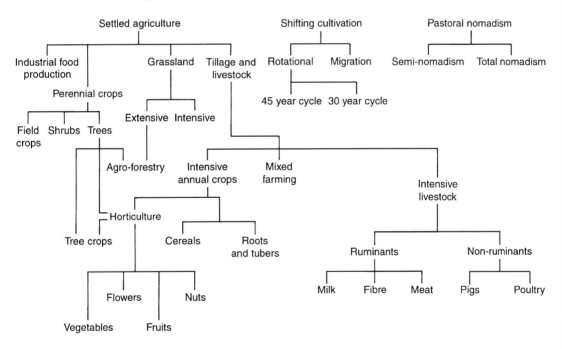

(a) Example of a classification of world farming systems

Basing the structure on existing, identifiable (even named) systems avoids duplication, which otherwise can hardly have been avoided. For example, the structure shown in figure (b) risks having to repeat subdivisions such as 'temperature' or 'tropical' many times, but it also shows how any part can be refined or given in greater detail.

Box 7.6 (cont.)

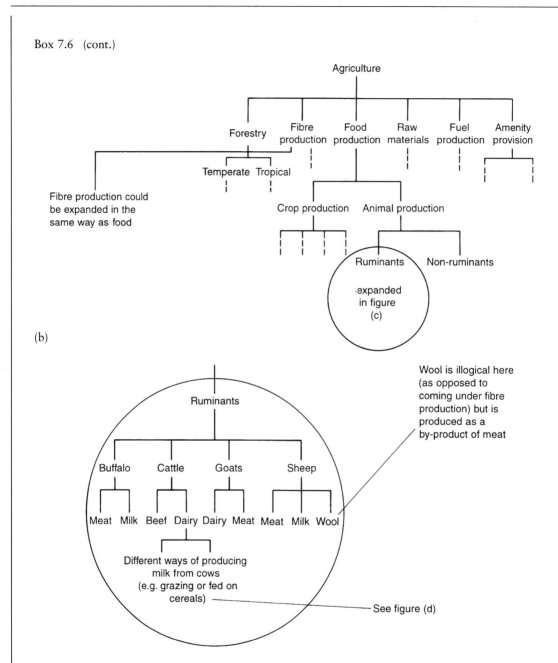

(c) The place of the dairy cow in world agriculture.

Such hierarchical diagrams usually finish up by identifying, at the base, the actual systems that exist (or could exist) and their descriptions can be read upwards through the diagram.

Box 7.6 (cont.)

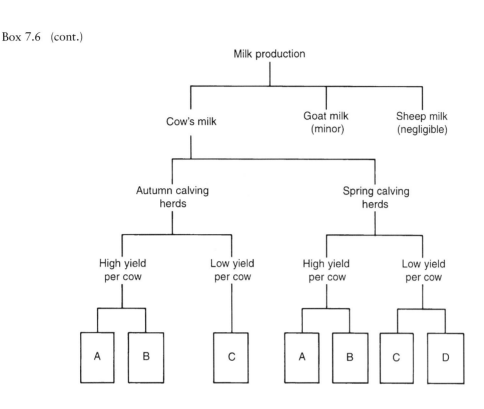

(d) A classification of milk production systems in the UK. A, intensive concentrate feeding; B, intensive forage with high yield per cow; C, intensive forage with high yield per hectare; D, extensive forage.

A different kind of example is given below for mixtures of agriculture and forestry.

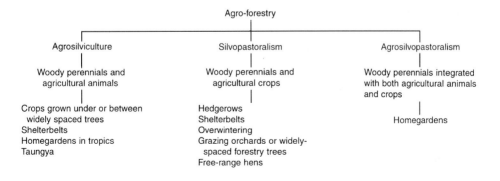

(e) Classification of agro-forestry systems.

Questions

1. Are the farmers part of the agricultural system they operate? What about their spouses and their children?
2. What do you mean by a picture of yourself? How much does it have to show? (one ear?).
3. Is it only production systems that have purposes?
4. What is the purpose of a squirrel?
5. Does a 'picture' have to be **seen**? In the mind? Does a bat see the picture formed by its 'radar'? Do birds see worms in the mud beneath them? Can touch produce a picture?

Energy and water in agriculture

> If you do not change direction, you will finish up where you are going.
>
> *Chinese proverb*

Both energy and water are required in agriculture, usually on a massive scale. All biological processes – all life – depend upon adequate supplies of both: this is equally true of all the essential nutrients, of course, but not on the same scale. It is convenient to treat them in the same chapter largely because the water cycles depend upon energy and, without water, energy supply will produce nothing.

8.1 Energy use in agriculture

8.1.1 Solar radiation

The biggest source of energy is the sun and agriculture is the biggest user of solar radiation, except where forestry occupies a greater area of land.

As described in Chapter 5, plants use solar radiation to fix carbon in the process of photosynthesis. However, not all of the solar radiation can be used in this process, only *c.* 45% of the energy received being in the photosynthetically useful range of 0.4–0.7 µm. In fact, very little of the total light received by plants is ever used and about a quarter of the visible light is reflected. Of that absorbed, most is converted to longwave radiation and remitted or used in the evaporation of water

in transpiration. On average, *c.* 22 000 kW of energy are received per hectare each day but less than 2% of this is generally used for photosynthesis and for heating and cooling the soil.

Most leaves become light saturated at 25–30% of full noon daylight in midsummer (UK) and additional light makes no further contribution to photosynthesis.

Thus, the full exploitation of light requires a more complex receiving surface than a single horizontal layer and plants exhibit a great variety of ways of arranging their leaves to best advantage.

Since the angle of the sun also varies, there is some merit in near-vertical leaves (as in many grasses) and in leaves which move to orient themselves to the changing (relative) position of the sun.

The area of leaf per unit area of land is called the leaf area index (commonly denoted by L) and, under British conditions, most of the light will be intercepted at values between 4 and 7.

Since plants and leaves vary in their photosynthetic efficiency, there is scope for research aimed at improving it. All crop production for human consumption and the production of food for animals is (virtually) entirely dependent on photosynthesis, directly or indirectly.

As the total amount of solar radiation received

in a year is about the same for all points on the earth's surface (though cloud and dust may interfere with this), the rate of production is usually limited by other factors, such as temperature, water supply, soil fertility, and incidence of pests and diseases. These can be manipulated by the farmer to a varying degree, often involving the use of additional support energy.

8.1.2 Support energy

As described above, agriculture is the biggest user of solar radiation, except where forestry occupies a greater area of land. But, in developed countries agriculture also uses substantial amounts of 'support' energy (usually fossil fuels), mainly in the production and use of fertilizers and machinery (Table 8.1). Even so, it uses a relatively small proportion of the national total (Table 8.2) up to the farm gate: the proportion used over the whole food chain may be much larger, however (Table 8.3).

The situation is complex and the use of fertilizer, whilst decreasing the efficiency of support energy use, may increase the efficiency with which solar energy is used, simply by greatly increasing output per unit area of land (see Chapter 9). Where nitrogen is fixed by plants, as in legumes, there is also a substantial energy cost, a minimum of 6 g CH_2O being required per g N_2 fixed (Hardy, Heytler and Rainbird, 1983).

In considering support energy, largely based on fossil fuels (themselves derived from past solar radiation), it is important to recognize that, although all energy can be expressed in the same terms [usually joules (J) or MJ], not all forms of energy can be converted from one form to another.

For example, fossil fuel energy cannot usually be converted into food energy, although it can influence the rate of food production. Thus, fertilizer may promote greater crop growth and thence food production, but the energy in or used to manufacture the fertilizer does not appear in the product. That is fixed solar energy.

Consequently (see Chapter 9), efficiency expressed as energy produced in the food per unit of support energy used is not an **energetic** efficiency, since the two sorts are not interchangeable.

What such a ratio expresses is how much food energy (from sunlight) is obtained per unit of support energy employed in the process. That is why it is called **support** energy: it supports and facilitates production but it is not part of the energy production conversion process. The ratio is similar to that which could be calculated for the energy produced per unit of water applied.

The inputs of support energy vary with the product: examples are shown in Table 8.4, but bear in mind that the energy output (in the product) is not derived from the support energy used. The latter is in addition to the solar radiation used, which is itself used with very low efficiency (see

TABLE 8.1 Energy usage (MJ ha^{-1}) in the production of winter wheat (after Spedding and Walsingham, 1975)

	Energy usage (MJ ha^{-1})
Manufacture of fertilizer	
N	7987
P	747
K	394
Herbicides	155
Tractor fuel	1240
Labour	21

TABLE 8.2 Use of support energy in agriculture (based on Spedding, 1989)

Agriculture	% of total used in the country
UK (1973)	3.9
USA (1980)	3.0
Europe (1980)	3.4
Australia (1975)	2.0

TABLE 8.3 Support energy use in the food chain as % of national (1981) total (based on Wilson, 1992)

Energy use in the whole food chain	% of total
UK	15.8
Australia	15.0
USA	12.0–15.0

Table 8.5). It will be noted that only the 'usable' solar radiation has been put into these calculations: when interpreting the efficiency of solar energy use, it is necessary to know whether it is based on 'total' or 'usable' radiation.

Support energy is now recognized as being of enormous importance because it is non-renewable, limited in quantity and likely to rise in price. If fresh supplies are discovered it only alters the time-scale and, because the cost of extraction tends to increase disproportionately (since new sources are generally in less accessible sites), may not greatly affect the rate of price increase. Since obvious alternatives are difficult to envisage on the necessary scale, it follows that the efficiency with which such a resource is used must be carefully considered.

There are some other sources of energy used in agriculture (such as hydro- or wind power) but the most important are animal power and human labour.

TABLE 8.4 Relative efficiency (E) of crop and animal production (from Spedding and Walsingham, 1975)

	E[a]
Wheat	2.2–4.6
Maize	2.8
Peas	3.2
Potatoes	1.1–3.5
Milk	0.62
Beef	0.18
Eggs	0.16

[a] E = gross energy in product/support energy input, up to farm gate.

TABLE 8.5 Proportion of energy harvested – an example for meat production from sheep[a] (source: Spedding, 1992)

Energetic efficiency calculation	%
$\dfrac{\text{Gross energy in boneless carcase}}{\text{Gross energy in herbage produced}} \times 100$	2.5
$\dfrac{\text{Gross energy in herbage}}{\text{Usable incident solar radiation}} \times 100$	0.73
$\dfrac{\text{Gross energy in boneless carcase}}{\text{Usable incident solar radiation}} \times 100$	0.02

[a] Mean of an experiment over five years in South England.
Note: Basic data ha^{-1} yr^{-1}.

8.1.3 Animal power and human labour

Both are really derived from solar radiation since they get their energy from their food.

Support energy may represent machinery and fuel that displace human labour or animal power and the effect may be quite different in these two cases. Where human labour is displaced, output per unit of land may be no higher – it may even be lower. This rather depends upon how much labour is available, especially at critical times when tasks such as weeding, cultivation and harvesting have to be accomplished rapidly. Such 'timeliness of operations' (Dr John Pearce, then at Reading University, when asked what the difference was between a good and a bad farmer, used to reply 'about a fortnight'), in relation to crop physiology or the weather, may be easier to manage with machines, although individual selectivity may still benefit from human labour: this can be seen in the selection of appropriate leaves in the harvesting of tea by hand. Of course, the use of machinery will naturally lead to higher output per person, although it is hard to be sure that this is always so when the labour involved in making the machinery and supplying the fuel is all taken into account.

In the case of animal power, it seems fairly clear that net output per unit of land is lower but that output per unit of support energy is higher (Table 8.6) where animals are used instead of tractors, but it is hard to make sensible comparisons with human labour.

Animal production is always based on plant production of some kind and adds its own losses to those incurred in the initial crop phase. The conversion of crops that could be consumed directly can never avoid a decrease in biological efficiency, therefore: this is not so for economic efficiency, of course, or animal production would not be undertaken.

The real problems with human labour are the rather small areas that one person can cultivate, weed, harvest, etc. Animal power greatly increases the area that can be worked, part of which has to be used to feed the animals (see Box 8.1), but

TABLE 8.6   Energy budgets for animal power versus tractors

Proportion of food energy* produced per hectare that is required to supply:	
A tractor	*A horse*
c. 3%	*c.* 7%
Support energy required to produce and run a tractor on a cereal farm = 2.8 GJ ha^{-1} yr^{-1}	Feed energy required per horse per year = 68 GJ yr^{-1}
Gross energy produced by cereals = 96.8 GJ ha^{-1} yr^{-1}	Support energy required by the horse (harness, stable, shoeing, etc.) = *c.* 15 GJ yr^{-1}
	Gross energy produced per horse per year as cereals[b] = 1162 GJ

[a]   Obviously, 'feed energy' could not, in fact be used directly for the tractor: gross energy is used throughout the calculation.
[b]   Assuming the same cereal output as in the 'tractor' column.

tractors and associated machinery can cope with very much larger areas.

In addition to using energy, agriculture can also be used to produce it, as a product and not merely in human food or animal feed.

8.1.4   Energy production by agriculture

Energy can be obtained from agricultural wastes (straw, excreta) but it may also be a main product, as in vegetable oils, dry material for burning, wet vegetation for digestion (to produce methane) or fermentation (to produce ethanol).

There are energy costs in these production processes, particularly where distillation is needed (e.g. ethanol). This is because it costs a lot of energy to evaporate water: consider the time taken to boil water and then the much greater time needed to boil it all off.

Production per hectare can be quite high (Table 8.7) but, economically, it is generally difficult to compete with the current (low) price of oil, which (mostly) only has to be harvested, not produced.

In many developing countries, where oil cannot be afforded, fuel for cooking comes from animal dung (reducing its availability for use as a fertilizer), by burning or methane generation, or crop by-products, but mainly from wood – with devastating consequences for the natural vegetation as well as for the labour requirement (see Figure 8.1).

The possibilities of generating significant quan-

TABLE 8.7   Fuel crop production (after Slesser and Lewis, 1979)

Crop	*Short-term high yield* [*g m^{-2} day^{-1} (dry wt)*]
Tall fescue	43
Potato	23
Wheat	18
Maize	17–52
Sugar cane	31–37
Sorghum	51
Algae	24

tities of renewable energy from agriculture depend mainly on vegetable oils or total (usually above-ground) biomass. These sources of fuel energy are renewable and they reduce the need to use fossil fuels. This has a beneficial effect in reducing the total output of carbon dioxide to the atmosphere (see Chapter 10).

Apart from forestry, many crops are potentially usable as a fuel source (see Box 8.2).

8.1.5   Energy use in forestry

Forestry is normally considered as a subject quite separate from that of agriculture but is an extremely important form of land use. The concentration here on only those aspects of forestry that impinge on agriculture should not be seen as in any way underestimating its importance in producing a vast array of goods, to the protection of soils, to its role in hydrological cycles, the energy budget of the earth and biodiversity. Furthermore, it is claimed

Box 8.1 Animal power (see Starkey and Ndiamé, 1988; Starkey, 1995)

Sometimes described as the oldest form of renewable energy (Löwe, 1986), though this ignores manpower, which is also based on food based on solar radiation. Certainly, the use of animals for transport and traction is very ancient but they have been harnessed for a great variety of purposes in addition to these, mainly:

1. pulling wheeled tool-carriers for ploughing, seeding, applying fertilizer, weeding and harvesting (for a review see Starkey, 1988);
2. pulling tool-bars without wheels (much lighter);
3. raising water: a great variety of methods have been used, some using animals walking up and down a slope, others where they walk round in circles (see Starkey, 1989); in Egypt, for example, Löwe (1986) estimated that 3000–4000 animals powered water-raising systems;
4. in forestry for logging (on wheels or skids);
5. in road building, mainly using bovines, equines and camels and carts able to tip their loads easily;
6. as pack animals, carrying loads on their backs.

The main species used as sources of animal power are given below.

Species	World numbers	Attributes
Oxen (including buffaloes and cattle)	300m	Staying power (sustaining a heavy draft force over time)
Horses Mules Donkeys	80m	Speed, can work longer Slower and work for less hours
Camels		Adapted to hot, dry climates

In total, there are probably c. 400m draught animals in the world, used to cultivate some half of the land farmed in developing countries, and c. 80m animal-drawn carts.

The efficiency with which animals can operate is much influenced by the method of harnessing (Starkey, 1989) and this is not always satisfactory. The work achieved (Barwell and Ayre, 1982) varies with the species. For examples, bullocks (weighing 500–900 kg) have a speed of 0.6–0.85 m s^{-1}, whereas light horses (400–700 kg) achieve 1.0 m s^{-1}. Cows, donkeys and mules are intermediate for speed. The power developed varies from 0.25 kW for donkeys to 0.75 kW for the light horse. Animals in good condition generate a draught force of about 10% of their weight. Animals harnessed together exert greater power but less as individuals: the increase is therefore not proportionate.

In the past, countries like Britain used oxen and then horses extensively in agriculture but these were displaced by tractors. In developing countries, however, the situation is quite different, with tractors making relatively little contribution.

Power used for crop and livestock production in developing countries (1974–1976) (from Goldenberg et al., 1988)

	Total power (%)	
Human	Animals	Tractors
66	27	7

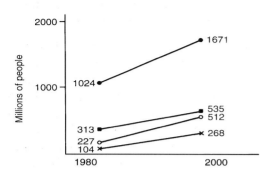

FIGURE 8.1 Number of people facing acute scarcity of firewood (after Myers, 1995). ●, East Africa; ■, Sub-Saharan Africa; O, Latin America; X, Near East and North Africa.

that well over half of all tropical deforestation is due to slash-and-burn agriculture by displaced landless peasants (Myers, 1995).

The annual increment of wood per hectare varies with the type of forest but has been estimated to be, on average, *c.* 4.7 m^3 with an energy content of 12.54 GJ m^{-3} (Goldenberg *et al.*, 1988). Biomass plantations (e.g. coppiced willow, poplar) are estimated to have a potential yield of 10–15 tonnes ha^{-1} year^{-1}, but this can be as high as 61 tonnes ha^{-1} year^{-1} for the most productive eucalyptus stands in Brazil. It has been suggested that the energy equivalent of the 1980 world oil production could be grown on 380m ha, or 10% of the world's forest area.

So, the potential contribution is very high and the economics look better than for agricultural energy crops.

Support energy use in fuel wood production is mainly for machinery, transport, chipping and fire protection. It has been estimated (Cousins, 1975) that the total support energy usage for a 10 000 ha forest (where 360 ha are clear-felled each year) is of the order of 36.52 × 10^6 MJ year^{-1}. The yield of wood energy, at 1.84 × 10^9 MJ year^{-1}, gives a support energy input of 0.02 MJ MJ^{-1} of wood energy produced. There are a great many species of shrub and tree species which can be used as firewood crops (NAS, 1980).

Box 8.2 Biofuels

The potential for biofuels in the UK has recently been reviewed by Carruthers, Miller and Vaughan (1994). There are four main sources:

Liquid biofuels	Bioethanol
	Biodiesel
Solid biofuels	Energy coppice
	Fuel crops

Liquid biofuels
Bioethanol technology is well established and such liquid fuel can be derived from the fermentation of starch and sugar crops, such as cereals, potatoes and sugar beet, or from relatively novel crops (e.g. Jerusalem artichoke). Bioethanol can replace petrol or be blended with it. Currently the process is not economically viable.

Biodiesel is also well established as a technology and also currently not viable without a subsidy. It is usually derived as RME (Rape methyl ester) from oil-seed rape crops. It is less polluting than diesel but can be used in similar ways.

Solid biofuels
Energy coppice The technology of growing, harvesting and using energy coppice is still developing and recent advances have been substantial. Poplars and willows appear to be the most promising species, yielding 8–12 tonnes DMha^{-1} yr^{-1}. When harvested, the wood has rather a high water content (*c.* 55%) and drying is therefore necessary. Probably approaching economic viability, but currently needs a subsidy.
Fuel crops The main options currently appear to be whole-crop cereals and Miscanthus. Fuel crops do not require the development of different methods of production or machinery and the harvested crops are relatively dry. Economically, similar to energy coppice and still needing a subsidy.

8.1.6 Agroforestry systems

Agroforestry systems are mixtures of agricultural crops and animals with trees grown for fuel, fertilizer (e.g. using the leguminous *Leucaena*), fodder or shelter. A great many different combinations are possible.

8.2 Water use in agriculture (see Box 8.3)

Apart from drinking water for animals (see Table 8.8), which is quantitatively minor but of great importance in arid areas, the water required for agriculture is for crops. Some of these are fed to animals and then the water required per unit of product may be very high (Table 8.9). These amounts vary, of course, with the environment and with the productivity of the animal, but they dwarf the quantities of drinking water. This is because plants need vast amounts of water, mainly for the following reasons:

1. they evaporate water by transpiration;
2. transpiration may keep them cool;
3. the nutrients taken in by the roots are transported in water and are in very dilute solution.

The water required by pasture grasses and legumes (Spedding, Walsingham and Hoxey, 1981) varies from *c.* 300 g of water g^{-1} of plant dry matter, produced, to over 1000 g. Much the same range applies to root, tuber and grain crops, although some of the values quoted for cereals go up to 1400 g. However, the global ratio of water loss to yield of dry matter is given by Stanhill (1985) as 5.6 kg g^{-1}.

Transpiration occurs mainly through stomata in the leaves (see Chapter 5 and Box 8.4) and the size of these can be varied by the plant. Where the water supply from the soil is not limiting, transpiration is dominated by the meteorological

---

Box 8.3   The hydrological cycle

About 97% of the world's water is contained in the oceans and nearly three-quarters of the rest is represented by the ice sheets of Antarctica, Greenland and the arctic ocean (Gilliland, 1979). The proportion of the total held in ice and snow is about 2% and less than 1% is found in rivers, lakes and groundwater: the atmosphere contains *c.* 0.001%.

Some 420 000 km^3 (9.26 x 10^{16} gallons) is evaporated annually and approximately the same quantity falls as rain. This is the hydrological cycle and it is powered by solar radiation: in fact it is the largest single user of the radiation reaching the earth. Only some of the rain falls on land and much of this disappears back to the oceans as run-off to rivers.

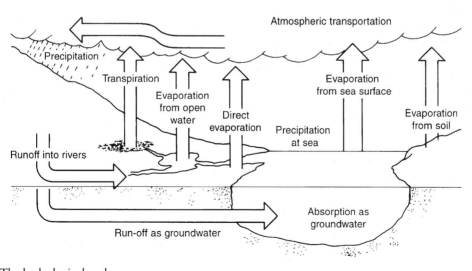

(a) The hydrological cycle

TABLE 8.8 Water requirements of farmed animals (after Spedding, Walsingham and Hoxey, 1981)

Species	Liveweight (L/wt) (kg)	Environment	Water (1/100 kg L/wt day^{-1})
Beef cattle	400	15–21°C	7
Sheep	40	15–21°C	9
Goats	15–20	Desert	4.2
Pigs	90	Housed	7
Horses	600	Housed	4–14
Rabbits	2.5	Housed: dry diet	6
Hens	2–3	Housed: dry diet	7
Musk oxen	365	Arctic (April)	3.5

TABLE 8.9 Water requirements for animal production (after Spedding, Walsingham and Hoxey, 1981)

Product	Total water use per kg of product (l)
Wool	420 000
Beef	42 110

TABLE 8.10 Energy used by plants in photosynthesis and evapo-transpiration (a range of values measured for different crops in different parts of the world)

Gross daily insolation (A) (KJ cm^{-2})	Percentage of A used in photosynthesis	Percentage of A used in transpiration
2.5–2.9	1.2–2.8	34–58

conditions, especially incident solar radiation. Under these conditions, when the soil is completely covered by short, green plants – as in a grass sward or lawn – the amount of water transpired is much the same as the evaporation from a free water surface of the same area. This is known as the 'potential transpiration' and maximum plant growth occurs when the actual transpiration equals the potential (see Chapter 5). When the supply of water from the soil is limiting, irrigation can greatly increase production.

Rather more energy is usually employed in transpiration than in photosynthesis but this may also come from heat in the air and sources other than direct solar radiation. Even for plant growth, the amount used in photosynthesis is small relative to that used by plants in evapo–transpiration (Table 8.10). This enormous evaporation of water from warmed leaves is absolutely essential to plant growth and the energy used in this way is just as much a part of the real energy cost of production as is that employed in photosynthesis. It may be argued that this emphasizes the inefficiency still

further but, if this did not happen, the hydrological cycle would still have to be powered and the environment as a whole kept warm enough for plants, animals and people to tolerate.

So, plants use rather more of the solar radiation they receive than at first appears and only a small proportion of the total is completely available for photosynthesis. Even so, it has been calculated that at normal light intensities the theoretical maximum conversion of available visible radiation is of the order of 5–6%, at times when temperature is adequate and water and nutrients are non-limiting.

Irrigation is quite widespread throughout the world, although probably less than half the potentially irrigable area (c. 200m ha) is actually irrigated, and it is worth remembering that many parts of the world suffer from acute water shortage – for example, some three-quarters of Australia.

Water is not only important as such but, since it is the medium for transport of nutrients, a lack of water leads to a reduced intake of minerals as well. It can even happen that roots obtain adequate water from deep down in the soil, where nutrient concentration is very poor, and fail to grow due, in particular, to lack of nitrogen, whilst continuing to evaporate water. Normally, both nutrients and roots are concentrated in the upper profile, e.g. the top 20 cm.

Of course, excess water also has damaging effects, by preventing oxygen reaching the roots (as in water-logging) and leaching nutrients from the soil (see Chapter 10). Irrigation by overhead sprays can be very wasteful of water, by evaporation and, especially, if used just before rain. Modern methods (e.g. trickle irrigation) are much more effective in the use of water, trickling the supply through perforated pipes on, or in, the soil.

Box 8.4 Stomata

Stomata are essentially pores in the leaf epidermis, through which water vapour escapes and CO_2 enters to be taken up by the chloroplasts: both processes are essential for plant growth. They are normally no larger than *c.* 15 cm by 35 cm and number about 10 000 cm^{-2}.

If the plant is short of water, the leaf will wilt and the stomata will close, thus reducing water loss: they thus function as a kind of hydrostat. Some plants (e.g. Marram grass) have leaves which roll up to protect the stomata: this is achieved by loss of turgor in the 'hinge' cells [see figure (a)]

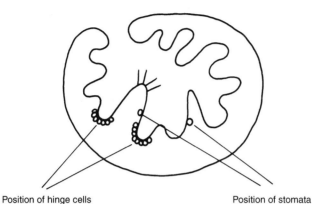

Position of hinge cells   Position of stomata

(a) Transverse section of a rolled-up leaf of Marram grass.

Guard cells

(b) Transverse section of stomata.

In many parts of the world, irrigation is by periodic flooding using open channels and in arid regions faulty irrigation can lead to saline soils, in which very little can grow. The reason for this is that irrigation water always contains some dissolved salts and these steadily accumulate as the water evaporates. In these circumstances, excess water may have to be used to leach the unwanted salts to the drainage system.

Irrigation need is based on the amount being evaporated being less than the supply (from rainfall etc.) and it is not always easy to calculate. Since some 90% of the earth's surface is covered by water, much of it to enormous depths, it is somewhat odd that so much of the land should be so arid. The problem, of course, is that the vast majority of the water has a very high salt content. Desalination (see Box 8.5) is quite possible and is

Box 8.5 Desalination

Desalination simply means removing the salts from sea–water in order to produce usable fresh water. It is carried out in the Middle East (e.g. in Saudi Arabia), for example, where water is greatly needed and oil money is available for the substantial capital costs. It has also been done in Hong Kong, Israel and Oklahoma, USA (Arad and Glueckstern, 1981).
Methods of desalination have included:

distillation (problems with scale formation)
  multi-stage flash evaporation
  multi-effect distillation
  vapour compression;
electrodialysis (mainly used for brackish water);
reverse osmosis (membranes become fouled with particulate matter);
humidification;
freezing;
chemical processes (use of solvents, hydration).

Distillation (the most commonly used process) also includes the use of solar stills but these tend to operate very slowly: the problems of scale deposition can be solved by feedwater pretreatment.
In general, desalination methods are now reliable but involve substantial capital costs and use large amounts of energy. It is worth noting that brackish water can be used, without desalination, to grow some plant species and genetically-engineered salt tolerance is now a possibility.

practised on a considerable scale, but it is expensive and depends upon energy, mainly from fossil fuels.

Finally, it is worth noting that, eventually, it may be expected that a safe, plentiful and cheap source of energy may become available. After all, the natural forces and sources of energy are immense – it is mainly a question of devising safe and cheap ways of tapping into them. The world would then face enormous problems in controlling activities, such as war, that are currently, at least to an extent, constrained by energy-supply limitations.

Chapter nine

The concept of efficiency

Foolish is the man who tries to pick up two water melons with one hand.

Anon.

It seems so obvious that increased efficiency is better, that it is quite a shock to realize that it is not necessarily so at all. This is not, however, a moral argument: the fact that being a more efficient murderer is not a good thing is mainly because murder itself is not a good thing: different circumstances can be imagined where greater efficiency made it better (in the sense that suffering might be less) and where it made it worse (e.g. by decreasing the chance of detection).

A more useful starting point is to pose a question of your own efficiency, such as 'Are you more or less efficient than your neighbour (father, daughter, etc.)?' You would immediately reply 'At what?' because you would recognize that you are not efficient or inefficient, except in relation to some specified activity.

It is also necessary to distinguish between efficiency and effectiveness. To be effective means to achieve some objective, without reference necessarily to the effort exerted. It might be more efficient to achieve half the objective but only using one-tenth of the effort – but it would not be effective. This can be seen in cost-cutting exercises, which use money more efficiently (in terms of what you get per £) but may fail to achieve the objective.

Of course, it is perfectly legitimate to seek the most efficient way of achieving the objective, but is also necessary to ask 'efficient use of what?'. Money, people, time?

The fact is that efficiency can refer to many different processes or activities and, for each one, can be expressed in many different ways.

9.1 The meaning of efficiency

Efficiency can most usefully be defined as the ratio of output to input. Biologists have tended to denote efficiency by E (Spedding, 1973), although physicists and engineers, who have employed the concept for much longer, have adopted η, reserving E for energy. E will be used in this book and the definition expressed as $E = O/I$, O representing a chosen output and I a chosen input.

Such a ratio can apply to innumerable combinations of output and input and each of these can be expressed in many different terms, some of them allowing several different outputs (or inputs) to be combined. The fact that efficiency can mean many different particular things does not reduce its usefulness: it is no different, in this sense, from words like 'animal', 'plant', 'aeroplane' or 'book'. The general concept is clear and obviously of value, and part of its value lies in its applicability to a great many, different, particular cases.

Particular versions are often named, such as energetic efficiency, relating energy output to energy input, and it is sometimes thought that outputs and inputs should be expressed in the same terms. This is not essential, however, and unnecessarily limits the usefulness of the concept. Furthermore, everyday usage of the term already includes ratios in which output and input are of a quite different kind, quite apart from differences in the terms employed. Familiar examples are 'miles per gallon', 'words per minute', 'miles per hour', 'gallons per cow' and 'protein per ha'.

There is no question of efficiency being good or bad, therefore, or of greater efficiency being beneficial or disadvantageous, nor of one ratio being right or another wrong, except for specified purposes. If we are interested in milk production, then miles per gallon and words per minute are clearly irrelevant, but several different ratios may still be relevant and these can be expressed in several different terms. Once our interest has been rigorously specified, however, the most relevant ratio will be fairly obvious.

Specification of interest has to flow from a definition of purpose, why we are interested and what for, how the ratio is to be used and what do we think it is going to tell us. Confusion often occurs here, simply because there are usually several purposes involved and one may not clearly override all others. So it may be necessary to choose the most useful of the various possible and relevant expressions of efficiency, or to accept that several have to be considered at the same time. Very often, judgement then has to be exercised in order to weight each ratio, giving greater importance to one or another, and also to take account of other factors, constraints and limitations.

9.2 Biological efficiency

There is no essential difference in applying the concept of efficiency to biology, agriculture or to other applications. The simplest view is that biological efficiency represents the efficiency of a biological process and, since the latter could use physical or chemical inputs, a biological efficiency ratio could involve non-biological inputs. Similarly, the results of biological processes may also be non-biological (e.g. a pearl, a territory, a house-martin's nest, a footprint, a desert): this notion has to be even further expanded if human activities are included.

However, biological efficiency has also been used to describe the biological success of organisms and this generally has to be expressed in biological units. For example, the success of a species is usually described in terms of its numbers, its rate of population increase, or the number of different habitats or niches it can occupy, or its competitive success relative to other species. These are, of course, legitimate uses but they are not excluded by the wider view and there are a great many ratios of interest that could not be included if **biological success** was made an essential part of biological efficiency.

The wider view, then, would simply define biological efficiency as the efficiency of a biological process or processes. This means that we can assess the efficiencies of combined processes, even including such complicated combinations as ecosystems. The latter include agricultural systems and thus allow the determination of biological efficiency within agriculture. There is nothing fundamentally different in the calculation of the efficiency of a biological system and that of one that is not biological. In many cases, the processes may be very similar. It is also worth remembering that biological systems commonly use non-biological inputs, such as solar radiation, and that some of the most relevant efficiency ratios will be based on them. Indeed, as soon as one considers solar radiation, it is clear that efficiency is already a long-established concept in biology. It is certainly true that agriculture is greatly concerned with efficiency, but not only that of biological processes.

Clearly, the possible number of interesting and important ratios is very large indeed and cannot be dealt with comprehensively here. Two examples have therefore been chosen to illustrate the kinds of efficiency that are measured.

9.2.1 Fertilizer use by crops

Fertilizer has to be purchased by the farmer and, even if natural manure is used, work is required to apply it. These costs have to be justified by increased output, so the relationship between output of crop and input of fertilizer is very important. The three major fertilizers are based on nitrogen, potash and phosphate (see also Chapter 10). To non-leguminous crops, the supply of nitrogen is usually the most important quantitatively, although such a statement is clearly nonsense where another plant nutrient is in limiting supply. The response curve to nitrogen supply is shown for wheat in Figure 9.1. In most cases, applying fertilizer nitrogen to a legume merely reduces the legume's own ability to fix nitrogen in its root nodules and may not result in any greater production than before.

The response of many grass pastures that contain various proportions of clover is therefore partly due to greater grass growth and partly to reduced nitrogen fixation by the clover (see Figure 9.2): at very high levels of nitrogen application, the legumes may actually die out.

9.2.2 Feed conversion efficiency by animals

The second example takes feed as its starting point since the cost of feed usually represents a very high proportion of the total costs of animal production. The use made of feed is therefore of great biological and economic significance.

If we measure the amount of feed (energy or protein) consumed by an animal and then measure the energy and protein produced in its products we find considerable differences between species (Table 9.1) but, of course, not all these species could live on the same feeds or in the same environments. So it may be more useful, at times, to have sheep producing protein with an efficiency of 4% from grass, which we cannot eat, than to have

a hen producing with an efficiency of 20% from grain that we could eat (although it is worth noting that the comparable figure for a cow is about 24%).

However, individual animal efficiencies are only part of the study, since breeding populations always have to be maintained by someone somewhere.

In most cases, the efficiency of feed conversion is reduced when expressed per unit of a breeding population (Table 9.2) since the feed required to support breeding females and the necessary proportion of males has to be included (see Chapter 6), and for the whole time, not merely when they are being productive. The minimum breeding unit must include the necessary males and females but these animals do not live for ever and so have to be

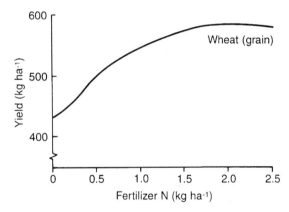

FIGURE 9.1 The response of wheat to fertilizer nitrogen (after Norman and Coote, 1971).

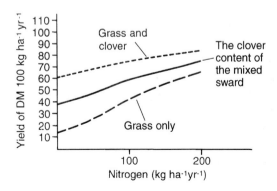

FIGURE 9.2 The response of grass and grass/clover swards to applied nitrogenous fertilizer (after Spedding, 1970).

TABLE 9.1 Ceiling values for efficiency of feed conversion[a] (*E*) by whole animal populations (after Large, 1973)

	E% (approx.[b])
Rabbit	42
Dairy cow (for milk)	39
Broiler fowl	25
Hen (for eggs)	18
Sheep (including wool)	17
Sheep	14

[a] E = N output in product(s)/N input in feed × 100 (per annum)
[b] The values depend greatly on the levels of performance assumed for each species. In this calculation, the levels were as follows: rabbit, 50 progeny per doe per year; dairy cow, 10 000 kg milk per cow per year; broilers, 120 eggs per parent hen; hen (for eggs), 250 eggs per hen per year; sheep, 6 lambs per ewe per year (based on a highly theoretical possibility of a ewe lambing twice a year and having triplets each time).

TABLE 9.2 Efficiency of production by individual animals (sources: Large, 1973; Spedding and Hoxey, 1975)

Animal product	Efficiency	
	Energy (energy in product/ energy in feed) × 100	Protein (N in product/ N in feed) × 100
Cow's milk	20	17–42
Rabbit meat	12.5–17.5	34
Beef	5.2–7.8	8
Lamb	11.0–14.6	16.4
Hen's eggs	10–11	16–29
Broilers	16	30
Pig meat	35	25–32

TABLE 9.3 Feed conversion efficiency by animal populations (calculated for breeding units of one female plus progeny and including the relevant proportion of the feed intake by the male) (sources: Large, 1973; Spedding and Hoxey, 1975)

Animal product	Efficiency (ceiling values)	
	Energy (energy in product energy in feed) × 100	Protein (N in product/ N in feed) × 100
Cow's milk	12–16	40
Rabbit meat	8.0	23–40[a]
Beef (suckler)	3.2	9
Lamb	2.4–4.2[a]	6–14[a]
Hen's eggs	11–12	24
Broilers	14.6	25–26
Pig meat	23–27	17–22

[a] Depending upon the prolificacy of the dam.

regularly replaced. Any assessment of the efficiency with which feed is used by an animal population therefore has to take into account the feed used to produce replacement breeding stock and output in the form of old or impaired breeding stock (culls) being replaced and losses due to disease. Population efficiencies thus tend to differ somewhat from those of simple breeding units (Table 9.3).

However, as mentioned earlier, different species vary in their capacity to utilize different feeds and not all feeds can be grown on all soil types. Animals and crops also differ in the climatic conditions under which they can be kept or grown. The relative efficiency with which land is used may therefore be quite different from that calculated for feed use.

On arable land, for example, crops or grass could be grown and converted by animals. In some cases (e.g. barley) the same animals and in other cases (e.g. grass) only a restricted range of animal species (those able to digest fibrous plants, such as grass, for instance) can be used. Comparisons can be made in several ways, therefore, and the relative efficiency per unit of land expressed accordingly.

However, it will be clear from the first part of this chapter and from earlier chapters that the efficiency with which one resource is used usually influences the efficiency with which all the others are used and some of the latter may be as important as the one first examined.

Not only are there many possible ratios to express efficiency for different purposes, it is often impossible to increase them all simultaneously. This is easily illustrated by a familiar example: a car. Consider two measures of the efficiency of a car: (1) miles per hour (how fast it gets you to your destination) and (2) miles per gallon (how efficiently it uses fuel). In any one car, if either one of the ratios is substantially increased, the other will generally decline. It is not possible to maximize both simultaneously and usually some compromise is achieved, which involves not maximizing either (except in special circumstances). It would be possible to design a new car that was better in both regards, but the same proposition would hold for that car.

In most operations, including those in agriculture, there are many more than two ratios of importance, so decisions are quite complex. What, then, can it possibly mean to propose that you are going to increase the efficiency of agriculture? Yet, this is a commonly accepted objective of research establishments, although strictly (and more sensibly) they aim to obtain the knowledge that makes possible increases in efficiency that are desired and selected by those operating agricultural systems.

9.3 The calculation of efficiency

It is obvious that efficiency cannot be calculated without specifying both the inputs and outputs of interest. There are three other important features that must be specified in addition. They are: (1) the process or system whose efficiency is being assessed; (2) the context or environment in which it is assumed to be operating; and (3) the period of time to which the calculation refers.

1. Where a simple biological process is being considered, there may seem to be little problem in specifying it. It is not usually a simple matter to be so specific, however, as the following example shows.

   Consider a plant growing in soil, producing protein in its leaves. Nitrogenous fertilizer may be added to the soil and the efficiency of nitrogen output calculated per unit of nitrogen input. But the soil may contain nitrogen already and the additional quantities may be large or small relative to that initial amount. Or the plant may be a legume or be associated with other plants able to fix atmospheric nitrogen. Clearly, the meaning of the efficiency ratio is going to differ according to exactly where the boundaries are drawn around the system considered.

   Another example illustrates further the problem of the system content of the particular input selected. Suppose we wish to calculate the efficiency with which feed energy is converted to milk energy by cows. Clearly, we have to specify the kind of cow (its breed, age, weight, stage of lactation) and the kind of feed (digestibility, energy content, toxicity, dry matter content), but the fat reserves of the cow may be difficult to assess. Yet a significant proportion of the milk energy may be derived from these reserves and not from the feed, at least for a time (such as the peak of lactation), in a high-yielding or a poorly-fed cow.

2. No system or process operates in a vacuum and it is unlikely that the entire context will be included in the system specification. Most descriptions of agricultural systems and processes, for example, will probably not mention oxygen, not because it is not an essential component but because it is assumed that all relevant environments will include it in non-limiting concentrations.

   Similarly, features of the environment, such as temperature, humidity and topography, may not be specifically mentioned but will be assumed to be suitable. Thus, the calculation of efficiency of feed conversion by fish may not actually include any reference to water. Yet it is obvious that the efficiencies of cacti and pondweed will depend on the environmental context, as will the efficiencies of sheep, goats, cows, reindeer and pigs. No efficiency ratio can have much meaning if the environmental context is not specified.

3. Efficiencies can legitimately be calculated over any period of time and comparisons are not always best made over the same absolute period. It does not really make sense to compare the reproductive efficiencies of mice and elephants over exactly the same period but, if this is to be done, the period chosen must be long enough for the elephant. Very often, lifetime performances may be appropriate and these will vary with the species.

Where land use is being considered, it is often sensible to take into account natural periodicities of growth: seasonal and annual cycles can be used. With animals, breeding cycles may mark off periods of time that are comparable for different species.

Whatever period of time is chosen, it is clearly necessary to specify it, since interpretation depends upon this.

9.4 The interpretation of biological efficiency

Given a sufficiently completely specified calculation, interpretation should be straightforward: the

results of the calculation should serve the purpose for which it was made.

Purposes will tend to be of two main types. The first is simply to decide which process or system is more (or most) efficient at doing a specified thing in a specific context. The object is to make it possible to choose the most efficient of those considered. Frequently, however, it will be desirable to know what effect on efficiency will follow changes in some of the most important attributes of the system or the environment.

The second purpose is to assess efficiency in such a way that it can be improved if it is not satisfactory. A simple ratio does not allow this,

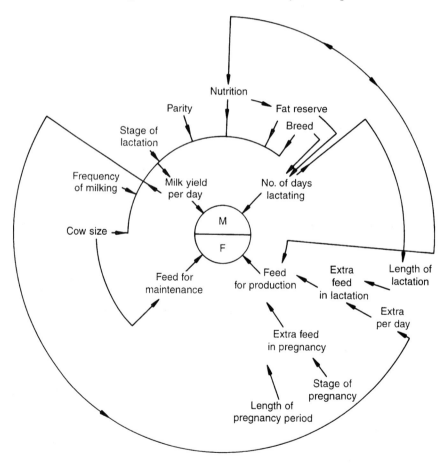

FIGURE 9.3 The expansion of an efficiency ratio, illustrated by the factors affecting milk output (M) per unit of feed (F), for an individual cow over a period of one year.

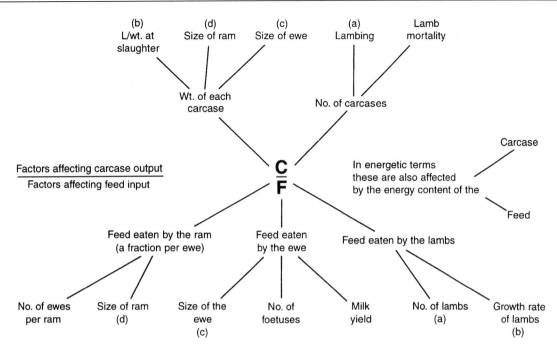

FIGURE 9.4 The expansion of an efficiency ratio. In this example, energetic efficiency of feed use by sheep (over one year) is calculated as caracase energy output (C)/feed energy input (F) per ewe (letters in parentheses indicate the more obvious connection).

as it cannot indicate how either component can be changed.

For these further purposes it is necessary to expand the ratio, to show what factors are important determinants of it and to determine the consequences of varying them over relevant ranges of values. Figures 9.3 and 9.4 illustrate how an efficiency ratio may be expanded and indicate how variables may interact with each other (for a systematic treatment of this, see Spedding, 1975).

There are two main, important conclusions to be drawn from these considerations. The first is that a single figure for the efficiency of a process, or a system, can only apply to a highly specific situation. The figure will be different for changes in a whole range of variables and efficiency can be expressed as a series of response curves to the most relevant changes in factors that are likely to change or can be controlled.

The second is that, if efficiency calculations are to be used as a basis for change and improvement, the process or system has to be described in sufficient detail and in such a way that its sensitivity to change in important variables can be adequately expressed. Furthermore, this has to be in sufficiently quantitative terms. To do this requires some form of modelling, ultimately in mathematical terms.

The advantages of modelling in the study of agricultural systems have been discussed at considerable length over many years (Dalton, 1975; Spedding and Brockington, 1976) and they apply here just as much as anywhere else (see Chapter 7).

Indeed, if agriculture is nearly always concerned with efficiency, since we are rarely interested in any output except in terms of how much is produced per unit of some input, it follows that calculating efficiency is simply one way of describing the study of agricultural processes and systems.

As any one of these processes and systems can be considered in so many different ways, in different environments and for different periods of time, the calculation of all possible efficiency values for each system is quite impossible.

9.5 The improvement of efficiency

Since it was argued earlier that efficiency (*E*) can be represented by the simple ratio of outputs (*O*) over inputs (*I*):

$$E = O/I$$

it seems deceptively obvious that the ratio can be improved (i.e. increased) in one of two ways: (1) increasing *O*; or (2) decreasing *I* (or both together).

Unfortunately, as described in Chapter 7, few real situations approach this degree of simplicity and agricultural units are best envisaged as complex systems. The fact is that one cannot usually change either *O* or *I* without affecting the other, often by an indirect route. This is illustrated for the cost of lamb production in Figure 9.5. This expanded ratio forms a circular diagram (see Spedding, 1975) with the starting ratio (cost per kg of meat) at the centre and radiating from it those factors which affect each part of the ratio.

Thus, one obvious way of increasing the meat output of the system is to increase the number of sheep (ewes and lambs) per unit of land, but this is bound to increase the costs of food, labour and veterinary services, and so it is for any change and it is quite possible to increase an output factor, and, as a result, disproportionately increase costs.

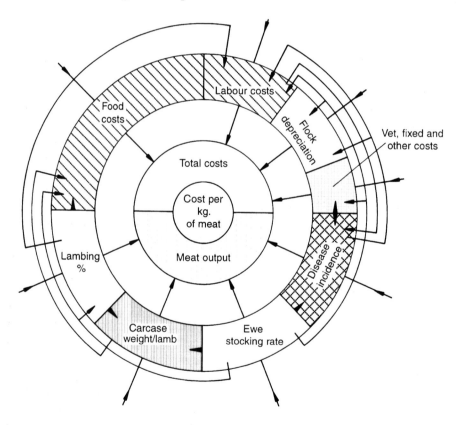

FIGURE 9.5 Lamb production and costs per hectare.

The improvement of even one efficiency ratio is thus not a simple matter and really has to be worked out on some kind of a model of the system – as suggested in Chapter 7. All this is related to one of the reasons for calculating efficiency – in order to improve it.

The second reason given earlier was in order to make comparisons and thus choose the more efficient process, practice, animal, crop or system.

9.6 The relative efficiency of production systems

Comparisons of the efficiency of different systems can only be made where both the products produced and the resources used are sufficiently similar or where they can be expressed in comparable terms.

Economic comparisons are the most universally applicable since all products and resources can be expressed in monetary terms and these can be interpreted internationally. Such comparisons rank systems in order of profitability, for example, or return on capital for the particular costs and prices used in the calculation. Since the latter do not remain constant, this may be a poor guide to the future: it may be an essential guide to the farmer in the short term, however. Similar considerations may govern the outlook of a nation, particularly if agricultural products form a significant proportion of its exports.

If we are concerned with agriculture in terms of the efficiency with which resources are used in feeding people (see Chapter 3) it makes sense to compare systems as energy or protein producers, or to assess the number of people they can feed. Even then, comparisons have to be based on the use of common resources in common environments. Since there are many different ways of looking at systems, there are bound to be at least as many ways of comparing them. A few of the most useful will be considered here.

TABLE 9.4 Output of protein and energy per unit of land (from Spedding and Hoxey, 1975)

Product harvested	Protein ($kg\ ha^{-1}\ yr^{-1}$)	Energy ($MJ\ ha\ yr^{-1}$)
Crop		
Dried grass[a]	200–2 200	92 000–218 000
LPC[b]	2 000	–
Cabbage	1 100	33 500
Maize	430	83 700
Wheat	350	58 600
Rice	320	87 900
Potato	420	100 400
Animal		
Rabbit	180	7 400
Chicken	92	4 600
Lamb	23–62	2 100–7 500
Beef	27–57	3 100–4 600
Pig meat	50	7 900
Milk	115	10 460

[a] Not directly consumable by humans.
[b] Leaf protein concentrate.

9.7 Crops versus animals

There is a range of efficiency to be found within any production system, so it is important to base comparisons between them on comparable versions of each. Similarly, there is a wide range of efficiencies within crop production systems and within animal production systems. Nevertheless, crop production is generally more efficient than animal production in the output of dietary protein and energy on land that will grow crops (Table 9.4). This is so for the use of land area, incident solar radiation, water, fertilizer and support energy.

On land that cannot grow crops, or from which crops could not economically be harvested for direct human consumption, it may still be possible to use animals. On such land, animal production is, of course, vastly more efficient in the use of all resources.

Since many farming systems are practised on quite different kinds of land and in different climates, it is not always possible to make direct comparisons between them. Within these con-

TABLE 9.5 Relative efficiency (*E*) of energy and protein production in agricultural systems (source: Spedding and Walsingham, 1975)

Production system		Land[b] (ha)	*E*[a] for the use of:		
			Solar radiation[c] (MJ)	Support energy[d] (MJ)	Fertilizer N (kg)
Milk[e]:	Energy	15 576	0.00047	0.54	129
	Protein	210 × 10³	0.0064	7.30	1 736
Beef[f]:	Energy	5 772	0.00017	0.11	22
	Protein	90 × 10³	0.0027	1.66	339
Sheep meat[g]	Energy	4 929	0.00015	0.23	38
	Protein	53 × 10³	0.0016	2.50	404
Wheat[h]:	Energy	58 600	0.00178	5.4	586
	Protein	350 × 10³	0.0106	32.0	3 500
Potatoes:	Energy	100 400	0.003	5.1	619
	Protein	420 × 10³	0.013	21.1	2 593

[a] *E* is expressed as energy (MJ) or protein (g) output of milk, carcases, grain or tubers, per unit of resource used, on an annual basis.

[b] Assumed to be receiving nitrogenous fertilizer at the annual rate of: 121 kg ha^{-1} for milk; 265 kg ha^{-1} for beef; 131 kg ha^{-1} for sheep meat; 98 kg ha^{-1} for wheat; 125 kg ha^{-1} for potatoes.

[c] Based on annual radiation receipt of 33×10^6 MJ ha^{-1} yr^{-1}.

[d] Support energy is defined here as the additional energy (labour, fuel and electricity) used on the farm, plus the 'upstream' energy costs, i.e. those used to manufacture the major inputs (fertilizers, machinery, herbicides, etc.; human labour is excluded) and the 'downstream' energy costs of processing and distribution.

[e] Permanent pasture dairy farm, *c*. 76 ha and 100 cows averaging 900 gallons (4217 kg yr^{-1}).

[f] Suckler herd, intensive grassland system.

[g] Lowland fat-lamb system.

[h] Winter wheat.

straints, however, it is useful to examine their productivities per unit of land, solar radiation and support energy (Table 9.5) because this indicates the size of human population that can be associated with these systems in the areas where they are found.

9.8 Efficiency in the food chain

Since a high proportion of food is now **processed** in developed countries (see Box 13.1), it is important to consider efficiency over the whole food chain. An example will illustrate the reasons for this (see Table 9.6).

Wheat is more efficiently produced than milk, if the calculation is up to the farm gate, for reasons already given, for the use of support energy, as for other resources. But milk can be drunk virtually as it is and usually only incurs rather small additional energy costs in transport and treatment. Wheat, on the other hand, is normally ground and cooked, involving processes that incur very heavy energy costs, quite apart from the costs of storage, transport and packaging.

If the two products are compared at the point of consumption, therefore, the large difference evident at the farm gate may completely disappear. Thus, conclusions may be different for calculations made up to the farm gate, within the food industry, or over the whole food chain.

TABLE 9.6 Efficiency of energy use in the food chain

	MJ of energy in product per MJ support energy used
Wheat at farm gate	3.2
Bread – white, sliced, wrapped	0.5
Milk at farm gate	0.65
Milk bottled and delivered	0.595

Questions

1. Which is better; a knife or a fork?; an apple or a pear?; high efficiency or high effectiveness?
2. How would you measure your **own** efficiency?
3. Which is more efficient, a sea-lion or a cow? (e.g. at milk production).
4. Which is the more efficient athlete, a sprinter or a long-distance runner?

Chapter ten

Agriculture and the environment

The higher the baboon
climbs, the more he shows
his less attractive features.

Ethiopian proverb

There seems little need to define the environment (Einstein claimed that 'the environment is everything that isn't me'), but it will be different according to where you stand and who you are. It has always been the case that the environment affected agriculture, indeed, historically, it determined what kind of agriculture could be practised.

Early agriculturalists grew the plants that the environment allowed or encouraged and kept the animals that could live and be fed in that environment. Both plants and animals had to fit, not only the soil and climate of the region, but had to be sufficiently resistant to, or tolerant of, the pests, parasites, predators and diseases of the area.

All this is fairly obvious and is still the case in many parts of the world, for the major crops and for ruminant animals. However, as agriculture has developed, it has tended to become more intensive and to exert more control over the environment. Pigs and poultry are now mainly kept in houses with varying degrees of control over day length, temperature and humidity, but, in any event, protected from wind, rain, insolation, etc.

Crops may be grown in glasshouses, also with highly controlled environments, independent of the local soil, water and nutrient supply. But, even in the field, irrigation may be used on a massive scale and the nutrient supply may be dominated by the application of manufactured fertilizers.

This does not happen all over the world, largely because the inputs cannot be afforded, so, much of the developing world (see Chapter 4) is still heavily dependent on the local environment as well as on major climatic events, such as floods, gales, hurricanes, cyclones, incursion by the sea, drought, excessively high and low temperatures, frost and snow.

Even without these hazards, the normal annual and seasonal variations in temperature and rainfall pose a continuing problem for most of agriculture and cause enormous variation in agricultural output.

Increasingly, human activity affects agricultural operations. Some of these activities are only just being realized, such as the possible effect of fossil fuel use on 'global warming' (see later in this chapter). Others are hardly recognized or taken seriously yet, including the possible effects of nuclear accidents (such as that at Chernobyl) in the future. The effects of Chernobyl are still being felt, even in the UK, and had the wind been stronger or in a different direction at the time, the effects could have been even more devastating.

It is quite possible to imagine nuclear accidents in the future, effectively putting out of action major grain-growing areas of the world for many years. Most of the natural disasters have a limited effect in terms of time, though volcanic eruptions may be one exception, but radiation can blight an area for tens, perhaps hundreds, of years.

It may be, therefore, that we currently underestimate the effect of the environment on agriculture. Certainly, in developed countries, partly because of worries about food surpluses – mainly because of the cost of producing, storing and disposing of them – public concern increasingly focuses on the effects of agriculture on the environment.

10.1   The effect of agriculture on the environment

Of concern, but somewhat tangential to the main issue, is the effect of change of land use from, say, rain forest to agriculture, on a massive scale. Such changes can affect the water balance of a region, precipitating flooding, and expose the soil to both wind and water erosion. This is not strictly an effect of established agriculture but of changed land use to what turns out to be an inappropriate form of agriculture.

Changes in the form of agriculture may also have similar effects, as well as changing the landscape. The sort of changes that have occurred in the UK – towards larger fields, the removal of trees and hedgerows, new crops (e.g. oil-seed rape), even set-aside (see Box 10.1) – have altered the landscape substantially.

Such changes also have implications for wildlife, and particularly for conservation and biodiversity (see later in this chapter). The main effects of agriculture on the environment can be separated into: (1) local impacts and (2) those beyond the farm.

10.1.1   Local effects

Whether such effects remain local depends on the degree of isolation and separateness of the farm. No farm is isolated in terms of atmospheric interchange (affecting gases, smells, small particles of

---

Box 10.1   Set-aside

The object of set-aside (i.e. putting land out of production) is to reduce output of products currently in surplus within the EU. This EU-wide scheme gives farmers payments (per ha) for putting out of production a minimum percentage (e.g. 15%) of their farmed land. There are various forms, related to whether rotation of the area round the farm is allowed or not and to whether public access is included.

Other, similar schemes are in operation in the USA and their effect is often regarded as negligible. This is sometimes because farmers put aside their poorest land and endeavour to increase production on the remainder.

In general, the public has difficulty with the concept, which appears to be paying farmers for doing nothing. The payments are intended to compensate farmers for loss of income but some consider that some use (e.g. for research or demonstration related to alternative crops, especially energy crops) could have been required.

The problem is how to reduce surpluses and the methods available all have disadvantages. For example:

1. **Quotas**, as used to control milk production, prevent new entrants unless they can purchase quota. In time, quotas can become excessively valuable and farmers can go out of milk production themselves and live on the rent charged for quotas used by others.
2. **Price reduction** would be the normal economic answer but this might eliminate the producers in one country and greatly damage the farming industry. In some cases, farm incomes might be so reduced that the less productive areas would cease to be farmed at all.
3. **Input restriction** has also been proposed, especially for nitrogen, but it is argued that this would build in inefficiency (using less than the economically optimum level of input) and ossify practice.

---

spray, etc.) but farms vary greatly in the extent to which water and soil movement occur between them and surrounding areas.

Even if an effect is confined within the farm it does not mean that it is unimportant. It may affect the health of the farm family or workers (see Chapter 13), it may build up to a point where it does have a wider impact, or it may accumulate as a future hazard.

In this last category is the accumulation of 'heavy' metals, such as lead, cadmium and copper, which are highly toxic and do not degrade or disappear. Once accumulated, they are difficult to remove, although there are some plant species that take them up without being adversely affected: whether this method could be used to decontaminate land (or even to harvest valuable metals) remains to be seen (Box 10.2).

Heavy metals are probably the most extreme of local effects, eventually affecting whether, or what kind of, agriculture will be possible on that land in the future. Such a problem may put land out of action for a very long time. Other local effects may also have quite long-term consequences, mainly within the soil.

For example, if heavy applications of fertilizer, herbicides and pesticides (collectively known as 'agrochemicals') are used for many years, the population of soil plants and animals may be greatly altered, generally being drastically reduced. A landmark in the history of concern over the effect of agrochemical use on wildlife was the publication of *Silent Spring* by Rachel Carson in 1962.

Many of these organisms play a major role in maintaining and improving soil condition and fertility, as well as influencing soil-borne diseases, pests and beneficial insects, earthworms and other invertebrates (see Chapter 6). One of the purposes of 'organic' farming (see Chapter 17) is to maintain balanced populations of micro-flora and -fauna in the soil: this is a major reason why the use of agrochemicals is not allowed in organic systems.

Such a situation is not immediately corrected by a cessation of agrochemical inputs: it may take years (5–7) to restore the 'natural' balance. The length of time may also depend upon the scale of the enterprise. The question has to be asked 'Where are all these organisms to come from and will they thrive when they arrive?'

In some cases, the smaller the organism, the less

Box 10.2 Heavy metals

Land may become seriously polluted with heavy metals, such as lead, cadmium, zinc, copper and nickel, from contamination with industrial wastes, car exhausts or even long-term fertilizer application. In general, these metals accumulate in soils because most crops remove so little and may become so concentrated that such crops no longer thrive.

Physical or chemical methods of removing them are prohibitively expensive. Scraping 1 m of topsoil from polluted land, for example, would result in *c.* 3000 tons of contaminated soil per hectare, containing only a few kilogrammes of metal.

A major source of heavy metal pollution of agricultural land is sewage sludge. About half of the 1m tonnes of sewage sludge produced in the UK each year is spread on agricultural land. This sludge often contains heavy metals which accumulate in the surface soil. If the concentration is high enough they can poison crop plants but, even at lower levels, the vital processes by which the Rhizobium in clover fixes atmospheric nitrogen may be adversely affected (McGrath, Hirsch and Giller, 1988).

However, in the last few years, it has been found that certain plant species, known as 'hyper-accumulators', are able to take up thousands of times more metal than most species. There is thus the possibility: (1) of using such plants to decontaminate soil and (2) even to recover useful amounts of the heavy metals in this way.

For example, it has been reported (Watts, 1993) that a tree called *Sebertina acuminata*, a native of metal-rich soils, may contain up to 11% of nickel in its latex. Since latex can be tapped from the living tree (as with rubber) this offers the intriguing possibility of regular tapping. Even more recently, it has been reported (Rajan, 1995) that the Indian mustard (*Brassica juncea*) accumulated up to 60% of its dry weight as lead.

---

the problem: however, earthworms, for example, spread very slowly (Box 10.3). However, clearly creatures that fly (butterflies, beetles, flies, birds) are not so constrained. Creatures that run, such as rabbits, hares, foxes, moles, mice, obviously could spread but may need 'corridors', such as hedgerows, in order to provide protective cover. Such animals, incidentally, are also a means of spreading effects (e.g. diseases) beyond the farm.

10.1.2   Effects beyond the farm

Farm operations can have effects on the surrounding countryside: noise from machinery, smells from slurry and silage, traffic congestion due to slow tractors, mud on roads from farm vehicles and the spread of chemical sprays or smoke on the wind, are all causes of complaint. The burning of stubble was banned in the UK (in 1992) for this reason.

Some of these complaints come from non-country people whose picture of rural life did not contain such features: others may have consequences for human health (see Chapter 13).

In the main, however, the most important impacts of agriculture outside the farm are: (1) contributions to global warming; (2) use of agrochemicals; (3) effect on conservation; and (4) effect on biodiversity. These will be dealt with in turn.

10.2   Global warming (see Box 10.4)

This is essential to life on earth and without it the planet would be some 30°C cooler. 'Enhanced' global warming is the current concern, that human activity is generating 'greenhouse' gases that result in higher than normal concentrations of them in the atmosphere, thus trapping more heat from the sun, resulting in an increase in the world's temperature.

Box 10.3   The spread of earthworms

Since there are *c.* 1 800 known species of earthworms (see Minnich, 1977) and they are found virtually all over the world, it is obviously rather dangerous to generalize about them.

They all belong to the phylum Annelida ('ringed'), the Class Chaetopoda (bristle-footed) and the Order Oligochaeta (bristles in groups). The commonest species in Europe and North America belong to the Family Lumbricidae and the largest common species is *Lumbricus terrestris*, which burrows deeply (commonly down to 1 m) and makes casts on the surface. There are about 220 species of these. They can live for 4–8 years but most probably only survive for a few months (Edwards and Lofty, 1972). They do not travel far horizontally and, indeed, many species have permanent burrows. Nevertheless, populations do expand at the margins and Hamblyn and Dingwall (1945) reckoned that *Allolobophora caliginosa* spread at the rate of 10m per year. More rapid dispersal was thought to be associated with cocoons in soil carried on agricultural implements.

The main 'greenhouse' gases are shown in Table 10.1, from which it can be seen that they have different effects per molecule and last for different periods in the atmosphere. In fact, as Emsley (1992) pointed out, water vapour is actually the principal 'greenhouse' gas. Air contains between 1 and 4% of water vapour (this is 10–40 000 ppm, compared with CO_2 at 350 ppm). Furthermore, different gases tend to absorb energy at different wavelengths.

TABLE 10.1   The 'greenhouse' gases

	CO_2	CH_4 (methane)	CFCs (chlorofluoro carbons)	N_2O (nitrous oxides)
Lifetime in the atmosphere (years)	50–200	10	65–130	150
Relative effect per molecule	1	30	17 000	160
Sources	Fossil fuel burning	Livestock, rice paddies, landfill, termites	Refrigeration propellants	Fuel burning, agriculture

Box 10.4 Global warming (see also Box 5.2)

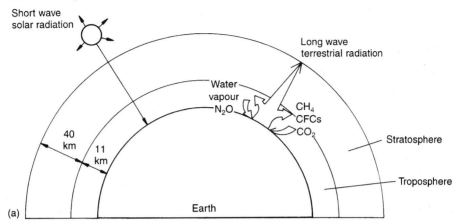

(a)

As described in the text (section 10.2), an 'enhanced' greenhouse effect could result in an increase in the average global temperature (estimates vary from 0.5 to 4°C over x years, x relating to how fast CO_2 concentrations rise and the levels they reach). Commonly, estimates relate to a period of 40–100 years.

The recent increase in CO_2 content of the atmosphere is shown in figure (b) and relates to estimated past figures – based on the composition of air bubbles trapped in Antarctic ice (WMO/ICSU, 1990).

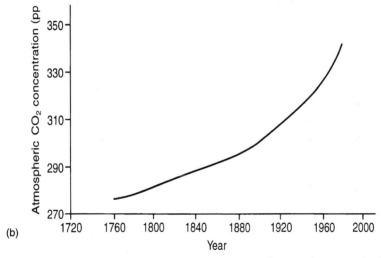

(b)

It is a big step from these increases in CO_2 concentration to prediction of increases in global temperature and even if the average rose, it is by no means obvious what would happen in particular countries or even regions. Weather patterns could change, as could ocean currents: the consequences to the UK of a major change in the path of the gulf stream might actually be a reduction in temperature.

Amongst all the imponderables, there is a growing conviction that significant warming could occur and probably will if current levels of greenhouse gas emissions are not reduced.

The reason why all this is taken so seriously is because the consequences could be so serious and, by the time the matter was proven, it would be too late to do anything about it. One of the most devastating consequences could be an appreciable rise in sea level, due to the melting of polar ice caps (notably the ice covering Antarctica and Greenland), the melting of glaciers and an expansion of the warmer oceans.

Agriculture contributes significantly to three of these gases, CO_2, methane (CH_4) and nitrous oxides (N_2O).

10.2.1 CO_2 emissions

Plants and animals (including ourselves) emit CO_2 as part of respiration but this only releases the carbon that has been taken in during life. In general, such activities are CO_2 neutral.

Trees accumulate carbon in their wood (as explained in Chapter 5) and, when this is burnt, this is released as CO_2. But no more is emitted than was taken in during the life of the tree.

The burning of fossil fuels (coal, oil, gas), however, releases the CO_2 that was absorbed millions of years before and that has accumulated over a long period. It is this massive additional quantity that is raising the CO_2 content of the atmosphere. However, there are still uncertainties about the rates at which such CO_2 may be absorbed by current forest growth (Pearce, 1992) and by the oceans.

It is also possible that algae in the oceans produce a sulphur compound which may influence cloud formation (Fell and Liss, 1993). Clouds have a profound effect on climate and it is not known how they will be affected by global warming.

The situation is complex and full of uncertainties and the reason it is taken so seriously is that, by the time we can be sure about what climate changes may occur, it will be too late to take any corrective action.

Agriculture's contribution to the increase in 'greenhouse' gases comes from methane and nitrous oxides.

10.2.2 Methane (CH_4)

Methane is emitted by agricultural animals, chiefly ruminants belching during the process of rumination, and by bacteria in rice paddy fields.

It is thought that less than half of the methane emissions come from rice-growing, ruminants and biomass burning combined (in descending order of magnitude), with about a quarter coming from marshes, wetlands and oceans, and more than a quarter derived from mining, gas drilling, landfills, termites and oceans, with termites, surprisingly, generating more than the oceans.

At this point, it is worth asking how anyone can possibly know this: the problems of estimating the methane output of termites, bogs or herds of wild game are obvious. But, as stressed elsewhere, nearly all world statistics suffer from almost impossible problems in collection, measuring and estimating them.

Recently, serious doubts have been cast on the accuracy of estimates for methane production from Asian paddy fields (Menon, 1994), based on samples taken in other countries.

10.2.3 Nitrous oxides (N_2O)

Although they are produced in the burning of fossil fuels, it is thought that most such gases originate from agriculture and the burning of biomass. The latter is causing increasing concern, particularly about major regional fires. The nitrous oxide emitted from soils is commonly associated with the use of fertilizers.

10.2.4 Impact on agriculture

Should global warming occur as many predict, average temperatures would rise by what may appear to be very small amounts, between 0.7 and 1.5°C (some estimates give higher figures) by 2030. But this could result in significant changes in the world climate, sea levels, hurricanes and flooding, which could cause the disappearance of islands and low-lying regions, with consequent human migration.

None of this can be predicted with precision and it is not known whether a particular country would necessarily become warmer or what would happen to cloud cover, rainfall distribution, incidence of frosts and droughts. The UK, for example, is warmed by the Gulf Stream and if changes in ocean currents altered this, the country could become cooler.

Any significant changes in temperature (which would be moderated by all the other related changes) could affect the incidence of pests and diseases, alter the plants we could grow and the animals we could keep.

In fact, an increase in CO_2 content increases growth rates in some plants (so-called C_3 plants because they use 3-carbon compounds in photosynthesis) and is deliberately arranged for some greenhouse crops. So there could also be changes in crop yields, but these would probably vary from C_3 crops, such as wheat, barley and oats, to C_4 crops, such as maize and sugar cane, which seem unaffected by CO_2 concentration. Of course, all this would also apply to weeds.

10.3 Use of agrochemicals

The main agrochemicals used are fertilizers, herbicides and pesticides.

10.3.1 Fertilizers

By far the greatest quantities of fertilizer used are nitrogenous but significant amounts of potash and phosphate are also applied.

The purpose of applying fertilizers is to replace plant nutrients lost in the removal of the crop (see Table 10.2), by leaching and as volatile gases from the soil and animal excreta. Very little of the nitrogen applied may actually finish up in the final product, however (Table 10.3). The losses incurred in removal of the crop are inevitable unless human waste is also returned to the soil and if they are not replaced will result in a fall in soil fertility. Nitrogen is quantitatively the most important element: the nitrogen cycle (Figure 10.1) is therefore of great importance.

Nitrogen is mainly absorbed by plant roots as nitrate (with some ammonium) and used to form amino acids and proteins. It is worth noting that this nitrate is apparently identical whether it is derived from organic manures (animal excreta, compost or green manure) or from manufactured fertilizers. As described in Chapter 5, atmospheric

TABLE 10.2 Main nutrients removed in crops (from Jollans, 1985)

	N	P	K	N	P	K
	(kg tonne)			*(kg removed per ha[a])*		
Wheat grain	18	3.5	4.2	115	22	26
Wheat straw	3	0.6	12.4	7	1.5	31
Field beans	38	3.6	11.3	114	11	34
Potato tubers	3	0.4	4	109	14	133
Grass	6	0.7	5	128	14	100

[a] These figures, of course, depend upon the yield of the crop.

TABLE 10.3 Examples of nitrogen recovery in crops and animals[a]

	Amount recovered in the plant or animal as a percentage of the nitrogen applied as a fertilizer
Pangola grass	79–103
Ryegrass	75
Beef (individual cattle)	8
Milk (individual cattle)	9.5

Animal populations	Amount recovered in animal products as a percentage of that in feed *(from calculations by Wilson, 1973)*
Beef	6
Milk	24
Hen eggs	20
Broiler chickens	20
Rabbit	17
Lamb	4

[a] Taken from actual experiments reported in the literature.

nitrogen is also fixed as ammonia by legumes within root nodules colonized by bacteria. So there are other ways of replacing the nitrogen removed in crops: potash and phosphate, however, have generally to be imported from outside the system.

The availability of soil nutrients is affected by acidity and in many areas of the UK lime is applied to correct the pH. The response to applied fertilizer nitrogen is well illustrated by grass in the UK (Figure 10.2). This also illustrates the law of diminishing returns as inputs are increased: as soon as the needs for one input are met, something else (another nutrient, temperature, light, genetic capacity of the plant) limits growth. The response to applied nitrogen being so great, it is usually profitable to use enough to reach point X in Figure 10.2,

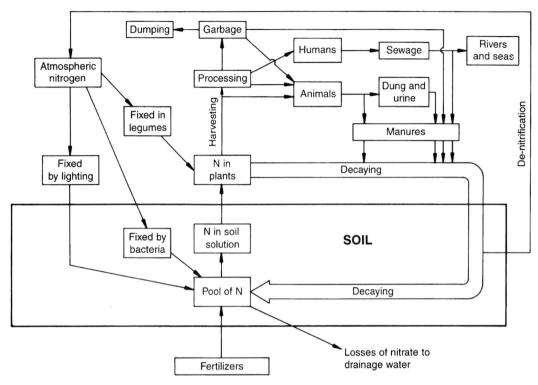

FIGURE 10.1 The nitrogen cycle.

since the other costs of growing the crop (land, labour, other inputs) do not rise as fast as the productivity.

It is generally assumed that 50% of the applied nitrogen is removed in the harvested parts of the crop (Tinker, 1985). However, fertilizer nitrogen has other effects. Those on soil fauna have been mentioned already: two others cause concern. One is the effect on the composition and quality of the crops and their products. For example, the nitrate content is often raised. The possible consequences to human health are discussed in Chapter 13. The other is the leakage of nitrate (particularly) out of the field (called 'leaching') into water courses. As Whittemore (1995) has recently pointed out, 'the polluting power of animal excreta (faeces plus urine) is prodigious'.

In the EU, with some 12m sows, there are 240m tonnes of excreta to dispose of annually from the pig sector alone. Of course, some of this can be

used as fertilizer but the intensive concentration of pigs in some areas makes this difficult. This is a major problem in northern Europe.

When nitrogen gets into lakes it can cause an excessive content of nutrients ('eutrophication'), resulting in too much vegetation growth and hence too much rotting vegetation. This deprives the water of oxygen and can result in the death of fish. However, eutrophication is usually caused by excess phosphates rather than nitrogen and these do not usually come from soil phosphate but, in industrialized countries, from detergents contained in sewage. The extent of nitrogen losses from agriculture is indicated in Box 10.5.

Since substantially more nitrogen is applied than is removed in the crop, it must either accumulate or be lost to the atmosphere or leached into water. The latter represents one of the major current concerns about intensive farming in Western Europe. The contamination of water supplies has

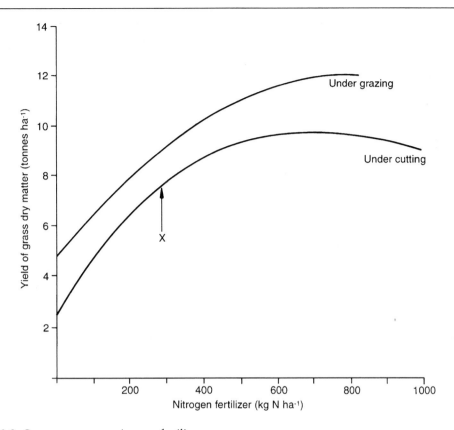

FIGURE 10.2 Grass response to nitrogen fertilizer.

potential health implications (see Chapter 13) but it is not a simple matter to establish the link between fertilizer application and the nitrate content of drinking water. A UK example of this is described in Box 10.6.

As a result of rising levels of NO_3 in drinking water, the UK has established nitrogen-sensitive zones (see Box 10.7) in which the use of fertilizer is restricted.

10.4 Conservation

The term 'conservation' is used in many different ways, some of them denoting **preservation** (as in jam-making and silage-making) but the latter is not possible for living things. Preservation can be applied to artefacts, like buildings, but, even here, it may not be possible to preserve anything in its original state unless it is totally enclosed in something (a glass case, plastic film, inert gas, sugar, salt, etc.).

Preservation of individual life is clearly not possible, except for limited periods. Species can be sustained, on the other hand, as can ecosystems and habitats, because they involve reproductive and other essential processes without which they would cease to exist.

'Conservation' is usually preferred to 'preservation' as a more appropriate term to describe biological sustainability and it is instructive to ask why the latter term is necessary: how does it differ from 'conservation'? (see Chapter 11).

It is sometimes the case that 'conservation' relates to external action to maintain a system without regard to its external effects (e.g. pollution, disease consequences, etc.). Thus, a malarial swamp could

Box 10.5 Loss of nitrogen in farming

The whole point of applying fertilizer nitrogen is to increase output: there is thus a loss of nitrogen in the product removed, as illustrated below:

Crop	Amount applied	Amount removed at harvest $(kg\ N\ ha^{-1}\ yr^{-1})$	Loss
Winter wheat (Biscoe and Dawson, 1983)	200	160	40
		Amount present in live animal	
Grass (grazed)	33	7	26
(Spedding, 1976)	66	7.6	58.4

However, there are also losses from:

1. decaying plant material, as nitrate or ammonia that is not then taken up by plants;
2. leaching of nitrate (which is very soluble in soil water) to ditches and rivers, or to groundwater by leaching downwards;
3. volatile gases released to the atmosphere from soil and animal excreta.

The amounts applied and lost seem almost insignificant in relation to the nitrogen content of the atmosphere ($c.$ 80 000 tonnes over each hectare of land surface) and of the soil, estimated by the Royal Society (1983) as: 20.5 tonnes ha^{-1} under forested land; 12.5 tonnes ha^{-1} under established grassland.

be 'conserved' as an ecosystem without regard to effects on a nearby human population.

Conservation would seem to be impossible if the system to be conserved is not sustainable – in whatever sense this is used. But a system cannot be said to be 'conserved' if it is altered, hence the notion that conservation can only require external action. However, questions of sustainability can be addressed to any system – new, altered or existing. So the following relationships may be useful in distinguishing preservation, conservation and sustainability:

Preservation relates to **static** systems (i.e. non-living)
Conservation relates to **dynamic** systems controlled by **external** influence only
Sustainability relates to dynamic systems whether subject to **internal** manipulation or **external** control (or both)

Conservation would thus be regarded as one means of achieving sustainability, relevant to one class of system: there may be an almost infinite number of means for achieving sustainability for other sorts of system, but they have to do with the **design and operation** of systems.

It has been suggested (IOB, 1992) that conservation can be considered in three categories:

1. conservation of habitat with no other form of land use;
2. conservation of habitat in balance with other forms of land use, such as agriculture and forestry;
3. conservation of habitat demanding changes in agricultural or silvicultural practices.

There are specific habitats which have arisen because of agricultural practices, sometimes with long

Box 10.6   Leaching of nitrogen (based on Jollans, 1985)

Only about 10% of the NO_3 leached in a year comes from the fertilizer applied in that year. Leached nitrate reaches a river quite soon. But, for an underground aquifer, the nitrate in water may move down through pores in the rock at less than 1 m per year.

Thus, in the UK, measurements by Young and Gray (1978) showed that high NO_3 levels are associated with times when little nitrogen was applied but when grassland was ploughed up in World War Two.

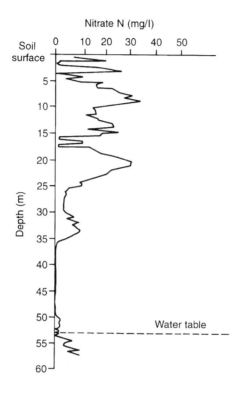

(a)   The progress of leached nitrate to an aquifer (1975).
Note. These measurements were taken in chalk below fertilized arable land with temporary grass leys.

traditions. Conservation then implies continuation of certain traditional agricultural practices which may be under economic pressure, such as profit levels, increasing human population leading to demands for food, etc.

This emphasis on habitats is partly because they represent one major objective of conservation (as judged by people concerned) and partly because it appears to be the most effective way of conserving species (of plants and animals). Obviously, individ-

ual species cannot be conserved in isolation from the habitats that sustain them, but it does not follow that only one habitat can do so.

In the UK, urban foxes are an excellent example where pressure on their original habitat results in them adapting quite easily to suburbia. Rats and mice are equally adaptable, so are sparrows, starlings and even kestrels.

Thus, conservation is not needed for many species and not desired for many others. People tend

Box 10.7   Nitrate sensitive areas (NSAs)

Following the designation of 10 pilot areas, a further 22 were added to the scheme in 1994. These are spread widely throughout England and Wales, on sandstone, limestone or chalk. The aim is to help meet the rules of an EC directive on the maximum permitted level of nitrates in drinking water.

There is a **basic scheme** limiting nitrogen applications, whether from organic or inorganic sources, to 150 kg ha^{-1} in any one 12-month period. It applies to arable cropping, with some exceptions (e.g. potatoes and vegetable brassicas), with compensatory payments that vary from one area to another.

There is also a **premium scheme**, with much higher payments per hectare, in four different forms. All involve conversion of arable land to grass, with limited application of nitrogen per year in one case but no nitrogen fertilization at all for the other three. Grazing is limited and cutting regimes are controlled in one way or another.

It has yet to be demonstrated what effect such schemes will have on the nitrate content of water.

to be rather hypocritical – or at least not objective – about such matters. If people in this country were asked about mice, they would probably wish to conserve the harvest mouse but not the house mouse. If asked about squirrels, they would be in favour of red but not grey ones. If asked about buttercups, they would want them to be maintained in farmers' fields but not in their own lawns.

It is reported to be similar in Australia, where koala bears attract enormous public support (with an Australian Koala Foundation) but the more endangered hairy-nosed wombat and the bilby raise little interest. And this in spite of the fact that the Koala is said to be 'nocturnal, slow, lumbering and easily upset'. Furthermore, it 'fails to fight back when attacked by dogs, despite having claws, is blinded by headlights, susceptible to chlamydia (an infection) and, because of its small adrenal gland, does not cope well with stress'!

This is not a trivial matter and someone has to

decide which species are to be conserved and which either ignored or positively eliminated. Several disease-causing organisms would come into the last category but what about dangerous or merely unpleasant animals? At this point, the judgement becomes subjective, yet many people would certainly choose candidates for conservation on the basis of appearance (beautiful, ugly, cuddly), feel (warm, cold, slimy), danger (poisonous, vicious, lethal), scarcity (rare, overnumerous) or damaging (to crops, animals, buildings).

Since all organisms can be viewed as resources, i.e. capable of conferring benefits on society – sometimes just by their presence, conservation can be interpreted as 'the wise, considered, planned, deliberate use of resources over time; a strategy for using resources' (IOB, 1992).

There are powerful reasons for concern. It has been estimated that, in the rain forests, at least 27 000 species are doomed each year, i.e. three every hour! Species are thus lost, not only before their potential usefulness is known but before they have even been described. The problems of estimating such losses are formidable and it is only recently (Lawton and May, 1995) that a major systematic attempt has been made to estimate 'extinction rates'.

Potential usefulness can be unimaginably high. For example, in surgery, cyclosporin, derived from an obscure Norwegian fungus, is a powerful immuno-suppressive agent which is today the basis of the entire organ-transplant industry.

Then consider that probably less than 1% of the species of fungi have been scientifically described: the figures for viruses and bacteria are estimated at $c.$ 1% and for nematodes $c.$ 3% (Pellew, 1995). Interestingly, Pellew (1995) points out that a species does not have to be named in order to be saved: conserving 1 ha of rain forest may protect thousands of species not yet known to science.

There is, then, a powerful argument for not allowing species to become extinct, or their populations to fall below some necessary threshold level, because their potential value (e.g. leading to knowledge that might help to cure dreadful diseases) has not been explored or exploited. There is, of course, a cost to conservation and, although many people call for it, it is usually far from clear who is to pay for it. If it is 'society', then it has to compete for funds with care of the aged and a host of other priorities. In very poor countries these problems are acute.

Pearce and Moran (1994) suggest that if only we could calculate the full value of a species, we would be better placed to persuade politicians that the species is worth preserving. But it is very difficult to do this, even for a large, well-known species like the elephant. Furthermore, the 'value' of a species may be quite unrecognized at any given time and its future value impossible to predict.

There is also the question of who should meet the costs of conservation. The extent of the difficulty is illustrated by, for example, the blue whale, which has 'existence value', with many people wishing to keep it out there, although they may never see one, and the cost is a global one, since only a global commitment can be effective. 'Existence value' goes beyond any utilitarian argument and leads to concepts of reverence for life – that life should not be needlessly destroyed – and stewardship.

It has to be recognized that any organism that is conserved successfully may become too numerous. If its numbers are not controlled it can become a serious nuisance or simply be allowed to be controlled by disease, fighting, predators or, most likely, starvation. If these outcomes are unacceptable, how should they be controlled and by whom?

The fact is that conservation usually requires population sizes that can sustain themselves and this may require action, such as captive breeding (see Box 10.8). By contrast, few people wish to conserve any organism that becomes too numerous. There is interest, however, in the number of species and this is what biodiversity is about.

10.5 Biodiversity

The term **biodiversity** was coined in 1985 (abbreviating 'biological diversity') and has been defined

---

Box 10.8 Captive breeding

If a wild population falls below the critical number needed to sustain itself in health or to increase in size, many people take the view that captive breeding may be able to correct the situation (IUDZG, 1993). Others argue that this is not dealing with the cause of the decline in population size and that, even if the captive-bred animals can be successfully released into the wild, the same causes of decline will operate.

Those in favour of captive breeding point out that at least it offers a chance and the alternative is to allow the species to disappear (or at least that population of it, which may be in a particular region of the world). Furthermore, there are cases where the cause of decline (e.g. hunting) has been stopped and the species has been successfully reintroduced: an example often quoted is that of the arabian oryx in Oman.

Other successful examples are reported to have been very costly: $400 000 per head for the Black-Footed Ferret in the US and $161 000 each for the Golden Lion Tamarin in Brazil (WSPA/BF, 1994). Amongst other successes are Conies in Jamaica, the Mauritian Kestrel and the Pink Pigeon (also Mauritian).

Behind these arguments lie some fundamental disagreements about the role of, and indeed the need and justification for, zoos. The reintroduction of captive bred mammals to the wild has been the subject of several studies and the International Academy of Animal Welfare Sciences (IAAWS, 1992) has produced a set of welfare guidelines.

The problems include all those of adaptation to the environment (such as resistance to diseases, ability to defend themselves and their young, ability to collect or catch their own food), but also relations with human beings, dependence and whether the environment has room for them.

Some consider that survival after release must be monitored but this may not be consistent with all the other requirements.

---

in many different ways. Holdgate (1991) regarded it as 'the total sum of life's variety on Earth, expressed at the genetic, species and ecosystem level' and suggested three reasons why it should matter:

1. All species deserve respect.
2. The living systems of the planet are our life support mechanism and we do not know the precise role that each species plays.
3. The economic importance for food, fibre, timber and medicines.

This has been recognized internationally by the World Charter for Nature, adopted by the United Nations General Assembly, and, in June 1992, some 120 Heads of Government attended the UN conference on Environment and Development. At this 'Earth Summit' in Rio de Janeiro, 153 countries (including the UK and the EC) signed the Convention on Biological Diversity. This conven-

tion defines biodiversity as 'The variability among living organisms from all sources including, *inter alia*, terrestrial, marine and other aquatic ecosystems and the ecological complexes of which they are part . . .'

There are three main levels of biodiversity:

1. between ecosystems and habitats;
2. between species;
3. genetic variation within species.

In 1994, the UK Biodiversity Action Plan (Cm 2428) was published with the goal of conserving and enhancing biological diversity within the UK and to contribute to the conservation of global diversity through all appropriate mechanisms.

Box 10.9 Numbers of species

[Based on Systematics Agenda 2000 (1994) and the Biodiversity UK Action Plan (HMSO, 1994)]
The numbers relate to terrestrial and freshwater species only.

	Known British spp.	Known world spp.	Estimated world no. of spp. to be discovered
Insects	22 500	*c.* 1m	8m–100m
Flowering plants	1 400	250 000	300 000–500 000
Arthropods (other than insects)	3 000	190 000	750 000–1m
Non-arthropod invertebrates	3 000	90 000	?
Lichens	1 500	17 000	?
Bryophytes	1 000	14 000	?
Ferns	80	12 000	?
Algae	15 000	40 000	200 000–10m
Protozoa	20 000	40 000	100 000–200 000
Fungi	15 000	70 000	1m–1.5m
Breeding birds	210	9 881	[a]
Wintering birds	180		[a]
Freshwater fish	38	8 500	[a]
Reptiles	6	6 500	[a]
Viruses	?	5 000	500 000
Mammals	48	4 327	[a]
Bacteria	?	4 000	400 000–3m
Amphibians	6	4 000	[a]
Total	88 000	1 770 000	

[a] It is estimated that the world total of known species for vertebrates is 45 000 with an estimated 50 000 to be discovered.

Box 10.10   Seeds banks

Seeds are often designed to last for a very long time and many can remain viable, if kept dry, almost indefinitely. Collections of seeds of genetic varieties of the main crops are held within the network of International Agricultural Research Centres (IARCs), which are distributed across the world [see figure (a)].

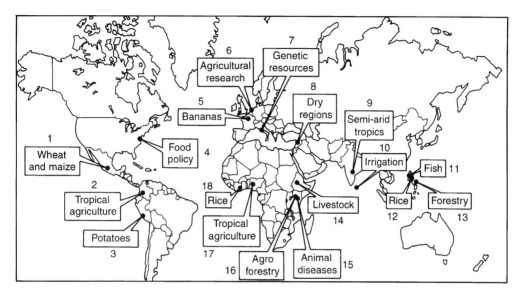

(a) 1, CYMMIT; 2, CIAT; 3, CIP; 4, IFPRI; 5, INIBAP; 6, ISNAR; 7, IPGRI; 8, ICARDA; 9, ICRISAT; 10, IIMI; 11, ICLARM; 12, IRRI; 13, CIFOR; 14, ILCA; 15, ILRAD; 16, ICRAF; 17, IITA; 18, WARDA.

Since 1983, the FAO Commission on Plant Genetic Resources has been the main forum for discussions on the control of plant genetic resources. In 1994, there was a major disagreement about who should control these valuable seed resources, mostly derived from developing countries which have often benefited little from their use (MacKenzie, 1994).

Seed production and sale is very big global business. In 1987, the 13 most important seed companies accounted for only 20% of the world seed market, but the total annual turnover was estimated at *c.* US$3bn (Gotsch and Rieder, 1995).

International Agricultural Research Centres (based on Tribe, 1994)

Centre	Date founded	Located	Subject area
IRRI (International Rice Research Institute)	1960	Los Baños, Philippines	Rice
CIMMYT (Centro Internacional de Mejoramiento Maiz y Trigo)	1966	Mexico City, Mexico	Wheat, maize
CIAT (Centro Internacional de Agricultura Tropical)	1967	Cali, Columbia	Rice, beans, cassava, tropical pastures

International Agricultural Research Centres (based on Tribe, 1994) (cont.)

Centre	Date founded	Located	Subject area
ICRISAT (International Crops Research Institute for the Semi-Arid Tropics)	1972	Hyderabad, India	Sorghum, pearl millet, chick-pea, pigeon pea, groundnut
CIP (Centro Internacional de la Papa)	1970	Lima, Peru	Potato, sweet potato
ILRAD[a] (International Laboratory for Research on Animal Diseases)	1973	Nairobi, Kenya	Animal diseases (e.g. Trypanosomiasis)
ILCA[a] (International Livestock Center for Africa)	1974	Addis Ababa, Ethiopia	Cattle, sheep, goats in sub-Saharan Africa
IPGRI (International Plant Genetics Resources Institute)	1974	Rome, Italy	Plant genetic resources
WARDA (West Africa Rice Development Association)	1970	Bouaké, Côte d'Ivoire	Rice
ICARDA (International Center for Agricultural Research in the Dry Areas)	1975	Aleppo, Syria	Wheat, barley, chick-pea, lentil, faba beans, pasture and forage legumes
ISNAR (International Service for National Agricultural Research)	1980	The Hague, the Netherlands	Research organization and management
IFPRI (International Food Policy Research Institute)	1975	Washington, DC, USA	Food policy
ICRAF (International Centre for Research in Agroforestry)	1977	Nairobi, Kenya	Agro-forestry
IIMI (International Irrigation Management Institute)	1984	Colombo, Sri Lanka	Irrigation
ICLARM (International Centre for Living Aquatic Research Management)	1977	Manila, Philippines	Fisheries management
INIBAP (International Network for the Improvement of Banana and Plantain)	1984	Montpellier, France	Bananas and plantains
CIFOR (Center for International Forestry Research)	1993	Bogor, Indonesia	Forestry management

[a] Combined in 1995 as the International Livestock Research Institute.

A comparison of the numbers of British species compared with world numbers is given in Box 10.9. However, it is well to remember that the numbers of known species is estimated to be only a fraction of those remaining to be discovered (also Box 10.9).

As applied to forests, biodiversity has been considered at three scales (Ratcliffe, 1993):

1. of biomass, biogeoclimatic zones and forest types;
2. of landscape;
3. at a localized scale, i.e. within a woodland.

In studying biodiversity, there are some special problems in marine ecosystems, because of their relative inaccessibility, but the oceans cover 70% of the earth's surface and most forms of life originated there. Furthermore, they may play a major role in the functioning of important global systems. However, our ignorance of the significance of biodiversity is substantial. It has been suggested (Cherfas, 1994) that terrestrial biodiversity affects productivity: the more species, the greater the plant production. Similarly, more diverse systems may be more stable and resilient, able to cope better with, for example, drought.

On the other hand, May (1989) has argued that more complex systems are, in theory, more likely to fall apart than simple ones.

In marine environments, there is apparently no evidence that species diversity is linked with sustainability or productivity: in fact, the ocean regions of high productivity are usually species poor with short, simple food chains (Whitfield, 1993).

Conway and Barbier (1990) almost define sustainability in terms of persistence and durability – the ability to maintain productivity against stress. Since modern agriculture tends to reduce complexity and diversity, and create rather simple systems, these relationships are of some importance.

In fact, as pointed out in Chapters 5 and 6, agriculture is heavily dependent on a very limited range of crop and animal species. About 25 species of plants and five species of animals provide 90% of all human sustenance and international commerce in foodstuffs (Solbrig, van Emden and van Oordt, 1992). It is worth noting that global warming could greatly affect this situation.

For crop plant species, seed banks have been created to ensure the survival of genotypes that might otherwise disappear (see Box 10.10).

10.6 Environmental auditing (See Box 10.11)

In view of the extent and range of effects that agriculture may have on the environment, it may be thought surprising that the 'environment auditing' being used in industry has not been applied to farming.

10.7 Consequences of land-use change

In addition to effects that agriculture may have on the environment, there are also important consequences of a change of land use to agriculture from

---

Box 10.11 Environmental auditing (see CSTI, 1994)

In 1990, 30% of UK companies had formal environmental policies and by 1993 65% had designated managerial responsibility for environmental issues: in 1994 71% had undertaken at least one environment audit. Germany, Ireland, Denmark, and the Netherlands were all ahead of the UK in 1990.

The definition of environmental auditing adopted by the International Chamber of Commerce in 1988 was:

A management tool comprising a systematic, documented, periodic and objective evaluation of how well environmental organisation, management and equipment are performing with the aim of helping to safeguard the environment by:
(i) facilitating management control of environmental practices;
and (ii) assessing compliance with company policies, which would include meeting regulatory requirements.

BS7750 (1994) defines environmental management audits as:

A systematic evaluation to determine whether or not the environmental management system and the environmental performance it achieves conform to the planned arrangements, and whether or not the system is implemented effectively, and is suitable to fulfil the organisation's environmental policy and objectives.

Box 10.12 Desertification (also sometimes called 'desertization')

The formation of deserts occurs in several ways, but always involves the loss of vegetation cover, essentially due to lack of available water. It can be progressive in the sense that loss of forests leads to poorer retention of such rainfall as occurs and, in some places, occurs by the encroachment of existing desert areas by blowing sand.

something else, such as forestry (see Chapter 8 and Myers, 1995). The net effect of such land-use changes may even result in desertification (see Box 10.12).

Questions

1. Should we reintroduce wolves and bears to the UK?
2. Who should pay for conservation?, for example, of tigers?
3. Are complex systems with many species better, more stable, more interesting?
4. Should we interfere with evolution?
5. Should the malarial parasite be conserved?
6. Should individual trees be preserved, even if they have to be propped up and treated for pests and diseases?
7. Should individual tortoises be preserved?, and tigers?, and mice?

Chapter eleven

Sustainability

> He who speaks the truth
> should have one foot in the
> stirrup.
>
> *Arabic proverb*

Even a cursory look at the world shows that poverty, cruelty and injustice are perfectly sustainable. No great effort is required to bring this about, so what does that tell us about sustainability? Clearly, it demonstrates that it is not necessarily good; and yet that is precisely what is implied by virtually all current usage. What then does it – or could it usefully – mean?

11.1 The meaning of sustainability

The most astonishing and currently most notable feature of the term is the apparent measure of agreement about it. In fact, it would appear that everyone is in favour of it and this alone is sufficient reason to doubt that it has a useful meaning and to suspect that it is being used by individuals to suit their own purposes. Its use is certainly widespread and this appears to be based on the feeling that some such term is needed to express an important concept. Unfortunately, because it is so loosely used, it accommodates a number of rather different concepts and it is hard to be sure which is intended at any one time. Because it tries to meet an important need, it cannot just be dismissed as useless, or even harmful: equally, it cannot just be accepted as if it had a sensible meaning, without further analysis. The number of definitions in use is very large (Conway and Barbier, 1990; Francis, Flora and King, 1990;

Holdgate, 1993; Harrington, 1995) and some of them are given in Box 11.1.

The most commonly quoted is the 'Brundtland' definition: this aims to be more comprehensive than most but it does not say how sustainability may be achieved and even its meaning depends upon being able to identify the needs of future generations (as if all of the world's future generations would want or need the same things). It is also somewhat negative in recommending that we should mainly avoid anything that would constrain this meeting of future needs.

A common gloss on this version is to emphasize the need to consider whether any action currently contemplated will put future generations at a disadvantage or saddle them with the costs of current action.

This kind of difficulty (in definition) is not unique to 'sustainability'. Indeed, the most important concepts carry the same problems, which is why the words used have been called 'umbrella' words.

11.2 'Umbrella' words

The best examples are 'freedom', 'truth' and 'beauty'. Take 'freedom' first and try to define it. Everybody knows what it means but no general definition is actually of much use. A useful definition can only relate to specific circumstances. If a lion is in a cage, it is quite clear what 'freedom'

Box 11.1 Definitions of sustainability

1. **The 'Brundtland' definition** The definition used in the Brundtland Report (Brundtland, 1987) is as follows:

   Sustainable development is development that meets the needs of the present without compromising the ability of future generations to meet their own needs. It contains within it two key concepts:

   – the concept of needs, in particular the essential needs of the world's poor, to which overriding priority should be given; and

   – the idea of limitations imposed by the state of technology and social organization on the environment's ability to meet present and future needs.

2. **Sustainable agriculture** is a system that 'can evolve indefinitely toward greater human utility, greater efficiency of resource use and a balance with the environment which is favourable to humans and most other species' (Harwood, 1990).

3. 'Sustainable development involves devising a social and economic system which ensures that these goals are sustained, i.e. that real incomes rise, that educational standards increase, that the health of the nation improves, that the general quality of life is advanced' (Pearce, Markandya and Barbier, 1989).

4. 'We thus define agricultural sustainability as the ability to maintain productivity whether of a field or farm or nation, in the face of stress or shock' (Conway and Barbier, 1990), where **productivity** is the output of valued product per unit of resource input.

5. Daly (1991) argued that 'Lack of a precise definition of the term "sustainable development" is not all bad. It has allowed a considerable consensus to evolve in support of the idea that it is both morally and economically wrong to treat the world as a business in liquidation.'

6. 'No single approach (to sustainable development) or framework is consistently useful, given the variety of scales inherent in different conservation programmes and different types of societies and institutional structures' (Heinen, 1994).

7. ' "Sustainable development", "sustainable growth" and "sustainable use" have been used interchangeably, as if their meanings were the same. They are not. "Sustainable growth" is a contradiction in terms: nothing physical can grow indefinitely. "Sustainable use" is applicable only to renewable resources: it means using them at rates within their capacity for renewal. "Sustainable development" is used in this strategy to mean: improving the quality of human life while living within the carrying capacity of supporting ecosystems' (IUCN/UNEP/WWF, 1991).

8. 'Development is about realising resource potential. Sustainable development of renewable natural resources implies respecting limits to the development process, even though these limits are adjustable by technology. The sustainability of technology may be judged by whether it increases production, but retains it within environmental and other limits' (Holdgate, 1993).

9. 'Sustainable development is concerned with the development of a society where the costs of development are not transferred to future generations or at least an attempt is made to compensate for such costs' (Pearce, 1993).

10. 'Most societies want to achieve economic development to secure higher standards of living, now and for future generations. They also seek to protect and enhance their environment, now and for their children. "Sustainable development" tries to reconcile these two objectives' (HMSO, 1994).

11. Riley (1992) pointed out the level of analysis of sustainability is important and quoted the following table from FAO (1989).

Box 11.1   (cont.)

Analysis of sustainability

Level of analysis	Typical characteristics of sustainability (cumulative)	Typical determinants
Field/ production unit	Productive crops and animals; conservation of soil and water; low levels of crop pests and animal diseases	Soil and water management; biological control of pests; use of organic manure, fertilizers, crop varieties and animal breeds
Farm	Awareness by farmers; economic and social needs satisfied; viable production systems	Access to knowledge, external inputs and markets
Country	Public awareness; sound development of agro-ecological potential; conservation of resources	Policies for agricultural development; population pressure; agricultural education, research and extension
Region/ continent/ world	Quality of the natural environment; human welfare and equity mechanisms; international agricultural research and development	Control of pollution; climatic stability; terms of trade; distribution

12. 'A sustainable agricultural system is one that can indefinitely meet demands for food and fiber at socially acceptable economic and environmental costs' (Crosson, 1992).

O'Riordan (1985) comments on the difficulty of defining sustainability, describing its definition as an 'exploration into a tangled conceptual jungle where watchful eyes lurk at every bend'. Perhaps this is the reason for the remarkable number of books, chapters and papers that even use 'sustainable' or 'sustainability' in the title but do not define either term.

means – at least initially, but clearly the lion and the bystander cannot both acquire significant freedom simultaneously.

Every example you can think of will illustrate the problems. We will all be in favour of it, in general terms, but in specific cases our views will differ. The same is true for all such words and 'sustainability' is of the same nature.

We should not, therefore, dismiss such terms simply because they cannot be pinned down by definition, but recognize that they are important, necessary and useful. Precise definitions have to be reserved for their use in specified conditions and for specified purposes. It is worth asking, then, what are the conditions/purposes to be specified for 'sustainability'.

11.3   What kinds of sustainability are there?

There are two common-sense propositions that would probably command general support:

1. A sustainable system (or process) must be based on resources that will not be exhausted over a

reasonable period (sometimes expressed as 'the long term');

2. A sustainable system (or process) must not generate unacceptable pollution (externally or internally).

It is quite straightforward to apply these to either physical or biological systems, but the latter are in some ways special.

11.4 Biological sustainability

No individual is 'sustainable' indefinitely, since all organisms die at some time. Preservation of individual life is therefore not possible except for limited periods. Species can be sustained, on the other hand, as can ecosystems and habitats, because they involve reproductive and other essential processes – without which they would cease to exist. However, many of these entities change and evolve as a result of such processes; sustainable processes, therefore, may not lead to sustainable entities (i.e. precisely as they were originally).

The implications of all this to issues of conservation were discussed in the last chapter (Chapter 10) but, of course, most biological systems have physical components, so there is considerable overlap between the use of physical and biological resources.

11.5 Sustainable use of resources

It sounds very plausible that a process should be sustainable in the sense that it can be continued (at least for the foreseeable future), without the resources needed running out. But if a resource is limited, what is the benefit of leaving it totally unused? There may be legitimate questions about the rate at which it should be used, the purposes for which its use is justified and by whom it should be exploited.

This question of leaving resources unused is of some importance (and has a conservation dimension, considered in Chapter 10). If, for example, we believe that we should leave a resource unused and our descendants take the same view, no one

will ever use it and we have created an 'unusable' resource by our arbitrary decision.

There are also resources which, if not used now, will not remain. The most obvious example is sunlight. The solar radiation received today, if not used or trapped, will not be there tomorrow. Nor will the rate at which it is used today affect the amount available tomorrow, although it may well affect the amount that can be used tomorrow. For example, if solar radiation is not used to grow a leaf today, the leaf will not be available for photosynthesis tomorrow. In fact, of course, solar radiation is not indefinitely sustainable and, as James Lovelock has said, if the sun was not running down, we would not exist.

Non-living resources may largely remain intact, though they may change greatly in form (e.g. radiation loss, soil erosion, weathering of rock) but living resources do not remain static. A dairy cow that is not milked presently, may provide no milk in the future and will eventually die and be totally unavailable for any purpose. All living things eventually die: thus a tree not used as a resource will eventually die and decay – liberating the same amount of CO_2 as if it had been burnt.

Even the products of living organisms, such as meat, milk and potatoes, have a limited storage life and even that may be very expensive (e.g. needing refrigeration) of both money and energy. Drying is often effective but is not suitable for all products and, unless carried out using the sun and/or the wind, requires a great deal of expensive support energy.

The use of living resources may either have to be considered over a relatively short period or be related to animal and plant populations, capable of reproduction. This has given rise to the concept of sustainable **harvesting**, taking only such proportion of the population as can be continued over time. It may, therefore, depend upon such factors as the reproductive rate in animals and the yield of seed in plants.

It is also sensible to distinguish between using a resource and using it up. Some forms of use (e.g. viewing the countryside) may not use it up at all:

other forms of use (e.g. walking in the countryside) may destroy it. Other resources, such as water and energy, can be used up but never destroyed. They may be changed in form, be recyclable or difficult to recapture. All of this, of course, depends upon the availability of energy in an appropriate form and at an affordable cost (see Chapter 8).

Many resources (e.g. fossil fuels) are totally changed in use and cannot be recreated on any reasonable time-scale. But they serve no purpose unused (except to take CO_2 out of circulation) and their use may allow other, substitute, resources to be found or devised.

Oil replaced coal as coal had replaced wood and, for all we know, may provide the energy needed to produce or discover new and better sources of energy. Many resources are only required as a means to doing without them. Fossil fuels can make possible the construction of dams (for hydropower) and windmills (for windpower).

Nor are these substitution and development processes necessarily foreseeable, so that what future generations will need is very hard to predict, quite apart from the difficulty of assuming that they will have the same values as us. And all this in an unpredictable world, full of uncertainties about climate, population growth, disease incidence, nuclear accidents and a host of 'Acts of God'.

The more one considers what should constrain the use of resources, the less important preserving, or even conserving, them seems to be, and the more important become any adverse consequences (pollution etc.) of using them. Soil erosion is an important exception to this (see Box 11.2).

This change in attitude is well illustrated by the way in which the use of fossil fuels was viewed in the 1970s, as a scarce non-renewable resource that must be husbanded, compared to the current concern that burning them may cause global warming by increasing the content of CO_2 in the atmosphere (see Chapter 10).

So far, the kinds of sustainability considered fit well within the common-sense usage referred to earlier but the concept has been used much more widely in recent years, to include **economic** and **social** sustainability.

> **Box 11.2   Soil erosion**
>
> Erosion is the loss of soil by action of rainfall, run-off or wind. The eroded soil may be deposited on other land, or in estuaries or be carried out to sea. Unfortunately, it tends to be the more fertile topsoil that is most readily lost.
>
> In world terms, the most vulnerable areas to water erosion are those between 40°N and 40°S which have semi-arid or wetter climates. The areas most affected by wind erosion are those with drier climates.
>
> Water erosion is judged to be the more serious of the two. It is greatly exacerbated by loss of ground cover and especially forest vegetation.
>
> Physical loss of the soil is not, however, the only way in which crop production may be rendered impossible – **salinization** is another. Salt-affected soils (saline, alkaline or sodic) constitute a major threat to irrigated lands and it has been estimated that over 50% of irrigated agriculture is damaged to some degree by salt-related problems (see Box 8.5). Salinization can follow forest clearance because the rising water table brings with it accumulated salt (Thompson, 1995).
>
> It has been estimated that 75bn tons of topsoil are eroded every year and c. 9m ha of productive land lost. Thompson (1995) stated that four-fifths of the world's agricultural soils are affected by erosion.

11.6   Economic sustainability

Economics is, of course, about the efficient use of resources, whether expressed in monetary terms or not. In this sense, economic sustainability is the same as the sustainable use of resources and can be thought about in the same way, except that, in monetary terms, one resource can substitute for another. If, however, inputs and outputs are expressed in monetary terms – and there are few alternative ways of aggregating different inputs and different outputs – then economic efficiency is dominated by the actual costs and prices obtained at the time. Since these cannot be predicted, there is no way of ensuring that any particular choice of process or system will remain economically efficient in the future. So it is not possible to choose an economically sustainable system for the future, except in relation to stable, predictable or 'arranged' (e.g. as with subsidies) costs and prices.

In general, what is meant by saying that systems must be economically sustainable is that if they do not remain profitable they will cease to be practised. This is fairly obvious and much clearer without the use of the word 'sustainable' at all.

The same applies to other extensions of the concept. In fact, meeting the needs of future generations, however desirable, has little to do with sustainability. Failing to meet needs has gone on for a very long time and could easily be continued in the long term – a kind of sustainability.

We do not meet the needs of current generations, a failure often due to war, greed and selfishness. It is perhaps too cynical then to observe that the destruction of millions of people, by war, famine or failure to control disease, might be said to increase the chances of future generations meeting their needs in a less crowded world.

It does focus attention, however, on what are legitimate 'needs' and who judges this. Most of us would define our needs in some relation to what we are accustomed to and what we hope for. The debate has thus already entered into the realms of speculation and subjective judgement.

11.7 Socio-economic sustainability

It is often argued that something is not sustainable because it depends upon unacceptable practices, such as near-slavery, sending small boys up chimneys or other conditions for workers that have been defended in the past on economic grounds. The trouble is that such judgements are bound to be subjective and we do not all agree on what is acceptable. In these cases, it is hard to see that the use of the term sustainable makes any sense. Words like 'responsible', 'acceptable', 'desirable' or 'mor-

ally right' seem both more accurate and more appropriate.

Acceptability introduces subjective judgement which may rest on moral, religious or social values but which may also be governed by fashion, custom or usage and may be quite capricious. Society is entitled to establish criteria of acceptability and these will change over time: hopefully, they will evolve in directions which are, in some way, 'better'.

Research and practice, alike, have to respond to these social criteria but both may help to change them by acquiring new knowledge, whether based on experiment or experience. Unfortunately, people differ in what they find acceptable, within and between societies, and, since the criteria change with time, there is no way of guaranteeing social sustainability into an unpredictable future.

Thus, the social sustainability of animal production, for example, is currently influenced by varying standards of farm animal welfare (see Chapter 12) but, in the future, would be transformed if society became generally vegetarian or vegan (see Chapter 17).

The question of how long a system has to be sustainable can only be answered by 'as long as it is needed/wanted' and this cannot be predicted.

11.8 General conclusions

Before considering the specific case of agriculture, some important general conclusions can be drawn from the foregoing discussion.

1. A straightforward meaning and definition of sustainability can be based on the long-term use of resources, i.e. susceptible to scientific measurement (of quantity of resource reserves, rate of usage, output of pollution and other important factors). The data measured can provide a basis for the design of improved resource-use systems.

2. The sustainability thus planned or achieved may be essential but it is not, of itself, sufficient. Sustainability can never be a sufficient end in itself. How long it is possible to sustain a

process or an activity is no indication of its value or effectiveness.

Life is full of examples of **necessary but not sufficient** criteria. Life requires that a human being breathes, this is an essential requirement but, by itself, it is not enough. No one would be satisfied with such a sustainable activity if that is all that was achieved.

3. This is just one example of the need argued by Handy (1994), to distinguish between **requirements** and **purposes**. As he pointed out, in order to live it is necessary (indeed, absolutely essential) to eat, but this requirement should not be confused with the purpose of living. In spite of some strong evidence to the contrary, the purpose of living is not eating.

Similar arguments can be used for all 'bottom-line' requirements, including profit: however essential, they are not the same as purposes or, when they are, they are only one within a multi-purpose activity.

4. When dealing with economic and social aspects of sustainability which are unpredictable or subjective, it is better to use self-evident terms, such as profitable, responsible, acceptable or desirable.

5. It is desired to describe a system (activity, process, society) as sustainable, it has to relate either to a specified aspect or to a package of such specified aspects.

Following up this last point for agriculture, it ought to be possible to compile a package of the most essential and/or desirable facets.

11.9   A sustainable agriculture

Such a package can be based on the six essential attributes of future agricultural systems given in Box 11.3.

The first point is an essential prerequisite or the purpose is lost. It should be noted that the products referred to are not confined to food and that they must be wanted sufficiently for consumers to pay a satisfactory price (i.e. there must be an economic demand for them, however created). The first three

Box 11.3   Essential attributes of future agricultural systems (after Spedding, 1994)

1. they should be highly productive, of safe, high quality products (within identified constraints, such as those listed below);
2. they should be physically sustainable (i.e. use physical resources at rates or in ways which allow adequate long-term development);
3. they should be biologically sustainable (i.e. the biological organisms and processes on which they depend must be sustainable in the long term) – this could encompass the avoidance of 'internal' pollution, such as the build-up of heavy metals;
4. they should satisfy agreed standards for human and animal welfare;
5. they must not give rise to unacceptable pollution, by-products or effects (including visual);
6. they must be profitable (since they will not be practised if they are not) – this also assumes that the products are wanted (otherwise there will be no demand and the business will collapse).

of these points are susceptible to scientific research and assessment. The last three are different.

The word 'agreed' in (4) is subjective and it has to be recognized that what is agreed will vary with culture, ethics and custom.

Similarly, the word 'unacceptable' in (5) is subject to non-scientific judgements, but this subjectivity does not mean that research cannot be carried out. It merely means that the criteria to be satisfied have to be laid down by others (e.g. society). There is no logical way in which such judgements can be predicted and, for this reason, it is probably better to recognize that (5) deals with 'acceptability' rather than 'sustainability'.

In (6), future profit depends upon quite unpredictable costs and prices. A profitable system can only be sought for given values of these.

Research can (as it does now) establish the relationships from which financial outcomes can be calculated but there is really no scientific way of predicting whether too many farmers will choose to produce a given product and the resulting over-production can render any system unprofitable. Such a 'package' may well be improvable but it seems likely that it will consist of a number of 'sustainability' attributes plus several that are essentially social and economic.

Others have taken a different approach. Conway and Barbier (1990), for example, have defined agricultural sustainability as 'the ability to maintain productivity, whether of a field or farm or nation, in the face of stress or shock' (such as increasing salinity, or erosion, or debt, or a new pest, or a rare drought or a sudden massive increase in input prices) (see Box 11.1). This 'resilience' is certainly a factor in determining whether a system survives (or is sustainable). Definitions of this kind tend to relate to a **development** context.

11.10   Sustainable development

It has been argued (e.g. by Conway and Barbier, 1990) that international development policies are now incorporating the concept of sustainability. Thus, the World Bank (1988) listed three criteria for agricultural development: 'First, it must be sustainable, by ensuring conservation and proper use of renewable resources. Second, it must promote economic efficiency. Third, its benefits must be distributed equitably.' This has the advantage of keeping resource-use sustainability separate from the economic and social goals but it leaves a subjective element in sustainability by referring to the 'proper' use of resources. Who determines what is proper?

The debate should, of course, be open to all those affected by the development but this is the basic problem of development, i.e. the purposeful or economic use of land and other resources whilst satisfying those who, in some cases, do not really want it to be used at all (or as they might put it 'exploited for profit').

Yet the social and economic objectives can only be achieved by the use of resources and whether this use is proper will be judged differently by different people. For example, Holdgate (1994) puts one

view of the situation. ' "In 1991 IUCN – the World Conservation Union, the United Nations Environment Programme and the World Wide Fund for Nature published a second World Conservation Strategy entitled *Caring for the Earth. A Strategy for Sustainable Living* (IUCN/UNEP/WWF, 1991). It has a simple message. We need development, to ease the problems of the billion or more people who live in poverty and to cater for the 3 to 5 billion we are warned will be added to the world population during the next 50 years. But that development must both provide real improvement in the quality of life and conserve the vitality and diversity of the Earth. To quote: 'We need development that is both people-centred, concentrating on improving the human condition, and conservation based, maintaining the variety and productivity of nature. We have to stop talking about conservation and development as if they were in opposition, and recognize that they are essential parts of one indispensable process." '

Other authors argue that sustainability may not always be possible, unless human population growth is controlled. For example, Wilson (1992) holds that 'The raging monster upon the land is population growth. In its presence, sustainability is but a fragile theoretical construct.' However, he also describes specific systems of forestry that are designed to be sustainable in a perfectly understandable way.

Strip logging involves clearing a corridor along the contours of the land, narrow enough to allow natural regeneration within a few years. Another corridor is then cut above the first, and so on, through a cycle lasting decades. This he describes as allowing a sustainable timber yield from forests, including the relatively fragile rain forests. The problem is that more money may be made by doing it some other way. The concept in relation to agro-forestry is discussed by Park (1993).

Questions

1. Should everyone live as long as possible?
2. What is a 'free' society and would you wish to live in one?

3. Is our responsibility to future generations greater than our responsibility to those currently in need?

4. Is running a sustainable activity?

5. Is a garden sustainable? Does it produce all its inputs, including the food of the gardener?

Chapter twelve

Animal welfare

It is false economy to burn down your house in order to inconvenience your mother-in-law.

Chinese proverb

Do animals have rights? To many people this is the ethical question on which attitudes to animal welfare have to be based. If they do, is this an innate property or have they been conferred upon them? The problems are obvious. If they have rights, what are they and who should protect them?

Consider an owl and a mouse. Both have the right, one supposes, to fight for life but their interests are in opposition and cannot both be supported or defended – as individuals (species would be another matter).

The same questions can be (and are) posed of people, babies and foetuses: yet millions of people are, in practice, at the mercy of events and other people and their rights cannot be protected. If we **confer** rights on others, it seems inevitable that we must accept some responsibility for defending them. What do rights mean, if they cannot be defended?

An alternative starting point is to accept that if we use or own an animal, we acquire a duty of care, a responsibility for its welfare. Does this apply to the entire animal kingdom? The numbers of individuals and species, are enormous (see Box 12.1) and most of them are invertebrates.

Quite apart from the impracticability of such a notion, it is hard to justify concern for the welfare of very lowly forms of animal life, mainly because they do not appear to be capable of suffering. Most concerns therefore focus on sentient animals.

12.1 Sentient animals

In the main, it is generally accepted that sentient animals include all the vertebrates. The distinction is based on the degree of development of the nervous system and it is difficult to delineate boundaries: it has recently (Bateson, 1992) been recognized that octopods can feel pain and suffer and should be regarded as sentient.

It is difficult to agree on the most suitable words to describe the characteristics of a sentient animal. 'Feeling pain' suffers from the problem that we may assume that this means in the same way that we feel pain, and 'suffering' implies pain. We would be less concerned at, say, suffering the loss of a leg, if this involved no pain. [It is interesting that most of us would object to a fly having its wings pulled off (by a lizard?) even if no pain was involved. Our objection appears to be to the action itself and the sort of person who is doing it.]

The animal, of course, is not at all concerned about our thoughts and attitudes, it is concerned about what we do (or fail to do) that affects its welfare.

12.2 Does it matter?

Attitudes to cruelty and animal welfare vary greatly between people, nations and cultures and they change with time.

Box 12.1 Number of species of animals (after Wilson, 1992)

Total living animal spp. currently known = **1 032 000**.

Descending order of magnitude Phyla	Example	Numbers
Coleoptera	Beetles	290 000
Lepidoptera	Butterflies and moths	112 000
Hymenoptera	Ants, bees, wasps	103 000
Diptera	Flies	98 500
Homoptera and Hemiptera	Bugs	82 000
Arachnida	Spiders and mites	73 400
Small insect orders	Earwigs, mayflies, grasshoppers	65 500
Other arthropods (not included above)		50 000
Molluscs	Snails, mussels	50 000
Non-mammalian chordates	Fish, frogs, reptiles	38 300
Platyhelminths	Flatworms	12 200
Nematodes	Roundworms	12 000
Annelids	Earthworms	12 000
Minor Phyla		9 300
Cnidaria and Ctenophera		9 000
Echinoderms	Starfish	6 100
Porifera	Sponges	5 000
Mammals	Humans, mice, whales	4 000

In the UK, attitudes to horses have changed greatly over 100 years and animals such as dogs are held to have a special place in our affections. However, the statistics for the gross neglect of both dogs and horses are horrifying, and some of this cruelty stems from ignorance or thoughtlessness.

Bear-baiting, as a public spectacle, would no longer be tolerated in the UK but persists in other countries. Dog-fighting and badger-baiting, by contrast, are covertly (and illegally) practised. Even so, it is generally held that concern for animal welfare is higher in the UK than in most other countries. One thing is certain and that is that the total amount of animal suffering, over the world as a whole, is enormous and most would wish to see this reduced. Those who take this view believe that they are right and feel very strongly about it. Most of the suffering would be judged to be unnecessary and to have a bad effect on those causing it. None the less, it has to be recognized that others feel differently, have different priorities or simply don't care.

One of the criticisms of (other) people's attitudes to animal suffering is that it is anthropomorphic – it attributes to other animals the feelings that we have as humans. This is so often used to dismiss concerns that it is worth pausing to analyse it a little.

It is worth noting that some of the worst excesses, of cruelty by people to other people, have been associated with the conviction that the others are not fully human, that they are more like animals, that they are undesirable and should be ill-treated or even eliminated. You only have to think of attitudes to colour, race and religion to see the dreadful effect that can flow from refusing to attribute to others the feelings of one's own kind (group, race, football club, army, political persuasion, trade union, etc.)

So, there is some value in recognizing that others (including other animals?) may possibly suffer as we do. To recognize this possibility is not the same as attributing to other animals exactly the same sensitivity or susceptibility to pain or to perceptions of it. Any more than we have any right to assume that other people are as sensitive (or insensitive) to pain as we are ourselves. It is important to be clear about these issues because it is not a simple matter to improve the situation. The fact is that most people would put human beings first and some are faced with more acute choices than others.

Recently, Campbell (1995) has reviewed this debate and pointed out that 'welfarism does not place animal suffering on a par with human suffering as pain and distress are not ruled out totally. It merely requires that we seek to minimize them.' This attitude is often combined with a compulsion to avoid unnecessary suffering. The problem, of course, is who judges what is unnecessary?

12.3 Animals of most concern

There are different sorts of reasons for concern. For example, the sheer numbers of animals slaughtered for meat (see Table 12.1) make them a matter for concern but part of this has to do with the fact that they are being killed. Since no animal can live for ever and death is inevitable, the concern ought to focus on whether the methods are humane.

Other animals, such as companion animals (see Box 12.2), stand in a much closer relation to their owners and this gives rise to some quite different considerations.

TABLE 12.1 Numbers of livestock slaughtered annually in the UK, taken from selected examples for 1988 (based on Pickering, 1992)

	No. (m)
Cattle (beef and veal)	236.5
Sheep (mutton and lamb)	442.0
Goats (goat meat)	197.5
Pigs (pork)	870.5

Numbers are not available for other groups including poultry, which provided c. 22.5% of the total meat production in 1988. The total number of livestock slaughtered annually worldwide is probably of the order of 15bn (Spencer, 1995).

Very different attitudes apply to other people's activities (e.g. hunting, fishing) and this probably influences attitudes to livestock farming.

There are also animals (e.g. badgers, foxes) inadvertently affected by human activities, not directed at them, and it can be argued that this also incurs some responsibility.

Somewhat different arguments are applied to animals used for exhibition and entertainment, as in some zoos, aquaria, dolphinaria and the like. Here it can be argued that the whole activity is unnecessary and even degrading.

But some of the sharpest criticisms are directed at animals used in experiments, especially those that are judged unnecessary or related to the testing of, for example, cosmetics.

All these are ethical issues, either for the individual or for society and, historically, it has generally been the case that enlightened individuals are ahead of their societies and endeavour to persuade, lead, shame or coerce them into changing their ways. Whatever the ethical position adopted, as to whether some particular use of animals is acceptable, there is general agreement, amongst those who are concerned, that there is a duty of care and

Box 12.2 Companion animals	
Animal	Purposes (other than production) for which they are kept
Horses	Riding (recreation and sport), military (transport, traction)
Dogs	Guide dogs for the blind, sniffer dogs, guard dogs, pets, tracking
Cats	Pets, rodent control
Cage birds	Pets
Rabbits Guinea-pigs Mice Rats Hamsters	Pets, laboratory experiments
Ferrets	Hunting (especially rabbits)

a responsibility for ensuring good standards of welfare. So what is good welfare?

12.4 What is good welfare?

This question has been posed most acutely and answered more fully in the UK for farm animals. Following Ruth Harrison's book *Animal Machines* (1964), a study was carried out by Professor Rogers Brambell which was reported in 1965 (Brambell, 1965) and, as a result, the Government established a Farm Animal Welfare Advisory Committee in 1967 and this became the present Farm Animal Welfare Council (FAWC) in 1979.

Central to the thinking behind these developments was the idea that 'positive' welfare had to be more than the mere absence or avoidance of cruelty to animals (which is against the law in the UK anyway). Positive welfare embodies the notion of behavioural freedom, initially of a closely confined animal (see Box 12.3). This was extended to include the now internationally known and respected 'five freedoms' (Box 12.4). These represent directions in which provision for animals should move if positive welfare is to be achieved. It is quite possible to argue that none of these freedoms can be totally achieved and even, indeed, that they should not be: it can also be argued that they cannot all be achieved simultaneously. For example:

1. If an animal did not experience hunger and thirst, it would have no incentive to feed and drink, or to engage in the foraging and other activities associated with the satisfaction of nutritional needs.
2. Total protection or insulation from the environment – even harsh elements of it – could interfere with the proper development of an animal's own defences (e.g. hair and wool growth etc.).
3. Exposures to disease may be necessary in order to develop immunity and pain is an indispensable part of an animal's protective armoury. (Imagine the damage we might incur, if we could not feel pain.)

4. Normal behaviour is not always tolerable since animals are not necessarily well disposed towards each other – or us. The normal behaviour of predators is difficult to accommodate without some injustice to prey animals.

Box 12.3 Behavioural freedoms (Rogers Brambell, 1965)

Freedom for each animal:

1. **To turn round** Veal crates, for example, are too narrow for calves to turn round: this affects movement, obviously, but also grooming, etc.
2. **To groom itself** This is primarily a matter of space but may be affected by height as well as width of pens.
3. **To get up.**
4. **To lie down.**
5. **To stretch its limbs.**

Points (3)–(5) involve adequate space and this must have regard to the ways in which different species get to their feet and lie down. Point (5) must include freedom to flap wings in the case of birds.

These freedoms are clearly minimal and do not allow flight, running, jumping, rolling, etc. That is why they needed further elaboration (see Box 12.4).

Box 12.4 The 'five freedoms'

1. **Freedom from hunger and thirst** by ready access to fresh water and a diet to maintain full health and vigour.
2. **Freedom from discomfort** by providing an appropriate environment including shelter and a comfortable resting area.
3. **Freedom from pain, injury or disease** by prevention or rapid diagnosis and treatment.
4. **Freedom to express normal behaviour** by providing sufficient space, proper facilities and company of the animal's own kind.
5. **Freedom from fear and distress** by ensuring conditions and treatment which avoid mental suffering.

5. Fear is also a protective emotion and a total absence of it would generally be to the disadvantage of an animal.

All this has links with the discussion in Chapter 11 (on freedom, for example), where 'umbrella' words may be useful even though they cover a wide spectrum of more precise meanings which have to be specified in order to make further progress.

These 'five freedoms', therefore, are not just unachievable targets, they do not carry a sufficiently precise meaning to be achievable. But they are an invaluable guide to the directions in which we should move and should lead to a whole series of further questions. For example, what kinds of degrees of fear and anxiety are tolerable, valuable, or should be reduced or avoided? And the answer will vary with the animal, its species, age, size, degree of wildness/domesticity and a host of other attributes. It is also necessary to take account of duration: are we talking about acute or chronic conditions? 'Good' (or positive) welfare also requires answers to such questions as 'how can it be measured'.

Ordinary human experience illustrates the difficulties and most people would not find it sensible to apply a scientific test to assess suffering, quite apart from the possibility of the test adding to the suffering, only measuring one aspect or taking too long. In general, when we cannot ask the sufferer, we rely on observation of behaviour and this will be how most people would act in relation to animals. This is because the judgement is then left to the sufferer, who is the best judge, but will not always express it openly [to avoid attracting the attention of predators or because of limited capacity for, for example, facial expression (in birds)].

If good welfare is looked at from the animal's point of view, it accepts external and internal impacts as acceptable – up to a point. This is the point up to which it can **cope** with the stresses, strains and stimuli to which it is exposed. Beyond this point, when it cannot cope with the stress (and this could be due to lack of stimulation or boredom), it will usually exhibit behaviour patterns that are not 'normal' and signal that all is not well. This is a big subject, with a large literature, a selection of which is given in the bibliography for this chapter.

In a sense, therefore, it is easier to see signs of **poor welfare** but, even here, there are difficulties. For example, Webster (1995) has pointed out that grazing animals (such as cattle, sheep, antelope) have powerful reasons for not revealing pain, injury or sickness to watching predators and have presumably evolved to be stoical. So, the fact that an animal does not appear to be suffering may be no evidence that it is not doing so.

12.5 The causes of poor welfare

Poor welfare obviously occurs under natural conditions. Although the incidence of deliberate cruelty, with the intention of causing suffering, seems rare in nature (and even the most obvious examples may represent play, although not always associated with training of the young), there is no doubt that **suffering** is widespread.

Killing in the wild often appears to be savage but quick. However, this is not always so. For example, many animals that catch prey much smaller than themselves start eating it alive. Other prey is swallowed whole and alive, or simply torn apart. In addition, a great many animals are, as it were, eaten from the inside by parasites or, slowly, from the outside by pests (e.g. blowfly maggots on sheep).

None of this, however, should greatly affect our judgement of what is morally right for human beings who have control over other animals. The animals involved are in agriculture, sport and recreation (including zoos), as food producers, for transport and traction, as pets and companion animals generally. The greatest numbers are probably farmed livestock, whether for food or as sources of power. Pets and companion animals are very numerous in some countries: smaller numbers are used in sport and recreation. Hunting and shooting affect relatively small numbers worldwide but large numbers of small birds are shot or trapped in some countries and vast numbers of fish are caught at sea.

The larger the numbers, the less animals (as with people) are seen as individuals and the less individual attention they receive. This may not result in increased welfare problems – this may well be the case in intensive poultry keeping – but tends to do so when animals are handled in large numbers ('harvesting' of hens from battery cages, sea fishing, in transport and at slaughter).

The pressure of conveyor belts and other handling lines is not conducive to good welfare and it reduces the time operators have to think and observe. A great deal of animal suffering is caused by quite unnecessary poor handling and thoughtlessness.

There are, of course, wide cultural differences across the world, in attitudes to animals and their treatment and in beliefs as to whether animals can really feel pain or, indeed, whether it matters. In Northern Europe generally, and, it is often thought, in the UK in particular, the greatest concern is focused on farm animals: the main areas of concern are shown in Box 12.5. Before discussing these further, it is worth considering why farm livestock are highlighted in this way. Doubtless there are many reasons, amongst them that we nearly all consume livestock products but very few are involved in their production. So we can feel guilty at consuming a product unless we are assured that it was produced in a welfare-friendly fashion, yet neither we nor most people we know can be guilty of treating the animals badly.

Farm animal welfare is thus a much safer topic for concern and conversation than, say, pet keeping. Most pet keepers are as convinced that they treat their animals well as are farmers about their livestock. Yet cases of cruelty occur in all sectors and positive welfare is equally hard to ensure. No one keeps dogs and cats 'naturally' as many would believe cattle and hens should be kept. Dogs are often tethered in ways that it is held livestock should not be. However, there is no reason to suppose that our behaviour will be any more logical, rational or consistent in these matters than it is in most other areas

---

Box 12.5 Public concerns about farm animals

The main concerns are listed in Table 12.3.

In intensive systems, most concern focuses on overcrowding, space restriction, restraints, stress/fear and mutilations (see separate para).

In entensive systems, most concern focuses on malnutrition, exposure to adverse weather, disease, lack of care (especially at birth) and mutilations (see below).

The major mutilations (FAWC, 1981)

1. **Castration** is common practice for most male cattle and sheep (and was so for pigs), because:
   (a) very few males are needed for breeding (see Table 6.1 for examples of numbers required);
   (b) unwanted, uncontrolled reproduction has to be avoided;
   (c) uncastrated males may be dangerous (e.g. bulls).
2. **Desnooding of turkeys** Removal of the snood (Stevenson, 1994) is carried out to prevent damage by frostbite or fighting.
3. **'Dubbing'** This is the removal of combs from large-combed breeds of poultry (for the same reasons as for desnooding). Now generally not necessary, as a result of positive breeding.
4. **Tail-docking** Mainly of lambs to prevent accumulation of faeces on long, woolly tails attracting damaging flies to lay their eggs. Also done on cows in New Zealand, to avoid irritation when milking, and on dogs – mainly to conform to breed standards.
5. **Dehorning** and **disbudding** Removal of horns or preventing their appearance is carried out mainly on cattle but also, to a lesser extent, on sheep and goats, primarily to avoid damage due to fighting.
6. **Branding** Done with hot irons or by freeze-branding in cattle and horses.

In a country like the UK virtually all these practices are controlled, banned or only allowed to be carried out by a veterinary surgeon. This is not the case for the world as a whole.

---

of human behaviour. The concerns listed in Box 12.5 are entirely understandable.

Instinctive and anthropomorphic feelings are involved, in objections to crowding and mutilation, for example, but we actually need scientific or practical evidence about whether particular animals need particular levels of light or temperature. There is often an assumption that intensive farming is bad for welfare and that extensive farming is therefore good. Neither proposition is necessarily true. Intensity will usually mean that animals are well fed and watered, protected from the weather, pests, parasites and diseases, and looked after at and around birth.

Extensive livestock keeping sometimes means poor protection from the elements, pests and diseases, poor feeding in winter and neglect at critical times, such as parturition.

It is unwise to generalize about such things: indeed, even the terminology is suspect and some have proposed that the terms 'intensive' and 'extensive' should be abandoned as meaningless (Barber, 1991).

12.6 Farm animal welfare improvement

The range of animals farmed across the world is very wide (see Chapter 6) and it is sometimes hard to decide exactly what constitutes a farm animal. It can be argued that for a species to be farmed its reproduction must be under the control of human beings (Spedding, 1988) but this does not mean that this occurs on every farm or for every animal. The majority of meat animals are not bred from at all, milking animals usually are, and egg producers are in the same category as meat animals.

In the UK, for legislative purposes, farm animals are those kept on agricultural land; this includes fish that are farmed on such land, but not, for example, dogs bred in the farmhouse. However, once an animal is so defined, welfare standards apply to it during transport, at markets and in slaughterhouses.

The problems of improving welfare are quite different in these areas and what constitutes good or improved welfare has to be decided for each specific case. The main considerations in these areas are briefly listed in Boxes 12.6–12.9.

The most complicated situations relate to pro-

duction on the farm and there are increasing attempts to establish conditions that would warrant a 'welfare-friendly' label on the products. The credibility of any such scheme depends both upon the standards set and on the effectiveness with which they are implemented, inspected and policed. This is currently a particular problem for transport across national boundaries.

The situation on the farm is made more difficult because it is almost impossible to label a production system as good from a welfare point of view. Bad, certainly: there are battery cage systems, for example, where the cage dimensions are totally inadequate but, even here, it also depends upon the number of birds put into one cage.

This applies to every system, however well designed; if too many animals are held in it, it becomes a poor welfare system. The same is unfortunately true in relation to the operator: no system is invulnerable to the effects of a bad stockperson.

If a system cannot be simply assessed, how should one proceed? It is necessary to identify all the critical factors for the welfare of each species (age, sex, size, etc.) kept in a particular way and then to quantify the criteria that have to be satisfied. These criteria are summarized in Box 12.10 and an example given in Box 12.11 and Table 12.2. A good welfare system is then one in which all or most of the relevant criteria are satisfied.

How can improvement in welfare then be brought about?

12.7 The role of public opinion

As will be argued in a following chapter (Chapter 14), the citizen has both the right and the responsibility to say 'this activity/practice is not acceptable to me and should not be so in the society of which I am a citizen.' The citizen has no individual right to determine this but has the right and the responsibility to try to influence society in such matters. Most of us would prefer to live in a society where animal welfare is a matter of importance rather than in one where no one cared.

Furthermore, on the whole range of issues about

Box 12.6   Animal welfare on the farm (see FAWC, 1993)

This is a very large subject and cannot be easily summarized, since the needs of the different kinds of farm animals differ greatly, between species and over the course of their lives.
     The main areas of concern are:

1. **How to assess welfare** In particular, how to see the world as the animal sees it (perception) and how it thinks about it (cognition). This includes how the animal senses its environment, how it processes the information, how it compares it with, for example, relevant memories, and how it experiences its world as pleasant or unpleasant sensations.
2. **Maintenance of health and prevention of disease** Especially where animals experience pain and suffering that do not seriously affect performance or productivity or which arise as a direct consequence of breeding, feeding or housing.
3. **The avoidance of pain and injury** In particular, the need to recognize both and to relieve pain.
4. **The design of husbandry systems** This includes the design of housing and equipment, methods of moving animals, the supply of feed and water, removal of excreta, in order to improve 'positive' welfare.
5. **Management** This includes the appropriate use of husbandry systems but also the need to train operators.

An example of the detail that has to be covered is provided by the areas that FAWC (1991a) included in its recommendations for laying hens kept in colony systems:

space allowance;
provision of nest boxes (number, design, positioning);
provision of perches (design, length, positioning);
control and prevention of feather-pecking and cannibalism;
prevention of bone breakage;
minimum light intensity (and dark periods);
appropriate stocking densities;
limitations on beak-trimming;
number, size and distribution of popholes;
provision of litter (and nature of etc.);
arrangement of tiers in multi-tier systems;
provision of scattered whole grain (to encourage foraging);
management of waste, ventilation, water and feed supply;
refuge for bullied birds;
vegetation on free-range areas.

which a citizen of modern society is expected to have views, it is not feasible to be well informed. Thus (as I have remarked before), in a democracy, we are all expected to hold strong views on a range of topics about which we are almost wholly ignorant. It is no use complaining, therefore, that the concerned citizen is ill informed and does not fully understand the issues, and it is idle to suppose that all such issues can be safely left to the specialists. This means that those engaged in keeping animals, of whatever sort, should listen when fellow citizens express concern. The major current animal welfare concerns are listed in Table 12.3.

However, the right to identify something as unacceptable does not convey any right to say what *should* be done. This requires relevant knowledge, expertise and experience.

Liverstock keepers – of whatever kind – should look more critically at the ways in which they keep and treat animals and ensure that their standards are acceptable. If they do not, public opinion may suddenly gel and focus an uncomfortable spotlight on their activities. If the latter are commercial in any way, the consequences can be very serious and threaten the operator's livelihood.

Public opinion may work through public pressure, by influencing consumer behaviour or by bringing about changes in legislation. If it fails to

Box 12.7 Farm animal welfare in transport

Transport issues	Examples of FAWC recommendations (FAWC, 1991b)

1. Journey times
  (a) by road
  (b) by sea
  (c) by air
	Cattle and sheep should be rested, fed and watered at intervals not exceeding 12 h. Pigs, in addition, need water more frequently.
	A ring fence should be used to prevent escape during loading or unloading at an airport.

2. Lairages (approved secure premises close to an embarkation point where animals may be rested, fed and watered)	Lairages should be available where sea and air transportation may be delayed by adverse weather.
3. Resting intervals	At the end of a journey stage animals should be rested for at least 10 hours.
4. Staging posts and designated assembly points	These should be approved and regularly inspected by a competent authority.
5. Stocking densities (number of animals per unit of floor space – height, nature of floor, ventilation etc. also matter)	Optimum stocking densities should be calculated for all species, related to age, size etc., and controlled. Too many animals (overcrowding) lead to injuries and death: too few animals may lack support and be buffeted about.
6. Vehicle design	All dimensions of a vehicle need to be specified for all classes of stock: sharp edges should be avoided. Stock from different groups should not be mixed. Drivers should be trained.

Note:
1. 'Infirm, diseased, ill, injured or fatigued' animals should not be transported if this would cause unnecessary suffering (FAWC, 1988).
2. Special arrangements are necessary for the handling and transport of poultry (FAWC, 1990).

achieve change, the moderates may feel deep dissatisfaction and disappointment. Unfortunately, moderates are frequently ignored. In fact, one could almost define a 'moderate' as someone who can safely be ignored. This does not mean that they always are, especially if they are customers/consumers, and it is most unwise if it is the case.

Ignoring moderate opinion is a recipe for the creation of extremists, whose actions may not command the respect or support of the majority but who may be seen to be effective where moderation failed. So often, public opinion is amorphous, unfocused, ill informed, poorly articulated and, in any case, far from homogeneous. This is not a criticism, simply an observation, but when it is so, it is very easy to ignore or dismiss. That is why pressure groups have such an important role.

12.8 The role of pressure groups

Pressure groups are relatively well organized, often quite well resourced, focused on a limited number of issues and many carry out thorough research that puts them in a very well-informed position. They are sometimes tempted to go over the top, to exaggerate and never to admit that progress is being made. The reasons are obvious, they have to make themselves known and attract funds by persuading people that the issue is of major importance and that they (the pressure group) will get something

---

Box 12.8 Farm animal welfare at markets (FAWC, 1986)

Areas of welfare concern	Examples of requirements
Adequate lairage (see Box 12.7) and covered accommodation	Adequate shelter and ventilation
Penning and passageways (races) These should be designed to avoid overcrowding and injury	Non-slip floors (well-drained) Vertical rails are better than horizontal No dangerous projections
Bedding may be needed for all species	Calves, pigs and dairy cattle in milk or in calf have special needs
Separation of stock Different kinds of stock should not be mixed (but suckler cows and their calves should be kept together) **Facilities needed**	Bulls over 10 months of age should be penned individually (so should boars) Horned and polled cattle should be separated for milking For isolation (for sick or injured animals)
Lighting	Should be adequate after dark
Fire prevention	Availability of fire-fighting equipment

---

done about it. Their ability to achieve change, however, may be damaged if their actions in any way become extreme.

Naturally, on issues such as cruelty and welfare, people feel very strongly and can persuade themselves that only extremism will achieve results. This illustrates, again, that resistance to change may generate extremists.

Public debate is therefore of the greatest importance:

1. to demonstrate that the issue is being taken seriously;
2. to provide opportunities to listen to ordinary people;
3. to provide opportunities to clarify the issues and improve the level of information and understanding;

TABLE 12.2 Criteria which need to be satisfied (in great detail): an illustration of a comparison of welfare of laying hens in cages and on free range

Criteria derived from the 'five freedoms'[a]	Battery cage	Free range
Nutrition	Adequate	Adequate
Comfort		
Thermal	Good	Variable
Physical	Poor	Generally good
Hygiene and freedom from infectious disease	Usually good	Variable, parasitism (coccidiosis, etc.)
Freedom from pain and injury	Deformed feet, osteoporosis, bone fractures	generally good, predators
Fear	Rare	Predators, agraphobia
Behaviour		
Frustration	Common	Rare
Abnormalities	Common	Cannibalism

[a] These have to relate to the species, size, weight, age and sex of the animal; additional criteria may also need to be added in many cases in relation to: space requirements; temperature needs; light (intensity, night/day periods); scale of the enterprise; stocking density; provision of food; provision of water; 'furniture' needs (e.g. perches, dust baths and nest boxes), access to other animals of the same species; disease control.

Box 12.9 Farm animal welfare at slaughter (FAWC, 1984)

The slaughter sequence

1. Transport to the abattoir;
2. unloading;
3. lairage;
4. handling from lairage to stunning area;
5. stunning;
6. sticking.

1. Transport (see Box 12.7).

2. Unloading should be calm and unhurried (very difficult to guarantee with such large numbers of animals involved). This is affected by handling by staff (including the use of goads) and thus staff training, design of loading ramps, the use of hydraulic tail-lifts.

3. Lairage (see Box 12.7).

4. Handling from lairage to the stunning area should encourage unimpeded movement forward in a calm, unhurried atmosphere, with minimal risk of physical damage and stress. This is affected by design of pens and races, too narrow to allow animals to turn round but preferably curved. Noise should be minimized (e.g. metal gates clanging).

5. Stunning. In the UK there is a legal requirement that animals should be stunned before slaughter. The object is to render them 'instantaneously insensible to pain until death supervenes'. (This does not, however, apply to animals slaughtered by Jewish or Moslem methods. This is commonly but, in my view, unhelpfully, known as 'religious slaughter'.)
   Methods of stunning vary from:

(a) Electrical stunning. Mainly used for pigs, sheep and calves by devices which pass an electric current through the brain. Timing, correct voltage, maintenance of equipment, positioning of the 'tongs' are all of vital importance.
(b) Captive bolt stunning involves a pistol that fires a bolt that penetrates the skull and destroys part of the brain. Positioning is crucial and varies with the species.
(c) 'Gaseous' stunning is brought about by anaesthesia induced by an appropriate mixture of gases (mainly CO_2 and oxygen).

With all methods of stunning, especially with large animals, restraints may be necessary to hold the animal's head in an accessible position.

6. Sticking. The historical reason for separating stunning from actual death was to allow 'sticking' – the severing of main arteries – leading to 'bleeding out', which results in less blood in the muscles, and 'exsanguination' (blood loss leading to oxygen deprivation of the brain).
   Sticking also requires the animal to be suitably positioned (including for the collection of blood) and this commonly involves suspension by shackles. It is therefore of the greatest importance that animals are rendered absolutely insensitive to pain before any part of this process begins.

Poultry (FAWC, 1982)
Special consideration has to be given to the slaughter of poultry.
   Stunning is commonly carried out by passing the heads of suspended birds through an electrically-charged water bath. The birds are then subject to 'venesection' (cutting of the throat) before they pass through a 'scald tank' (in order to make plucking easier). It is clearly crucial that each stage is fully effective on all birds at all times. The slaughter of unwanted male day-old chicks presents a different problem, which still has not been solved to everyone's satisfaction.

Box 12.10 Criteria of positive welfare

A good animal production system has to satisfy the criteria listed below, spelt out in detail for every species, size, sex and age of animal. In the UK, such criteria are embodied in FAWC Reports and MAFF Codes of Practice based on them. There are few such codes and detailed criteria for pets, however, and perhaps there should be.
The main criteria relate to the following features of the environment.

1. **Light** must be sufficient for the animal to carry out its activities (eating, drinking, moving about) and for staff to inspect them regularly. Dark periods may also be required. Light intensity and even wavelength matter. Quite often (e.g. in poultry) low light is used to prevent fighting, feather-pecking and cannibalism.
2. **Temperature** must be controlled so that it is within a band appropriate for the animals housed.
3. **Space** has to be adequate in total for whatever stocking density is used but there are difficulties in setting standards. Number of animals m^{-2}, for example, takes no account of the size or weight of the animal. Weight and size do not increase in a simple linear way, just as a human of twice the weight of another may require no more height and more or less floor space, according to whether they are standing or lying down.
4. **Food and water sources** The points of supply have to be sufficient and not placed so that one animal can prevent access by others. Contamination of food, water and bedding have to be avoided.
5. **Furniture** Pens and cages should not be bare: most animals need to have an interesting environment that they can explore, but a degree of security is also important.

For poultry, such features as the presence of perches (and their dimensions), nest boxes, scratching area and dust baths are all examples of furnishings that may transform a bare cage into a welfare-friendly environment (this usually, of course, requires a larger cage).

4. to explore what change is necessary, possible or desirable; and
5. to consider the cost and consequences of change.

But public debate does not happen by itself, it has to be initiated, led, organized and informed.

Pressure groups can do this and thus create a climate in which change can at least be considered. Far better that pressure groups should unlock a log-jam than that extremist action should ever be thought necessary.

Opinion-formers in the media could also con-

Box 12.11 Specific criteria for laying hens in colony systems (a selection of 12 out of the 50 recommendations made in FAWC, 1991a)

1. A minimum space allocation of 1425 cm^2.
2. Nest boxes available.
3. A minimum perching space of 18 cm per hen.
4. Access to litter (30% of the area) with grain scattered in it.
5. A minimum light intensity of 10 1ux: plus a period of darkness each day.
6. Ban routine non-therapeutic beak trimming by 1996.
7. Sufficient pop-holes no smaller than 30 cm high and 75 cm wide.
8. Vegetation on outside area.
9. Stocking rate not to exceed seven hens per m^2 (except in multi-tier systems).
10. Stocking rate in rotational free-range systems not to exceed 1000 hens per ha.
11. 10 cm of feed trough side per hen.
12. Sufficient drinkers (bell type: one per 80 hens, nipples: one per 10 hens).

TABLE 12.3 Concerns about animal welfare

Close confinement
Overcrowding
Intensive feeding
Lack of food or water
Lack of attention to health
Physical ill-treatment
Very early weaning
Bullying
Lack of shelter
Bleak environment
Mutilations
'Unnatural' processes

tribute greatly to constructive public debate but they may not see this as their main job or as commercially rewarding.

The number of organizations devoted to animal welfare improvement is very large worldwide: this is illustrated in Box 12.12.

12.9. The responsibility for welfare improvement

Whose responsibility is it to bring about improvement in animal welfare? Although citizens have some responsibility for trying to influence how their society behaves, most cannot be held responsible for actually achieving the results.

If the central proposition is accepted, that anyone keeping, controlling or significantly affecting sentient animals should accept a duty of care that includes positive animal welfare, then clearly these same people must be held morally responsible. This would include, for example:

pet/keepers/sellers/breeders;
livestock farmers and their staff;
livestock transporters;
livestock market operators;
abattoir owners and staff;
companion animals owners/trainers/keepers;
zoo keepers/owners/visitors;

as well as those who hunt, shoot, fish, rear game and drive on roads that animals cross. The last illustrates the problem where there is no intention to affect an animal and the whole list could be expanded to include many more people. The scale of the issue is very large for a country such as the UK but, for the world as a whole, it is enormous.

As with so many problems facing humankind, the scale must not be used as an excuse for inaction. Few such problems can ever be solved: one has to

Box 12.12 Welfare organizations of the world (a small selection)

Organization	Acronym	
World Society for the Protection of Animals	WSPA	Has 323 members from 69 different countries including the following:
Royal Society for the Prevention of Cruelty to Animals	RSPCA	UK
Massachusetts SPCA		USA
American Humane Association	AHA	USA
American Society for the Prevention of Cruelty to Animals	ASPCA	USA
Humane Society of the United States	HSUS	USA
Deutscher Tierschutzbund		Germany
Schweizer Tierschutz	STS	Switzerland
Wiener Tierschutzverein		Austria
Ned. Ver. tot Bescherming van Dieren		Holland
Royal Society for the Prevention of Cruelty to Animals, Australia	RSPCA	Australia

determine the right direction in which to move and then forge ahead as fast and as far as is possible.

Government also has an important role, because society may insist that the behaviour of the people listed must be regulated. This may require legislation, inspection, monitoring, policing and enforcement. The difficulties of framing such legislation and the cost of implementing it should not be underestimated.

Sometimes, it is better to concentrate on voluntary improvements instituted by those involved and it should be remembered that voluntary actions may attract rewards or sanctions, quite apart from legislative constraints. For example, consumers and customers can refuse to purchase products that have not come from welfare-friendly production methods, and customers include processors and retailers. However, the pressure for improvement by any of these routes will vary greatly from country to country and between activities. No systematic improvement can be expected if there is no agreement on standards and this, too, varies between countries and, often, between states or regions within a country.

As mentioned earlier in this chapter, the UK has a Farm Animal Welfare Council (FAWC) which has the responsibility for advising Ministers in MAFF on what constitutes good welfare for farm animals and what changes are needed in legislation to achieve it. The Home Office regulates the use of animals used in experiments (Box 12.13).

However, there are no comparable bodies to recommend how pets and companion animals should be kept, although many organizations offer advice on such matters, directed mainly at good nutrition and maintenance of health. There are thus no welfare controls on the keeping of companion animals except that it is against the law to cause unnecessary suffering. Since it is quite infeasible to inspect all the hamsters, rabbits and guinea pigs kept in ones and twos all over the country, it is clear that much of the responsibility has to be taken by individuals, often with little guidance or even a recognition of the need for it.

There is obviously a role for education in all this

and, because of the difficulty of defining standards, the emphasis has to be on an increased appreciation of the needs of animals and how to assess their behaviour (which may indicate how well these needs are being satisfied).

Just because welfare standards exist, embodied in the UK in Codes of Practice (e.g. MAFF, 1994), for farm livestock, one should not underestimate the problems. For example, one of the most obviously needed standards is for the space allowed for each animal. This will clearly vary with the animal, its sex, weight, age and so on, but what is space? It is not just that there are three dimensions and that the material of the floor especially has a bearing on whether a space is adequate, it is that the quality of the space matters. Thus, for the battery hen, the provision of a scratching area, an adequate perch, a nest-box and a dust-bath, may be preferable to a big increase in environmentally-barren space.

The same is true for pigs and straw, and nesting facilities will be important to nearly all breeding

---

Box 12.13 Regulation of use of animals in experiments in the UK

The Animals (Scientific Procedures) Act 1986 regulates 'any experimental or other scientific procedure applied to a protected animal which may have the effect of causing that animal pain, suffering, distress or lasting harm': it came into effect on 1 January, 1987. The House of Commons has published a 'Guidance' to its operation (HMSO, 1990), which explains all aspects of the Act and the controls embodied in it.

A **protected animal** means 'any living vertebrate other than a human' but the definition can be extended to cover any invertebrate, by order of the relevant Secretary of State. It applies from halfway through gestation or incubation or when the animal becomes capable of independent feeding.

People carrying out regulated procedures have to be appropriately licenced. As might be expected, there is vigorous debate about all aspects of the use of animals in research, development, testing and education (see Smith and Boyd, 1991; POST, 1992; Anderson, Reiss and Campbell, 1993).

females (and males in the case of, for example, sticklebacks). It is amazing that such a point should ever need making, when you consider our own reactions to sitting in a room, however large, with no furnishings at all.

12.10 The cost of improvement

Resistance to improvements, especially amongst livestock producers, is often based on the assumed cost of achieving them. The argument runs as follows: what you are suggesting will cost more, so you will have to pay more for your eggs/milk/beef, etc: if you do not, but buy cheaper imported products, I will go out of business.

It is worth examining this argument in some detail, but first it must be couched in cost/benefit terms.

12.11 Cost/benefit assessments

McInerney (1991) examined the evidence available about cost increases associated with different egg-producing systems (see Table 12.4) and pointed to the need for carefully structured studies to establish the magnitude of the concern, because it is those concerned who would mainly benefit by their greater satisfaction at the introduction of better welfare systems.

However, Carruthers (1991) pointed to economic losses due to poor welfare conditions and Guise (1991) concluded that, in the UK, pig welfare improvements could result in increased profitability of £27.15m annually, or £1.80 per pig sold.

Relatively few cost/benefit analyses (Box 12.14) have been carried out and livestock producers have

TABLE 12.4 Costs of different egg-producing systems (after McInerney, 1991)

System	Index of production costs
Hens in battery cages	100
Perchery systems, 20 birds m^{-2}	108
Deep litter systems, 7–10 birds m^{-2}	118
Free-range systems, 400 birds ha^{-1}	170

tended to react defensively with the simple argument already stated. This ignores the fact that some welfare improvements cost very little, including those based on changed attitudes on the part of the people involved. It also reinforces in the public mind the view that the price of food in the shops is largely due to the cost of production. This is far from the case and food prices are generally about twice those received by the producer, or vastly more in the case of processed foods (Box 12.15). In fact, retailers are likely to charge what they can get, having regard to competition from other retailers.

However, consequent increases in capital costs may be very serious. For example, some 90% of UK hens' eggs are produced in battery cages and the investment in these cages (for many millions of birds) is very high. If such cages were banned they would be worth nothing and massive investment would be required in whatever kind of accommodation replaced them. If this were a different kind of cage (with perch, nest-box, etc.), the investment per cage would inevitably be even higher than the original and, with fewer birds per unit of space, the investment per bird would be higher still. It is not surprising, therefore, that such changes are resisted and that legislation tends to allow long transition periods from one system to another.

Some of the changes being sought would alter the structure of whole industries. Obviously, vegetarianism (see Chapter 17) would require radical change in the entire livestock industry. Acceptance of a vegan philosophy would have even more drastic effects.

But even changes to the existing industry, without altering its size, could require expensive restructuring. For example, a very reasonable proposition, that all livestock should be slaughtered as close to its point of origin as possible, would require a complete change in the size and distribution of abattoirs. The proposition is reasonable because so much suffering is associated with transport of animals to the point of slaughter. In some cases this involves extremely long, unsupervised, multiple journeys across several countries, under dreadful conditions and incurring substantial losses (including

Box 12.14 Cost/benefit analysis

It seems obvious that before any major planned change or development takes place, both the costs and the benefits should be calculated. The fact that this is frequently not done is, in part, a reflection of the difficulties in doing this.

It is not a simple matter to measure – or even identify – all the costs of a proposed action. Often the action is spread over a considerable period of time (raising problems of discounting etc.) and many of the disbenefits may not appear as monetary costs.

Similarly, some of the benefits may be extremely difficult to quantify and some may not occur for some time. As soon as time enters the equation, it is also clear that costs, benefits and disbenefits have to be predicted – with all the hazards involved in that.

Economists have developed techniques for dealing with the likely return and impact of contemplated investments: this is slightly easier to do retrospectively and important lessons can be learned from so doing. Cost/benefit analysis is one such technique.

Social cost/benefit analysis aims to identify and measure all the costs and all the benefits accruing to society as a result of an investment decision. It may not necessarily use current market values but take into account the 'real' scarcity of a resource or 'real' value to the community of a product (Ansell and Done, 1988).

In the case of research (see Chapter 16), it is often extremely difficult to see which piece of research led to precisely what benefit and it is important to recognize the dependence of all research on much that has gone before (sometimes many years before).

For a discussion of the application of such techniques to farm animal welfare, see McInerney (1991).

Box 12.15 Costs of production and prices of products

The price per kilogram varies with the product and its scarcity: world prices only reflect the volumes of product that are internationally traded. Examples are given in the following table (the actual prices are out of date but the relativities are illustrated by them).

Price per kilogram of protein for animal and plant foods

	Product	World price[a] (£)
Animal	Beef	10.08
	Poultry meat	3.28
	Milk (cond. and evap.)	6.87
	Cheese	6.44
Crop	Wheat	0.86
	Wheat flour	1.27
	Rice	3.26
	Soya beans	0.47
	Potatoes	5.33
	Cocoa beans	5.70

[a] Based on FAO (1985) world average export unit values.

Data on costs of production vary in similar ways, so the following figures have been calculated simply as the price received by the producer as a proportion (%) of the retail price. [Even these proportions may well have changed since these calculations (Spedding, 1982)].

Product	%
Milk	45
Eggs	58
Beef	56
Bacon and ham (uncooked)	40
Wheat – flour	42
Potatoes – prepacked	41
– crisped	2.6

actual deaths). Not only would abattoirs be affected but also the transport systems and, because carcases can be more efficiently moved than live animals, the number of refrigerated lorries would be less than those used for livestock. It is hardly surprising, therefore, that there is an important political dimension to welfare improvement.

12.12 The political dimension

Animals are commercially traded across national boundaries, products are imported from countries with different standards of production and exported to countries with different standards for transport and slaughter. Non-farmed animals, such as horses, for show or competition, are also moved from one country to another but, in general, such animals tend to have a high unit value and are relatively well looked after. This is not, however, the case for the trading of pets, where cruelty and heavy losses appear to be commonplace. These international interactions make welfare a political issue.

Within the EU, freedom of trade is a central part of the ruling philosophy and it is difficult to interfere with this on the grounds of differences in prevailing animal welfare. This raises severe political problems, since a government which bans a practice in its own country cannot prevent imports from other countries without such a ban. Thus, the effective banning of veal crates in the UK in 1990 prevented UK farmers from rearing veal calves in narrow, individual crates – the lowest-cost method, but could not stop veal being imported from other European countries (e.g. Holland) that continued to use that method. Farmers are understandably angered by what appears to them to be a hypocritical public, calling for a ban at home but buying imported products.

This has come to be known as 'exporting the welfare problem' since it does not solve or even reduce the welfare problem but merely changes where it exists. In the particular example above,

the extent of the problem was actually increased by the export of very young calves from the UK to be reared in veal crates elsewhere, veal then being imported back to this country.

None of these cases are quite as simple as they appear but they do illustrate the real possibility that 'home' producers may be put out of business by unilateral action without any real welfare benefit. It is understandable, therefore, that a government (such as that in the UK) should be reluctant to take any unilateral action that would put its own livestock producers at an economic disadvantage *vis-à-vis* their other European counterparts. But the effect is to make progress very unlikely, since no government will be in a position to initiate it. In short, if we do not act in unison, we shall not be able to act at all. What then are the chances of this?

12.13 European action

Not all European countries have bodies such as FAWC to advise Ministers on animal welfare matters and those that exist differ in many ways (see Box 12.16). Few, if any, are as well resourced as FAWC.

Theoretically, studies and advice could be coordinated, so that it would be much easier for the EU to act as one. But there are other considerations that

---

Box 12.16 Animal welfare organizations established in European countries to advise their Governments

Country	Name	Ministry responsible
Belgium	The Animal Welfare Council	Agriculture
Denmark	The Ethical Council Concerning Animals	Justice
France	No Council Several Commissions	Agriculture & Forestry
Germany	Federal Commission on Animal Welfare	Food, Agriculture & Forestry
Netherlands	Council on Animal Affairs	Agriculture & Public Health (two ministries)
Norway	Ethical Council Regarding Animals	Agriculture
Sweden	Standing National Committee for Animal Welfare	Agriculture
UK	The Farm Animal Welfare Council	Agriculture, Fisheries & Food

politicians have to take into account in addition to the welfare advice they receive before deciding when and how to act. However, there are also other bodies in Europe concerned with animal welfare (see Box 12.17) with important roles to play. Furthermore, since Europe imports and exports animals and animal products from countries outside Europe, similar arguments apply worldwide.

12.14 Worldwide welfare problems

The range and magnitude of world animal welfare problems are enormous, but then so are the human welfare problems. People suffering from the ravages of war, famine, disease and natural disaster (e.g. floods, hurricanes, earthquakes, volcanic eruptions) are in no position to think about animal

---

Box 12.17 European animal welfare bodies (based on Spedding, 1994)

The importance attached to animal welfare in Europe is illustrated by the fact that all EU Member States (and many non-EU countries) are contracting parties to the Council of Europe Convention for the Protection of Animals Kept for Farming Purposes. The general principles embodied in that convention therefore constitute a common welfare policy throughout Europe. This has been implemented by means of recommendations for individual species, laying down common welfare standards for the species concerned. The EU, being a Contracting Party in its own right, is obliged to implement the recommendations by means of Community Law.

There is also a Council of Europe convention for the Protection of Animals during International Transport, to which the EU became a party in 1979. Welfare at Slaughter is the subject of Council of Europe and EU policy. In fact, the Council Directive of 18 November, 1974, on stunning animals before slaughter, was the first Community law on farm animal welfare. The reasons for such legislation were: (1) that disparities in national legislation in the field of protection of animals may directly affect the functioning of the common market, and (2) that the Community ought to take action to avoid, in general, all forms of cruelty to animals.

There is now a Commission budget specifically voted for animal welfare studies, which are commissioned from consultant experts. In 1987 studies commissioned included the keeping of pigs and calves, transport, stunning and information exchange among EU countries in farm animal welfare, resulting in proposals to the Council in 1990. The Commission's objectives are clear: 'the improvement of the welfare of animals from which we profit and for which we bear the responsibility of care'.

Many other national animal welfare organizations contribute to the improvement of welfare within Europe. The more important ones are represented on the Eurogroup for Animal Welfare, which serves as a two-way channel of communication between these organizations, the Institutions of the EU (Council of Ministers, Commission, European Parliament, Economic and Social Committee) and the Council of Europe.

The policy of Eurogroup is to bring together in friendly cooperation the leading welfare organizations of the EU to participate in public policy decisions via the legislative process and to encourage humane treatment of animals by increasing public awareness and encouraging responsibility towards fellow creatures. Policy, of course, has finally to be formulated and effected by Governments, increasingly acting together in an EU context, and there has to be some mechanism for deciding what constitutes welfare improvement.

EU countries differ in the extent to which they have a designated body with responsibility in these matters. The UK Government, for example, does not have a single organization with overall responsibility for the welfare of all animals. The Agriculture Departments have the responsibility for farm animal welfare and are thus restricted to animals that are farmed, whether on the farm, in transit, at markets or at slaughter.

Increasingly, Governments' actions are harmonized throughout the EU member countries but there are concerns about the extent to which agreements, directives and regulations are implemented, monitored and enforced in different countries. A powerful European Inspectorate must be the ultimate answer to this kind of problem but, whilst discrepancies occur, there is resistance to controls being applied in one country, putting its livestock producers at a competitive disadvantage with those of countries not following the rules. In fact, this concern, that advances in animal welfare may impose costs not borne by competitors – whether within the EU or from outside it – is one of the major constraints on progress in this area.

---

welfare, never mind about acting to improve it. Even when life is 'normal', normality for many people is characterized by poverty, suffering and often hunger, and in these circumstances animal welfare will not rate a high priority, whether for individual action or government expenditure.

Examples abound of the acute dilemmas that all these circumstances pose. For instance, there have been cases of bear dancing, involving cruelty and degradation for the animal, being the only source of livelihood for an individual. Sometimes, when action is taken, it may not result in welfare improvements.

When the exhibition of dolphins was banned in Brazil in 1988, some dolphins were left in a worse position, in the same small pools but with no resources to look after them. (In this case, WSPA stepped in to remove the dolphins – but what do you then do with them? The problems associated

---

Box 12.18   Release of captive animals

Most captive animals (other than in agriculture) are thought of as being in zoos, but all pets are, in effect, captive, as are laboratory animals.

Zoos are generally defined as establishments which possess, or manage collections of, mainly wild animals which are confined for exhibition or study and which are displayed to the public (see IUDZG, 1993). However, many zoos regard themselves as centres for captive breeding of rare and endangered species. This raises two major issues:

1. how should such animals be kept?;
2. how should they be released into the wild, if breeding is successful?

Point 1.   The 'five freedoms' are a useful test for captive animals of all kinds but:

(a)   exhibited animals also require some protection of their privacy so that visitors do not intrude adversely;
(b)   enclosure should not be sterile leading to unnatural and stereotyped behaviour;
(c)   animals should not be isolated from their own species (or, indeed, other species);
(d)   ventilation with fresh air must be ensured;
(e)   the needs for animals to be active, including searching for food, must be met.

There is much debate about the desirability of zoos: criticisms of them and their activities have been recently embodied in a report (WSPA, 1994). Of course, there is enormous variation in the standards of zoos across the world and varying degrees of control of their activities.

Point 2.   The reintroduction of species, following successful captive breeding, usually involves the introduction of animals that have spent their whole lives in zoos or in protected areas. The main issues are:

(a)   where are they to be released and how gradually?;
(b)   will the habitat be suitable for the species and still have capacity to absorb new individuals? (in other words, if it is suitable, why isn't it full up?);
(c)   can the released animals
        feed themselves;
        avoid or cope with pests, parasites, predators and disease;
        avoid continued dependence on humans or attack by them?;
(d)   will released animals spread new diseases to the existing wild population?;
(e)   will they upset the existing ecological balance?

It should be noted that most of these problems are exaggerated if animals have been in captivity a long time – a problem that results from the closure of zoos, menageries, dolphinaria, etc.

There are also worries about disturbing the existing gene pool (in the wild population) and many people consider that released animals should only be returned to the areas and populations from which they came. This may not always be possible and, in any case, there may be no 'room'.

It is also felt that released animals should be monitored, but this also presents problems of such complexity that some hold the view that either release will rarely make sense or that continued captivity (let us assume, in improved conditions) will generally be preferable.

---

with the release of captive animals are illustrated in Box 12.18).

None of this provides reasons or excuses for giving up but it does suggest that progress may be slow and gradual, involving education and persuasion.

12.15 Achieving progress

Gradualism does not generally appeal to those greatly concerned about welfare, yet it has to be recognized that the more sweeping the recommendation, the less likely it is to be acted upon. Pressure groups have to decide whether to moderate their ideal solutions in order to increase the chance of achieving some improvement. In fact, it is not clear what an ideal solution would be.

Since farm animals only exist in an economic framework, there are real limits to the allocations of space, freedom and other resources that are feasible, i.e. will not result in unprofitable activity and failure of the enterprise. Reversion to a wild state is impracticable (and, in any case, not without welfare hazards) for domesticated species, so even abolishing the keeping of animals for profit would not solve any immediate problems.

Similarly for pets, the objective has to be improvement of welfare within some acceptable context, in which treatment is humane. This may mean that welfare standards should deal with the minimum that is acceptable or tolerable, rather than ideal.

12.16 Who sets standards?

The role of FAWC (and similar bodies) includes the stating of standards representing minimum, good or ideal welfare conditions. This is not quite the same as setting them in the sense that they must be followed and thus enforced.

Legislation clearly has a role but this is often largely negative, banning practices regarded as unacceptable. But legislation does not cover all countries and may thus have limited relevance to the general improvement of animal welfare.

Other avenues are available, however, and there is recent evidence that retailers may wish to set standards to which their suppliers must conform, wherever they are in the world. If retailers adopt different standards, the customer/consumer may well be confused and lose faith in all of them. It is again a question of trust but there would seem to be great merit in the adoption of common standards, independently established.

Questions

1. In countries where dogs are eaten, why shouldn't they be farmed?
2. Which has the more interesting life, a goldfish in a glass bowl or a battery hen?
3. Why don't we give as much freedom to our dogs as we do to our cats?
4. Is it natural for a horse to be ridden?
5. What behavioural freedoms does a gerbil need?
6. What rights does a rat have?
7. Should all animals be allowed to breed?
8. What makes an animal a pest?
9. Would you wish to lead a stress-free life?
10. Do plants have rights?
11. Does chronic pain lead to poorer welfare than acute pain?

Chapter thirteen

Agriculture and human health

Into the closed mouth, the fly does not enter.

Arabic proverb

The World Health Organization defines health as 'a state of complete physical, mental and social well-being and not merely the absence of disease or infirmity'. This is a very wide definition and leads to the three fundamental requirements for health:

1. healthy lifestyles;
2. healthy environment;
3. appropriate care.

There is some circularity in this idea and there is some subjectivity in the judgements employed, but agriculture has some impact on the first two.

Agriculture is commonly thought to have a major effect on human health, mainly because it produces the food we eat and thus affects our diet, which, we are continually told, has a major influence on our health. Consider then, the following:

1. Agriculture, in developed countries, such as the UK, produces very little of the food we consume; mostly it produces raw material for a well-developed food industry (see Box 13.1).
2. It is consumers who select their diets, choosing from what is available – and most of this is not bought directly from farmers.
3. A major factor in human health is the amount eaten, in total and of each dietary component. Agriculture has nothing whatever to do with this.

The food items purchased may be determined by price but this is set by the retailer and may bear little relation to cost of production (see Box 13.2).

In developing countries (Chapter 4), it is very often the case that people suffer hunger and malnutrition because they have insufficient money to buy the food they need. But this is not usually due to excessive prices: if less is paid to the farmers, they cannot meet their costs and will produce less.

Clearly, the issue of the effect of agriculture on diet and health is more complicated and, indeed, the issues of agriculture and health go way beyond diet, because agricultural practices may affect human health in many other ways.

13.1 Effect of agriculture on the health of farmers

Farmers and farm workers, and the families of both, may be affected in several ways. The stress and loneliness of farming, especially in remote rural areas, often have severe effects on the social lives of those involved and leave them very vulnerable to pressure. For example, farmers in the UK have been amongst the most likely to commit suicide (see Table 13.1) and the incidence of accidents on the farm may be very high (Box 13.3). They are also vulnerable to zoonoses (diseases transmissible from

---

Box 13.1 Proportion of food processed

It is often estimated that, in developed countries such as the UK, some 75–80% of all food sold is processed, depending upon whether butchering of meat is included. The proportion of food consumed that is processed is even higher, if cooking is included.

   The following table (Spedding, 1989) illustrates the situation for some of the major products.

Agriculture's production of raw materials for the food (and beverage) industry

Farm gate product	Processed [a] (%)	Examples of product
Raw cow's milk	97[b]	Pasteurized or sterilized liquid milk, yoghurt, cheese and butter
Sugar beet roots	100	Sugar, molasses and pulp
Cereals	100	Flour, breakfast food and malt
Live animals for slaughter	100	Carcase meat, cooked, cured and smoked meats, pies
Raw potatoes	51[b] (the remainder will be cooked before consumption)	Canned, crisped, dehydrated and frozen

[a] In addition to transporting and marketing.
[b] 1979/80.

TABLE 13.1 Suicide rates for farmers in England and Wales (1970–1972) expressed as standardized mortality ratios[a] (SMR)

Farmers Employers and managers (age 15–64)	Farmers Own account (age 15–64)	Agricultural workers (age 15–64)
242	73	190

[a]SMR = No. of observed deaths/No. of expected deaths × 100. Expected deaths are calculated from the death rate for the same national sex and age group (adjusted for numbers of each age).

animals to humans – the reverse may also happen) and to effects of the chemicals they use.

13.2. Zoonoses

Clearly, livestock farmers and their staff, who work closely with animals and handle them, are most at risk. Some of these diseases (see Box 13.4) are serious or unpleasant and some are extremely difficult to cure.

   But it is not only livestock producers who may

suffer from contact with the material they work with. Farmer's lung is caused by fungal spores associated with dusty hay and inhalation of dust from stored grain may cause irreversible changes in the respiratory system, the results of which can be disabling. More widespread is exposure to chemical substances now used in most forms of intensive agricultural systems. Some of these, in the form of wind-borne sprays, may also affect the public, but most have a more localized effect.

13.3 Exposure to chemicals

Chemicals that are injected into, or fed to, livestock are unlikely to cause problems to livestock keepers, although antibiotics could result in the development of resistant strains of bacteria to which human beings may then be exposed.

   The hazard to livestock keepers is well illustrated by the concerns about sheep dipping. The process consists of totally immersing each sheep in a tank or a large concrete trough filled

Box 13.2   Costs and prices of foods

When food is sold by the producer, for consumption by the purchaser, it is likely that the cost of production would be the base on which the price would be calculated. After all, the producer has to recover these costs and make a profit as well. But where the purchaser is not the consumer, but engages in processing, storing, distributing, packaging and retailing, substantial additional costs are incurred and the final consumer has to pay for these as well. Thus, in developed countries, where most food is processed and very little sold by the farmer for consumption, the price of food is commonly much higher (by a factor of two or three times) than the production costs.

Simply by way of illustration, the following retail prices were calculated (Spedding, 1993) and expressed relative to the costs of production (set at 100):

Commodity	Retail price	Product
Potatoes	184	Whole potatoes
Potatoes	2800	Crisps
Eggs	280	Eggs
Wheat	370	Bread
Milk	250	Milk

However, this addition of costs is a very simple way of looking at prices. We all know that scarcity is also a major factor in determining price (sometimes confused with value). But the dominant factor will reflect economic demand – how badly people want the product and how much they are prepared to pay for it (consider jewellery and works of art). So with foods: there is no necessary relationship between costs of production/processing etc. and prices charged to the consumer.

Box 13.3   Accidents on farms (UK) (Jollans, 1984)

Fatal accidents on farms have decreased over the years (e.g. the total for England and Wales in 1970 was 105 and only 50 in 1981). Most accidents involve motor vehicle traffic elsewhere than at work or at home, presumably on public roads. Death rates due to motor accidents are higher in rural rather than in urban areas. The reverse applies to suicides.

Thus, mortality due to accidents, poisoning and violence is high in the agricultural population and a quarter of the deaths are ascribed to suicide (this is an agricultural rather than a rural trend).

Mortality is well recorded, but morbidity is less so. The main health problems appear to be the same as those of the general population. Some diseases are more prevalent in farming, such as forms of pneumoconiosis (e.g. farmer's lung, caused by mouldy hay) and allergic responses to chemicals, pollen and mites.

with a solution of chemicals that kill the mites that cause severe itchiness (sheep scab) and loss of wool. The dip also kills off other ectoparasites and prevents subsequent fly strike (blow-flies laying their eggs on the wool: these hatch and feed on the flesh of the sheep). In the past, substances such as arsenic and DDT were used but these were banned because of the dangers considered to be involved in their use. More recently, organo-phosphorus compounds have been found to be very effective but some ill-effects on people have been attributed to their use. Protective clothing is very hot and uncomfortable, so it is not always used and, in any case, it is quite difficult to avoid any contact with the dip when hundreds of sheep have to be dealt with in one day. Opinion is divided on whether the dip causes the problems reported and it is a difficult choice between playing safe by banning it and risking disease and welfare problems in the sheep. This is an obvious case for research (See Chapter 16) but this is costly and takes time.

Other chemicals, such as those applied to crops, can be used in developed countries by operators in closed tractor cabs or wearing protective clothing. But, in developing countries, deaths have occurred. Boardman (1986), for example, stated that 'deaths from accidental or deliberate ingestion of paraquat are a regular occurrence in Papua New Guinea, Guyana and some other developing countries'. Paraquat is probably the only highly toxic herbicide in recent times.

Box 13.4 Zoonoses

Zoonoses are diseases that affect animals and are also transmissible to humans. Rabies is a spectacular example, but is not a risk in the UK. In the UK it is a **notifiable** disease; all suspected cases are reported to MAFF. Other notifiable diseases are:

anthrax;
tuberculosis;
brucellosis.

All cases of sudden death in farm stock are investigated for anthrax and currently there is routine testing of cattle for tuberculosis and brucellosis.

Other conditions of concern in the UK are:

orf (contagious pustular dermatitis);
ringworm;
salmonella;
leptospirosis;
listeria.

Many of these can cause abortion but enzootic abortion is very serious, caused by an organism called *Chlamydia*.

It is thus necessary to distinguish between hazards intrinsic to the chemical structure of the compounds and those which arise from the circumstances in which they are used.

Herbicides can be toxic to humans even though they are designed to kill plants. Pesticides are, of course, designed to kill animals and their effects depend greatly on the dosage. Conway and Pretty (1991), in an extensive review of the literature, concluded that the hazard from pesticides was 'less than might be expected'.

Even DDT, the most notable of the organochlorines, is reckoned to be about as hazardous an acute poison as aspirin, and it was used on a massive scale for the destruction of mosquitoes and thus as a control of malaria. The World Health Organization (WHO, 1979) concluded that 'the safety record of DDT is phenomenally good'. Yet it has been banned (in the USA in 1972 and in the UK in 1984) because it accumulates in the food chain, being concentrated in the tissues of the predators at the top of the chain, and in fat. These cumulative properties can give rise to higher concentrations and chronic poisoning.

However, much of the concern about pesticide use in agriculture rests on the fear that dangerous residues may occur in the food products.

13.4 Residues in foods

FAO/WHO lays down an acceptable daily intake (ADI) based on a figure of one hundredth of the amount of residue that would cause the least detectable effect if ingested regularly throughout life. There is thus a major safety margin. In general, actual intakes in developed countries are well below these levels but in developing countries this may not be so. Because dietary patterns vary, a measure of amounts in products has also been developed, based on the maximum residue levels (MRLs) likely to occur following application of a pesticide according to good agricultural practice. MRLs are sometimes exceeded but again the safety margin is considerable. This, of course, is only relevant to situations that are monitored and controlled.

In developing countries, leafy vegetables may be sprayed frequently and arrive at market with residues far in excess of human tolerance limits, and there are greater problems in finding and using clean water for food washing or preparation. Fish, also, may contain high residue levels. In some cases, fruit and meat may be imported with such levels but these are rejected after inspection at ports.

In developing countries, human milk may contain high levels of DDT, for example. Even so, the major hazard from pesticides lies in accidents, particularly in developing countries (Conway and Pretty, 1991), although it can always be argued that long-term effects (e.g. carcinogenicity) may yet be detected.

13.5 Residues in water

Heavy metals (illustrated in Box 10.2) can be very toxic and very persistent in soil, and water may become contaminated. Most of this is derived from mining and other industrial operations. The soil being cultivated may be rich in heavy metals and the latter may be applied in sewage sludge but the main source of heavy metals from agriculture is probably fertilizer, however cadmium (from phosphate rock) appears to be the only potential problem.

That is not to say that other toxic elements (e.g. lead and copper) cannot get into food and water but this is generally from non-agricultural sources, although, in the past, agriculture has used such substances as arsenic and copper to treat diseases (and the latter as an animal feed supplement).

Most of these metals are essential nutrients for plants and animals but usually in very small quantities. Some may be accumulated and cause problems, as with lead or selenium in forage.

In recent years, the main concern has focused on nitrates (NO_3) in water. As described in Chapter 10, high use of nitrogenous fertilizers can lead to leaching and water with a high NO_3 content getting into sources of drinking water, such as rivers and aquifers.

13.6 Nitrates

Here are some basic facts about nitrates:

1. they are a natural part of the living world – there is nothing new about them;
2. they are the form in which plant roots take in the nitrogen they need to form proteins and other essential tissues;
3. excess nitrate in our food is excreted by the kidneys and no known human physiological process changes its form;
4. the most important sources of dietary nitrate are actively growing green leaves and stems (e.g. cabbage, lettuce); it is in the leaves because it has been taken up by the roots and not yet made into protein;
5. our intake of nitrates in water is normally much less than that from food.

Why then the public concern? It is mainly because nitrate can be converted into nitrite by bacteria (this can be caused by bacteria in the mouth and nitrate in saliva) and nitrite may be absorbed into the blood stream where it combines with haemoglobin to form methaemoglobin, which cannot carry oxygen.

Nitrite poisoning is not seen in adult humans but has occasionally occurred in babies below the age of three months (a condition known as 'methaemoglobinaemia' or 'blue baby' syndrome). Only nine cases, with one death, have been reported in the UK and the last death in the USA was reported in 1964. The risk is therefore confined to bottle-fed babies fed on water with a high nitrate content and contaminated with bacteria: it is therefore surprising that Conway and Pretty (1991) found the condition to be almost unknown in developing countries, perhaps because it goes unrecognized.

There appears to be no evidence to associate stomach cancer with the intake of nitrates (and the formation of nitrosamines) and nitrate has been used as a treatment for patients who have had kidney stones (calcium carbonate) removed, to prevent the formation of new stones.

There is one other source of nitrate and nitrite intake and that is cured meat (e.g. bacon, ham), since both are used as preservatives. Table 13.2 illustrates intakes of NO_3 in the UK and the USA. Average nitrate intakes tend to be higher with vegetarians and heavy beer drinkers.

13.6.1 Legal controls on NO_3

The World Health Organization (WHO) suggested acceptable and maximum levels of NO_3 in drinking

TABLE 13.2 Estimates of intakes of NO_3 (in mg) by humans (after Jollans, 1985)

Source	UK mean	USA mean
Meat products	5.3	1.8
Milk	12.4	0.2
Cheese	0.7	0.2
Vegetables	32.0	65.0
Potatoes	8.6	65.0
Water	15.0	2.0
Other	–	5.5

water and the EU then adopted standards in a Directive to member States, reducing the WHO levels by half. The current maximum in Europe is set at 50 mg NO_3 per litre.

13.7 Effects of other farm practices

Quite apart from the application of fertilizers and other chemicals, other farm practices cause concern, even when they do not cause problems. Two examples are the use of bovine somatotropin (BST) and the occurrence of bovine spongiform encephalopathy (BSE).

13.7.1 BST

BST is a hormone that occurs naturally in milk and that can be injected into milking cows, inducing them to produce more milk than they otherwise would have done. Those wishing to do this claim that it is harmless to both cow and consumer and increases the efficiency of milk production.

It is not proposed to rehearse here the scientific arguments for and against the use of BST. These will change with time as more experimental evidence and practical experience emerge. However, the practice reflects the antithesis between scientific judgement and public perception and typifies a general problem. The opposing arguments are briefly listed in Box 13.5.

Whatever the outcome of these arguments, it has to be recognized that, if the public perception is that the use of BST is undesirable – for whatever reasons – demand for the product may be drastically reduced. This could be directed specifically at milk from herds where BST is used, if that is labelled, or reduce consumption generally, where it is not.

The difficulties are obvious. If milk is so labelled, the customer may play safe and avoid it, whether it is perfectly healthy or not. If it is not labelled, a whole industry could be blighted or producers be forced to assert on a label that BST is not used.

The implications for the milk and milk products

Box 13.5 Arguments for and against the use of BST

For:

1. it increases the efficiency of milk production by reducing the number of cows required to produce a given amount of milk;
2. it is a naturally-occurring hormone, so it is not introducing anything that the animal's body has no experience of;
3. it is harmless to the consumer, whose milk supply already contains it;
4. the method of injection does not hurt the cow;
5. there are no other ill-effects on the cow when BST is used.

Against:

1. Western European countries are over-producing milk and more extensive systems would be better anyway;
2. the presence of extra BST in milk may well be associated with the presence of other substances that may be harmful to consumers;
3. its use may increase animal diseases and disorders (e.g. mastitis);
4. continual injections may adversely affect the welfare of the cow;
5. adverse effects on both cows and people may take a long time to become manifest.

(cheese, butter, yoghurts) industry of having to keep the two sorts of milk separate at all stages, are horrendous.

13.7.2 BSE

BSE (so called because of its softening effect on brain and other nervous tissues) suddenly appeared in cattle in the UK in 1986. It has most unpleasant effects, causing cattle to stagger uncontrollably, and a great many cattle have died or had to be destroyed. It has some resemblance to a long-known disease in sheep, called 'scrapie'.

The causal agent is not completely known but can only survive in living tissue. One possible source of the causal agent is thought to have been inadequately treated offal that was subsequently incorporated into cattle feed.

Because no one can be absolutely sure what the risks are, nervous tissues from suspect animals are eliminated from the food chain and some countries refuse to import cattle from the UK. This is primarily an animal health problem but it has raised some fears that it might be capable of causing disease in humans. It is thus another unknown hazard where public perceptions can greatly influence agricultural practices and the demand for products.

In industrialized countries, with well-developed food industries, concerned with processing, preserving, storing, packaging, distributing and retailing food products, there are other concerns about food safety.

13.8 Food safety in the food industry

After food is harvested, it loses quality and may become unsafe, inedible or unpleasant. Where there is no food industry, simple measures of preservation are used but the losses in storage are often enormous (see Box 13.6). In the UK, historically, low temperature (e.g. in ice-houses), salt and sugar were widely used. Today, refrigeration is widespread.

In countries with hot sunshine, drying is used: drying is also used in developed countries, where it is costly in terms of energy. The causes of quality deterioration are largely physical, chemical, enzymic or microbial (Miller, 1990).

A priority in preservation is to limit the growth of micro-organisms, especially those that cause food poisoning, and the main methods are:

1. low temperature;
2. reduction in water activity (see Glossary);
3. vacuum and modified atmosphere packaging;
4. acidification;
5. heat;
6. irradiation.

The last one is less familiar and therefore more suspect to the general public, who worry that it may be used to disguise adverse changes that otherwise restrict the shelf-life of products.

The food industry in developed countries is rigorously controlled and, in any event, has to preserve its own reputation. But food is handled beyond the retailer, sometimes in a large catering industry (see Box 13.7) and, frequently, in the household kitchen. It is in the kitchen that major problems of food poisoning can occur. Salmonella contamination is a good example. The organisms themselves are very widespread and present in kitchens, as elsewhere. The result is that if food is undercooked or reheated there are very real risks.

Although the real problems are associated with micro-organisms and additives are bottom of the list for risk, public concern sometimes seems to focus more on them. Some of the reasons are that the additives (for flavour, colour etc) are often seen as unnecessary and artificial, not 'natural'. It is therefore rather startling to scan the list of naturally-occurring toxins to be found in our foods (Table 13.3).

13.8.1 Irradiation of foods

Ionizing radiation using, for example, cobalt gamma rays, X-rays or electrons, may be used to delay ripening, to kill moulds, pathogens and spoilage organisms. It thus has a role in promoting food safety and it is commonly used in hospital food preparation for this reason.

Such radiation can kill bacteria but there are concerns that survivors could give rise to resistant

TABLE 13.3 Some naturally-occurring toxins (after Waites, 1990)

Compounds	Common source
Alkaloids	Herbal tea
Psoralens	Carrots
Saponins	Legumes
Phytoestrogens	Legumes
Protease inhibitors	Legumes
Cyanogenic glycosides	Legumes
Lathyrogens	Chick-pea
Glucosinolates	Brassicas
Biogenic amines	Cheese, chocolate, wine

Box 13.6 Losses in storage

Very few products are available at all times of the year and large-scale production depends upon the mechanization of harvesting the entire crop at one time. Storage is therefore needed in nearly all circumstances. For example:

1. grass grown for feeding animals in winter is stored as hay (dried) or as silage (fermented by bacteria, with air excluded);
2. milk products are stored dried or as butter and cheese;
3. meat is refrigerated, or dried, smoked, preserved with salt or nitrite (to eliminate bacterial action);
4. grain is stored in containers (silos) or barns in huge heaps (sometimes with exclusion of birds, rats, mice, etc.);
5. fish are refrigerated, smoked or dried;
6. fruit is stored in controlled atmospheres (e.g. apples), or dried (e.g. figs, dates), or preserved with sugar (in jams etc.);
7. potatoes and root crops in earth-covered clamps or in barns.

For a smallholder, efficient storage affects available food supplies: losses may therefore be kept low. In these circumstances, large outlays on urban grain silos may result in larger losses during haulage (Mellor, Delgado and Blackie, 1987).

Even essential strategic food reserves can be decentralized but may also require central support. Since maize, for example, will not store for more than two years, an annual turnover is needed and this is expensive.

Storage of seed crops requires a low moisture content (between 13 and 15%, but 8% for oil crops such as oil-seed rape) in order to eliminate microbial growth. Temperature also influences fungal growth and insect activity. For root crops, with high moisture content (c. 80%) low temperatures are essential (Harris, 1992). Biologically, the major causes of deterioration in store are natural senescence, microbial and non-vertebrate degradation.

Of course, on occasions losses can be much higher. Yam losses up to 66.8% have been recorded in Nigeria;

The following are examples of post-harvest losses in selected crops (from Proctor, Goodliffe and Coursey, 1981, using 1977 FAO data)

Commodity	Country	Loss (%)
Cassava	Brazil	10
Cassava	Dominican Republic	24.4
Yams	Nigeria	10–50
Sweet potato	Indonesia	10
Potato	Brazil	1–30
Potato	Rwanda	20–40
Tomato	Venezuela	25
Tomato	Bolivia	30
Onions	Venezuela	9–12
Lettuce	Venezuela	11–14
Green beans	Dominican Republic	11.6
Carrots	Venezuela	5–6

sweet potatoes up to 65% in British Solomon Islands; onions in Guyana from 57 to 89% when stored dry from 82 days.

Box 13.7 Structure of the catering industry (based on Malik and West, 1989)

Organizations making direct profit	Indirect profit makers	Non-profit making organizations
E.g. Hotels Restaurants Food shops	E.g. Canteens (of schools, hospitals, workplaces)	E.g. Hospitals Social services Armed forces

Size of the UK catering industry:

Employs c. 2.5m people
10% of the workforce
about half being in the commercial sector

Sales (1986) £8.2bn

strains: however, this is not thought to present serious problems (Moseley, 1990). Although not yet permitted in the UK, it is allowed in more than 30 countries. Moseley (1990), after reviewing the evidence, concluded that, correctly applied, ionizing radiation 'provides an efficacious food preservation treatment'.

Since it is used to extend shelf-life of foods, some people suspect that risks, not yet known, may be taken purely for commercial gain. Other probable changes in technology in the future and the implications for human health will be further discussed in Chapter 17.

13.9 Effect of health advice on agriculture

It is worth considering whether the current recommendations about diet, designed to improve health, will have a major impact on the shape and structure of agriculture. The UK is taken as an example, because it is only in developed countries of this kind that the issue can be seen in this form.

13.9.1 Dietary change

There is a strong tide in favour of dietary change, based on the proposition that there are large numbers of people whose diet is 'unhealthy', and there has been some convergence of view in recent years as to what changes are desirable.

In developing countries, diets are also inadequate, mostly due to too low an intake of energy. On most diets, if the energy intake is high enough, so would be the protein intake.

In developed countries, the problem tends to be overconsumption and it has to be asked how much of the dietary problems would disappear if total food intakes were adjusted downwards (where intake is excessive). There are, of course, difficulties because human beings are so variable in so many important respects: we are not all alike. The treatment of individuals is unlikely to have a significant effect on agriculture but major changes in the diet of populations may do so. Persuading people to change is not an easy matter, particularly with those who remember what they would describe as 'earlier fashions'.

None of this alters the need to examine what the effects on agriculture would be, if given changes were actually made, but the agricultural industry can only adjust to what has happened or to probabilities of it happening (see Spedding, 1988). There is a separate argument about how the industry might assist desirable change by making choice in that direction easier, but this is more relevant to the food industry than to farming. There are three major aspects to this. First, the consumer cannot buy what is not on offer, so the availability of, for example, low fat foods affects choice. Secondly, the consumer cannot always tell what is contained in a food or interpret statements of content on, for example, the basis of dry matter or of total energy. Thirdly, the price of foodstuffs clearly influences the choice of some people or even, in the case of the very poor, may actually determine patterns of consumption. Education matters greatly, of course, to the ability of the consumer to exercise choice in an informed way.

However, it has to be remembered that there are other general trends affecting choice. One of them relates to 'organic' farming (see Chapter 17). Some believe that 'organically' produced foods are better for them, but there are others who wish to buy such produce because they prefer the ways in which it has been produced – in relation to soil fertility and notions of sustainability and ecological balance, to animal welfare, to pollution of the environment and to conservation of wildlife.

This illustrates the point that there will be other pressures for dietary change (including, for example, vegetarianism) in addition to those stemming from orthodox medical opinion.

In some cases, the impact on agriculture would be enormous and far greater than the consequences of the kinds of change caused by acceptance of the diet currently recommended for the prevention of coronary heart disease.

13.9.2 Dietary recommendations

These can be summarized as less fat, particularly less saturated fats, less red meat, less butter, less alcohol, less sugar, less cream, less salt; more fibre, more starch, fruit, vegetables, polyunsaturated seed oils, fish, chicken, turkey, rabbit and skimmed milk. These are recommendations about what people should eat: this is quite different from what should be produced.

The salient points here are:

1. most countries also export food;
2. most countries also import food;
3. a high proportion (75–80% in the UK) of farm output is processed by the food industry before purchase by consumers;
4. a high proportion of the food consumed is further processed and cooked, in the home, institutions, hotels and restaurants;
5. only a proportion of what is grown forms the product and only a proportion of the product is consumed.

Exports are obviously not consumed in the producing country and imports cannot be affected by changes in home agricultural practice. The food industry determines the nature and content of many foods and certainly most salt, for example, is added at (3) or (4) above. The proportions of products

that are discarded at various stages (including in the kitchen and on the plate) are very variable but are often significant.

All these features have a substantial bearing on the effect of the proposed dietary changes on the shape and size of the agricultural industry.

13.10 Farming

It is important to recognize that UK farming is not going to remain static anyway, quite apart from any reactions to dietary change. Overproduction in the EU is bound to be controlled, one way or another.

The methods by which overproduction is controlled, whether by set-aside, price restriction, quotas, input rationing or taxation of inputs, must, it they are to be effective, result in reductions in the total amount produced, notably of milk and cereals. Since these are such major outputs of farming at the present time, and since the reductions needed are substantial (e.g. *c.* 20% on cereals), the effect on the pattern of agricultural land use could be highly significant.

Within the food production sector there will be a greater awareness of the need to match supply to effective demand. Like any other industry, farming will doubtless also try to create demand for its products, old and new, but what is finally produced will have to equate to what is wanted or, more precisely, what people are prepared to pay for.

Of course, because of the size, power and influence of the food industry, the demands on the farmer – and the financial rewards for their produce – will rarely be determined directly by the ultimate consumer.

13.11 The power of the consumer

Whoever the consumer (or the purchaser) turns out to be, and this varies somewhat with the product, the power will lie there to determine what is produced. To some extent this has always been so but it will certainly be strengthened in the future, and this is probably the main way in which dietary changes will influence production.

For example, if the major demand is for cheap, uniform, blemish-free apples, the supply will tend to come from a limited number of productive varieties grown with substantial applications of agrochemicals. It may be argued that this limits the choice of minorities – as it does with all products from all industries – unless the specialist need is expressed and paid for. Even ancient varieties of apple would return if there were consumers prepared to pay for them. It might take time for such demands to be met but it would be likely to happen and more likely in conditions where the option of just growing what is easy or familiar will be greatly curtailed. So, any product which commands a sufficient premium will be supplied – eventually.

In general, if the consumers' needs can find expression as a significant economic demand, those needs will be met. Foodstuffs that do not meet these needs will remain unsold and, very rapidly, will no longer be produced. There are some constraints on this rather idealized view of how the food market works, however. First, it rests on consumers defining their own needs. In other words, the argument applies to what they want and are prepared to pay for.

The medical profession cannot control what is purchased, much less what is consumed, and will have to persuade people to want what they (the medical profession) think that they need. Others will also be engaged in persuasion, for quite different reasons. Whatever consumers finally decide that they want, they can only translate this into action if labelling is adequate to present them with an effective choice. This problem is much greater for choice on health grounds than for general advertising pressures. Advertisers only have to create an association between desirability and a brand name. Health advisers have to depend upon a prospective purchaser reading and interpreting the fine print on a label.

Any educational effort has therefore to embrace both the identification of needs and of the products that will satisfy them. But none of this exerts any control on the quantity consumed, in total, or of any one constituent of the diet. It therefore has little impact on the balance of constituents. Healthy eating is quite different from sensible purchasing and the situation is complicated by eating out and meals supplied by institutions.

If, as a result of educational programmes, the total demand for a product (e.g. milk) declines, then the amount produced will decline. If demand favours low-fat milk, more low-fat milk will be provided for sale.

In the case of meat, there may be additional problems because of differences in the standards of fatness required by intermediaries (e.g. butchers) or because fat serves other purposes related to juiciness, flavour or cooking. Such complications may reduce the impact of the consumer but eventually the latter must prevail. Again, it must be recognized that the consumer may be subject to opposing influences. For example, a move towards consumption of poultry meat, on health grounds, may be overshadowed by a move away from it on animal welfare grounds.

A more fundamental difficulty is that the degree of choice available to anyone must depend to a large extent on relative affluence. It is not inevitable that 'better' produce costs more but it may do so and is more likely to do so if it represents a minority demand.

In some cases, the reverse is true. Per kilogramme of potato, there is no more expensive way of buying them than as crisps. Yet this is the way that many poorer people buy them partly because of taste preference and partly for convenience. Even so, poorer people have less choice – in all kinds of ways – and may not be able to respond to persuasion to change their dietary habits.

Nor should we underestimate the pleasure derived from eating. This is substantial for most people, ranks very high with many and may be one of relatively few pleasures for the really poor.

The fact is that food serves many functions besides the provision of a healthy diet. This casts a slightly different light on the arguments that may have to be used, tending to emphasize the negative ones, which are by far the most powerful. Any real suspicion that a particular food is dangerous in some way is likely to result in sales falling dramatically.

13.12 Implications for agriculture

Major reductions in total fat consumption could result in major reductions in the farm output of the main foods:

1. directly, of meat fat, milk and butter;
2. indirectly, of reduced cereals used as animal feed to produce meat (also affected by reduced alcohol consumption).

Increased demand for millable wheat, potatoes, certain fruits and vegetables would make no great difference to agriculture, and an increased demand for some fruits would mainly increase imports. Reductions in sugar consumption could have the effect of reducing the area of sugar beet or of imports in particular countries. Increases in the demand for millable wheat would make little difference for those countries that cannot yet produce all their own needs and have to import the balance. Increased demand for potatoes could be met either from increased production (but this would not involve large areas) or by retailing the very small and very large grades from the present output.

There is no great difficulty in producing more vegetables, although they tend to be a high-cost source of nutrients. A penny spent on bread or potatoes will purchase some seven times as much energy and vastly more protein than a penny spent on carrots, for example (HEC, 1983). Equally, more vegetables could be produced in gardens and allotments, encouraging exercise as well.

The opportunities for reducing fat in meat are mainly:

1. to move consumption increasingly to poultry (this is happening already);
2. to cut it off – this can have very large effects (see Table 13.4);
3. to move towards leaner breeds of cattle, sheep and pigs (this takes some time but is perfectly feasible and has already occurred).

Changes in the direction of a higher ratio of poly-unsaturated to saturated fats also result from moving to leaner pigs and poultry.

Those who advocate change to achieve particular objectives, however, should accept some obligation to consider what other consequences those changes bring about. For example, greater consumption of fruit and vegetables may result in higher sugar intakes and thus may conflict with the recommendation to consume less sugar. Fruits often contain very high sugar contents [e.g. plums, c. 60% of the dry matter is sugar; peeled eating apples, c. 75%; oranges (flesh only), c. 61%; and vegetables do so to a lesser extent (boiled leeks, 50%; carrots, 50%; raw tomatoes, 42%]. This leads to questions as to whether constituents of fruit and vegetables are in some way nutritionally different from those not so embodied.

It is conceivable that sugar enclosed in cell walls behaves differently or that nitrate associated with other constituents has a different significance. Otherwise, the recommendation to eat more fruit and vegetables could increase nitrate intake more than by drinking water containing even high nitrate contents.

The point of principle is an important one, that even if the proposed changes will achieve their immediate objectives, it is vital to foresee what else they may also affect: this, of course, is another way of stating the need for a systems approach (see Chapter 7).

The effects on farming may not be the most serious or significant of the consequences of dietary change. Certainly, other changes (in the CAP, in the control of overproduction) and other dietary changes (for other than health reasons) may well have much greater effects on food production and the pattern of farming, and the influence of the food industry is of increasing significance.

Within the range of effects on farming of dietary changes brought about for health reasons, by far the largest consequences would flow from changes in consumption of animal products and particularly those derived from grass.

In this chapter, when considering diet, there has

TABLE 13.4 The effect of removing visible fat from meats (source: Paul and Southgate, 1978)

Meat	Cooking method	Fat in consumed product (by weight) (%)	
		Lean and fat	Lean only
Beef			
Rump steak	Fried	14.6	7.4
	Grilled	12.1	6.0
Sirloin	Roast	21.1	9.1
Topside	Roast	12.0	4.4
Lamb			
Chops	Grilled	29.0	12.3
Cutlets	Grilled	30.9	12.3
Leg	Roast	17.9	8.1
Scrag and neck	Stewed	28.2	15.7
Pork			
Chops	Grilled	24.2	10.7
Leg	Roast	19.8	6.9
Bacon			
Collar	Boiled	27.0	9.7
Gammon	Boiled	18.9	5.5
	Grilled	12.2	5.2
Rashers			
Back	Fried	40.6	
Middle	Fried	42.3	22.3
Streaky	Fried	44.8	

been a tendency to discuss mainly the negative aspects, i.e. those dietary constituents thought (by some) to have adverse effects on health. The positive aspects should not be ignored, however. It is hard to imagine how health could be sustained without adequate intakes of both the major and minor nutrients (see Table 13.5). Furthermore, there are many claims as to the value of particular constituents in relation to the prevention of ill-health. For example, most current dietary recommendations include encouragement to eat more fruit and fresh vegetables and some claims are made for nuts.

It is claimed that the risk of cancer can be reduced by improving diets. In some cases the idea of selenium as a protective agent against colon and breast cancers has to be balanced against the fact that it is toxic to humans in doses only slightly greater than the levels needed for good health.

Special claims are also made for allium compounds (in onions and garlic), glucosinolates and other compounds found in cruciferous vegetables (e.g. cabbage), limonenes (in citrus fruit oils) and folic acid (found in free-range eggs) for pregnant women. In general, diets that are adequate for the major nutrients, from varied sources tend to supply most of what we need.

In addition to foods containing 'protective' elements, plants were the original source of most medicines and are still widely used in many countries.

In 1636, a published list of medical uses included nearly 4000 plants (Phillips, 1992). The major constituents of medical importance are phenolic derivatives, isoprenoids, cardioactive glycosides and alkaloids.

Questions

1. How much of your food do you buy from a farmer?
2. Who determines your diet and how much you eat?
3. Are taste and flavour important in the meat you eat?

TABLE 13.5 Nutrient requirements for people (examples of daily allowances)

	Energy (MJ)	Protein (g)	Ca (g)	Fe (mg)	Vitamin A (iu)	Riboflavin (mg)	Niacin (mg)	Ascorbic acid (mg)
Men, depending on age and weight	10.9–13.4	65	0.8	12	5 000	1.6	13–16	75
Women, depending on pregnancy and lactation	7.4–13.8	55–100	0.8–2.0	12–15	5 000–8 000	1.4–2.5	10–15	70–150

Does fat meat taste better?

4. Do you have to eat the fat on meat?

5. How much effect does agriculture have on your intake of salt and sugar?

Chapter fourteen

The concerned citizen

He who takes the middle of
the road is likely to get
crushed by two rickshaws.

Chinese proverb

Who or what is a citizen? Many definitions have been proposed (see Box 14.1). As I argued some time ago (Spedding, 1979), the most useful definition of such a concept will surely include more than 'inhabitant or freeman of city', 'person holding some special status within the community', 'townsman' or 'member, native or naturalized, of a State or Commonwealth', which is what tends to appear in dictionaries. Nowadays, we would not confine citizenship to those in cities or, indeed, seek to debar anyone unless they lose their rights by their own acts.

Somewhat earlier, Aristotle (384–322 BC) defined a citizen as 'one who has the right to take part in the legislative proceedings of the state to which he belongs'. Clearly 'there has to be some element of participating membership of an organised, identifiable community, involving both rights and responsibilities, but, above all else, in relation to the community (city, state or world) as a whole. This last point imposes a responsibility on the citizen to be able to take part in debates on those issues which are in some sense central to the well-being of the state' (Spedding, 1979).

As Bernard Weatherill wrote, in his foreword to the Report of the Commission on Citizenship (HMSO, 1990), 'Citizenship, like anything else, has

to be learned'. The Commission recommended both a codification of the rights and duties of a citizen and adequate social entitlements to ensure that all citizens can exercise their rights. The encouragement of citizenship would be welcomed, judging from the school students surveyed, who were overwhelmingly keen to learn more about their citizenship rights and duties. The Commission recommended that 'The study and experience of citizenship should be part of every young person's education from the earliest years, and in further and higher education clearly an important dimension of education for capability.'

But it is also necessary to ask what kind of education is needed to enable citizens to discharge their responsibilities and exercise their rights. This question will be addressed in Chapter 15 but in relation to both agriculture and the environment it is clear that biology will be vital.

14.1 The importance of biology

The American Project 2061 Report on Biological and Health Sciences (Clarke, 1989) concluded that the study of biology was needed: (1) 'to understand oneself' and (2) because it 'teaches rules to live by'. These rules were envisaged as encompassing 'how

---

Box 14.1 Definitions of a citizen

The study and experience of citizenship should be part of every young person's education from the earliest years, and in further and higher education clearly an important dimension of education for capability.

No man qualifies as a statesman who is entirely ignorant of the problems of wheat.

Socrates

There has to be some element of participating membership of an organized, identifiable community, involving both rights and responsibilities, but, above all else, in relation to the community (city, state or world) as a whole. This last point imposes a responsibility on the citizen to be able to take part in debates on those issues which are in some sense central to the well-being of the State.

An inhabitant of a city or a person holding some special status within the community.

A citizen is one who has the right to take part in the legislative proceedings of the state to which he belongs.

Aristotle

Citizen (n.) Inhabitant or freeman of city; townsman; civilian; member, native, or naturalized, of a State or Commonwealth (usually **of; the world**, one who is at home everywhere, cosmopolitan); inhabitant of; hence -HOOD, -RY, -SHIP.

Concise Oxford Dictionary, 6th edn, 1976

A citizen, as defined by Aristotle (*Politics*, IIIi), is one who has the right to take part in the legislative proceedings of the state to which he belongs. He is a subject with particular privileges. In anct Rome there were 2 kinds of **cives**. The majority had certain private rights of citizenship, such as the right of inter-marriage (*jus connubii*) and right of trade intercourse (*jus commercii*) with the allies or friends of Rome. A few, however, had a special privilege of voting in the tribe, and were eligible for the higher offices of state. The rights of citizenship were generally acquired by birth, but both parents had to be Rom citizens. At a later period it merely denoted free birth as opposed to those who were born slaves. In Great Britain it has never been used to any great extent, and its meaning is indefinite.

Everyman's Encyclopaedia (1968) J.M. Dent and Sons Ltd, London

citizen n. an inhabitant of a city; a member of a state; a townsman; a freeman: a civilian (US): - *fem citizeness. adj (Shak)* like a citizen – *vt citizenise*, *-ize* to make a citizen of *-ns*. **citizenry** the general body of citizens; **citizenship** the state of being or of having rights and duties of a citizen: conduct in relation to these duties.

---

we treat our bodies and those of others and how we treat the environment we depend on'. It held that biological knowledge 'helps us to make sound decisions on health and diet practices as well as on economic issues'.

Clarke (1989) identified the principal theme running through their proposed biological sciences curriculum as 'the central role that knowledge of biology must play if the electorate of a democratic society is to make proper choices about the optimal ways of leading their own lives and directing the broader course of society.'

You could hardly wish for a better statement of the relevance and importance of biology to the

citizen. The fact that it was aimed at the American citizen emphasizes the universality of the proposition and the value of comparing the thinking going on in different countries.

14.2 The citizen's need for biology

Clearly, citizens have two rather distinct roles, even if one reinforces the other. The first is as individuals to understand their own bodies, how to look after them, feed and exercise them, and to understand their potential, possibilities and limitations – all of which change with growth, development and age. The second is in relation to their rights and responsibilities within society and the State.

The two roles obviously overlap because, for example, knowledge of human nutrition can be applied to others as well as to oneself. None the less there is a clear difference. The citizen's relations with society are not under private control, in the sense, for example, that citizens are free to eat what they wish, but are actually determined, in outline, by society. Thus, both the citizen's rights and responsibilities are laid down in terms of limits but the extent to which they are taken up depends greatly on the response of individuals. However, the most fundamental right and responsibility of a citizen is to contribute to the delineation and protection of both.

So, each citizen has both a right and a responsibility to help in formulating the rules by which society lives. The question then is what knowledge and understanding are needed in order to contribute wisely. It should be noted that wisdom implies some ability to use knowledge to good purpose and cannot be acquired in the same way as knowledge itself. William Cowper held that 'knowledge dwells in heads replete with the thoughts of other men: wisdom in minds attentive to their own.'

One way of approaching this question is to consider on what issues society will have to make decisions; then to consider what sort of knowledge and understanding are relevant (or vital) to those issues. This then has a bearing on what is needed in the education of a citizen, itself a subject on which citizens have to make decisions.

14.3 Issues of importance

Some of the biggest and more obvious issues are given in Table 14.1: those especially concerned with agriculture and the environment are discussed in Box 14.2. Every society has got to give thought to how its members are to be assured of an adequate supply of safe food of high quality. Since all food production is based on biological processes, the need for biological knowledge is clear.

It is also clear that, although food is essential, it

TABLE 14.1 Issues of public concern

Feeding the world's people
  Food production
  Food storage/processing/preservation
  Food safety
  Food quality
Prevention and control of:
  Pests
  Parasites
  Diseases of:
    crops
    animals
    people
Population control
Human genome project
Animal welfare
  Farm animals
  Zoo animals
  Companion animals
  Wild animals
Conservation
  Species
  Habitats
Biodiversity
Pollution and its control
Waste disposal
Agrochemical usage
  On farms
  In gardens
Biological control
Genetic manipulation
Biological warfare
Global warming and its consequences
Forestry
Non-food production from the land
  Biofuels
  Raw materials for industry
  Fibres

is not the only essential need (clothing, shelter and protection are others), and that producing food involves many inputs (e.g. implements, fertilizers, support energy) other than biological organisms and processes. Furthermore, simply producing food does not feed people who have no money to buy it, so there are socio-economic dimensions as well. Without law and order, no food supply system may work.

None the less, biology is relevant to plant and animal breeding, (including genetic engineering), crop and animal production, food processing, food

Box 14.2 Major concerns about agriculture and the environment

People's worries depend on where they live, how they live and what else they have to worry about. People in developing countries who have no other means of cooking, cannot worry about having to chop down a tree for fuel. Those whose small farms are devastated by elephants eventually turn against them, as those who are attacked by tigers have to defend themselves. Similarly, where food is scarce, food production may have to receive an overriding priority.

The affluent and well fed are free to worry more about the environment and are concerned about the topics listed in Table 14.1. Thus, in developed countries, the conservation of wild fauna and flora is a matter of concern, for their own sake. Strangely, when discussing the need to conserve species worldwide, the arguments tend to emphasize their potential value to science and medicine. But environment does not stop at national boundaries and the problems of ocean-going animals and migratory birds are difficult to solve.

Methods of farming also raise difficulties. This is most evident with animal welfare at the present time, where those concerned would wish to see standards raised worldwide and not just in their own countries. The use of agrochemicals may also have worldwide consequences that have to be balanced against an urgent need to produce food, and to control pests and diseases.

There are also major concerns about the reduction of the world's capacity to produce food, where soil is lost to erosion, rain forests are destroyed (80% for agriculture) and deserts created.

At present, 97% of the world's food comes from the land (rather than the oceans and other aquatic systems) and soil degradation is estimated to affect 30–50% of the land surface of the earth (Pimentel, 1993). In Europe, soil loss rates are said to vary from 10 to 20 tonnes ha^{-1} yr^{-1}; in the USA they average 16 tonnes ha^{-1} yr^{-1}. But in Asia, Africa and South America, the range is from 20 to 40 tonnes ha^{-1} yr^{-1}. Restoration is extremely costly.

Arising from this agricultural list, there are issues of food safety and quality throughout this 'food chain', as there are of energy use and pollution control.

During production, there are additional issues of animal welfare and environmental impact, the latter affecting wildlife and their habitats, biodiversity, pollution of land, air and water courses. Broader issues also emerge, including effects on landscape, atmospheric change, and land use. Table 14.1 illustrates why this book contains the chapters that it does, for virtually all the issues listed are dealt with in broad terms. Books could be (and have been) written on each of the topics mentioned and to understand any of them in detail is a considerable task, usually undertaken by specialists. The citizen cannot possibly be informed in detail about all these topics and the specialist is no better placed than the generalist citizen for all but one or two of them. How is the citizen then to understand all of them sufficiently to play their part in society? This brings us back to the education of the citizen (see Chapter 15, where the whole subject is further discussed) but some, rather obvious points are worth making here.

1. The problem for the citizen is no greater in these subject areas than for the wider range of topics that all of us are expected to have views about, at one time or another (see Table 14.2 for a minute selection).

2. No one can be well informed about such a range of topics.

3. Somehow, one must be able to form an overview of the wider picture and a sufficient understanding of how things work.

4. The latter can only be based on principles which, once grasped, can be applied to a number of different scenarios. (Thus, in this book, a chapter on 'Plants' cannot possibly even mention a significant number of species: it has to try and elucidate the principles governing what they do and how they do it.)

storage (including irradiation), food distribution, and the preparation and cooking of foodstuffs.

All this raises important questions about how the citizen can obtain the information needed, on an

TABLE 14.2 Wider issues of public concern

Primary school education (size of class and of school)
Nurses' pay
Hours worked by junior doctors
Northern Ireland
Deployment of troops in other countries
The best route for oil tankers round our coasts
The arming of the police
Dog licences
The running of prisons
Road building

almost day-to-day basis, quite apart from any educational process.

14.4 Sources of information

The citizen cannot be a researcher, in most cases, at all, and raw data may be unusable without knowledge, skills and experience that are probably unavailable. This means that someone or some body has to set out reasoned arguments for all the main points of view, without attempting to brainwash or persuade to a particular conclusion. Indeed, an analysis is needed that identifies the important questions to which answers are wanted and for which information may be required.

Consider the main sources of information available:

the media (TV, radio, newspapers, magazines);
consultative bodies;
advisory bodies;
pressure groups;
books, conferences, etc.;
Government;
commercial companies.

Clearly, the media, and the opinion-formers within the media, will have by far the biggest impact. But this is not, in the main, what the media is for. As we all know, television – whatever else it may do – is mainly for entertainment and newspapers have to aim at increasing circulation. That is not to say that they do not have a major effect, including informing people, or that there are not many responsible efforts to inform.

Furthermore, there is an enormous specialized literature, drawn upon by sectors of the population, and this may filter information through to many more people by informal contacts, conversation and so on.

The major problem for lay people, however, is not so much where to obtain information but who to believe. This is because the majority of people have little competence or time to devote to even major concerns. They therefore have to pick up preformed opinions and conclusions and they need to know whether the source can be trusted.

14.5 The need for trust

It would be easy to suggest that few people or organizations are, in fact, trusted, simply because they are perceived as interested parties. Government departments, commercial companies, even scientists, may well have a vested interest in what the public believes and, because disagreements are more entertaining, these are encouraged (or even invented) and the effect is to increase confusion.

That is not to say that listening to genuine disagreements, calmly and lucidly displayed, is not useful in allowing and, indeed, helping people to make up their own minds. There are special responsibilities, here, on bodies such as the Royal Agricultural Society of England (RASE), which not only contribute to public debate and interpretation of agricultural matters but also have remarkable opportunities to communicate with the public, through their involvement with the Royal Show.

But quite often, in matters of food safety, for example, it is very important for everyone to try and arrive at a soundly-based judgement. This will only be accepted from an institution or source that is judged to be authoritative, with access to the necessary information, independent and disinterested. In the UK, the Royal Society of Arts (RSA) is an example where open debates are organized for a wide range of people.

Since all bodies have to be funded in some

fashion, completely disinterested and independent sources may be rare but these attributes need careful judgement. The Royal Society is widely respected for its pronouncements, even though it is a scientific institution and those representing the professional scientists (e.g. the Institute of Biology and CSTI) also have a role.

There are bodies that are set up to be independent and funded accordingly, but, in general, these attributes have to be earned by an appropriate track record. The Centre for Agricultural Strategy (CAS) at Reading University is an example, now in its twentieth year (1995).

All this is only one side of the coin, however, because, having formed a view, the citizen must have some means of bringing it to bear on decision-makers.

14.6 Channels for expression of concern

Since individual views may carry little weight, it may be necessary to form groups, join societies or choose champions. The last route is often represented by pressure groups, who are not elected or even subject to the views of many of the individuals who fund them, except in the sense that if their actions dissatisfy supporters they will cease to contribute. This arrangement is bound to prove attractive if individuals feel that they are ineffective and not even listened to.

As was mentioned in the last chapter, individuals, especially well-behaved ones, can easily be ignored. If groups of moderate individuals are ignored, extremists will be created or extremism

---

Box 14.3   Consensus Conferences (CC)

The USA has used 'science courts' to assess publicly-funded science and the Danish Board of Technology adapted these in the 1980s in the form of CC. By 1995 there had been five CC in Denmark and this was followed by the Dutch: both found that holding a CC raised the general interest of the public in the scientific issues under discussion, increased public understanding of science and focused thoughts on the particular issue selected.

As described by Miles *et al.* (1994):

The CC format involves a group of people from the public (lay panel) interacting with a group of experts (the expert panel) in the presence of an audience, chaired by a facilitator. The lay panel is recruited through newspaper advertisements, and its members are selected to take part in the conference by a steering committee, which is also responsible for selecting the expert panel.

Lay panel members are given balanced briefings spread over two weekends about the subject area under discussion by a range of experts. Briefing is required to raise knowledge and understanding in order that the panel can ask a series of challenging questions to the expert panel. The expert panel is composed of scientists, industrialists, lawyers, economists, theologians and philosophers.

The climax of the conference is the final weekend where the questions devised by the lay panel are presented to the expert panel in the presence of an audience. The lay panel members, using the answers to the questions and the subsequent discussion, produce a report representing their views on the subject, which is read to the expert panel in the presence of the media.

The report is then published and made available to Government as an input to policy formulation.

The first to be held in the UK was a three-day meeting, held in 1994, organized by the then AFRC (Agriculture and Food Research Council), now the BBSRC (Biotechnology and Biological Sciences Research Council) and the Science Museum, on the subject of plant biotechnology. It was judged to be successful and produced a balanced report, highlighting three conclusions:

1. that all food products of genetic engineering should be so labelled;
2. that new crops produced by the biotechnology industry should not be patented;
3. that loopholes they perceived in the safety regulation of genetic experiments and commercial exploitation should be closed: they proposed that an independent ombudsman should monitor experiments.

---

encouraged amongst the (hitherto) moderates.

There have to be channels, therefore, for ordinary people to make their views known. Sometimes this can be combined with the involvement of such people in formulating views. Consensus conferences are an emerging example of this (see Box 14.3). In these various ways, people can become active citizens but the routes are not yet very well developed.

If the citizen needs, for example, biological understanding, it is also necessary to combine that with the understanding derived from other, including non-scientific, sources. There thus has to be very wide debate and dialogue between different sections of society. The importance of achieving this is illustrated by a quotation from Lopez (1992):

'What is at stake is so important that one of the novelties of the future will probably be the beginning of the dialogue which is needed between the scientific community and society as a whole.' We are all citizens.

Questions

1. Is everyone a citizen?
2. Is there a minimum age, degree of intelligence, amount of knowledge, a maximum age, to qualify as a citizen?
3. Should citizenship be lost, due, for example, to criminal activity? If so, does this involve both loss of rights and shedding of responsibility?

Chapter fifteen

Education of the citizen

Even the cleverest monkey
may drop a coconut.

Panamanian proverb

15.1 Education

Amongst the most important purposes of both education and training is to help people become better able to cope with the future. Training equips people for predictable needs and is generally fairly specific: it is therefore geared to the relatively short term. Education, on the other hand, must take a much longer view and equip people to adjust and adapt to change, most of it quite unpredictable, and indeed actually to bring about and control change. Both processes aim at fitting people for the future and ought to be complementary.

But relevance to the future has to take account not only of perceived specific needs and quite major uncertainties, but of the nature of the natural resources available. This includes recognizing that 'people' are not a homogeneous group, that individuals vary from region to region, and that there are important differences of sex, age, wealth and ability: such differences should not be regarded as inconvenient – they are part of the resource available.

The main feature of the 'non-people' resources it that most exist in the form of **systems**, either natural eco-systems or as more or less controlled agricultural systems (see Chapter 7). Changes in components of such systems, however beneficial they look to the specialist, are only relevant if they lead to better performance of the system as a whole.

Since such systems are always multi-disciplinary in character, this proposition has important implications for both education and training.

It also has considerable relevance to such questions as 'the role of technology'. New technology will (and should) only be applied where it makes economic sense but this has increasingly to be judged in human and environmental terms, even where these cannot be expressed as money (see Chapter 17).

Since education must equip people for real life and for playing a useful part in society, it always needs to contain a certain content of knowledge and this cannot cover all possible subjects. The distinctions between education and training should not be pushed too far, therefore, but education must prepare people for change and uncertainty (both of which are inevitable). Education, itself, has to change with time, partly because educational methods may advance and partly because the conditions for which it is a preparation can change out of all recognition within one lifetime. However, education should not just be about reactive responses to events: it should produce minds capable of **creating** the future (see Chapter 7).

The previous chapter (Chapter 14) argued that the citizen needed to be able to debate and hold views on the issues of the day. Many of these have been covered in this book and a major thrust has been that this expanded view of agriculture should be part of the education of a citizen. It also follows that **agricultural education** should take the same broad approach.

Before discussing these aspects, however, it is important to understand the ways in which the word 'education' is being used here.

15.2 The meaning of education

There are few satisfactory definitions but most contain two main elements: (1) instruction and (2) development. Like the 'umbrella' terms discussed in Chapter 11, there are within the general concept a host of specifics, such as religious education, cultural, physical and agricultural. Only when one of these is specified is it possible to discuss the content or subject matter of the appropriate instruction or the precise nature of the development envisaged.

Let us start, then, with the purposes of education. These include:

1. fitting for vocational needs;
2. acquisition of relevant knowledge;
3. developing the capacity to learn;
4. increasing the ability to think;
5. developing the ability to evaluate the thoughts of others (often thought of as a 'critical faculty');
6. developing the ability to communicate;
7. developing the capacity to solve problems.

The last is often neglected but problem-solvers are what we need and I often think that those who are not problem-solvers will probably be problems. Yet problem-solving is not always included explicitly in the educational process. However, it is worth asking whether people who can solve problems are usually, or necessarily, educated (it rather depends on what you mean by a problem, of course).

Can you tell when you encounter an educated person? Scott Fitzgerald claimed that the mark of an educated person was the ability to hold two opposing propositions in the mind simultaneously, without ceasing to function. It certainly seems to be the case that the desire for single, dogmatic, uncomplicated, 'headline' opinions is most marked in those least educated.

15.3 Educational methods

Methods designed to achieve all these purposes are very varied and quite properly so, since they need to take account of the subject, the teacher and the taught (both nature and numbers).

Some principles seem self-evident, however.

1. Lectures that could inspire students with a desire to learn should not shoulder the major burden of conveying information.
2. Students wishing to learn may still need instruction in how to do so.
3. Since the total quantity of information is endless, there have to be reasons to seek particular items.
4. Questions may not come naturally, especially to modest, shy or fearful students, but they provide incentives to learning.
5. It is often necessary to know how to start a process, such as asking questions or finding out the answers (see Box 15.1).
6. Discussion should include listening to the other person's views, as well as articulating one's own.
7. Practical involvement is an essential part of learning (as well as training). This has to involve discovery, experiment and problem-solving.

15.4 Problem-solving

The most important problems in life are multi-disciplinary: they usually involve money, people and several disciplines.

Very few real-world problems can be solved by the application of, say, chemistry or physics alone. Since the range of problems is infinite, it may seem unlikely that useful generalization will be possible. But some guidelines are helpful. For example, solv-

Box 15.1   Patterns of questions for exploring the world

Almost any question, if well answered, may lead to unexpected enlightenment and in suggesting that some questions are more relevant, useful or important than others, there is a risk of eliminating awkward, different or simply new ones. However, there is often a problem for those not familiar with a subject, to know what is a sensible question (or one that does not lead to embarrassing disclosure of ignorance or stupidity). It is no use saying that such fears should be ignored: in practice they do inhibit people.

A first obvious question is 'What is it?' (e.g. answer, 'A bird'). Next questions (Q) 'What sort of a bird?' and 'How can you tell?', are to do with identification. There then seem to be rather too many possible questions, which is why those suggested in the text are chosen to go right to the heart of the matter – from a biologist's point of view. Others might ask quite different questions. For example, confronted by a stone, a biologist might ask 'What is underneath it?': a geologist might ask 'What is it made of?' or 'How did it get there?'

The point is that some simple guidance can be provided (for the very young, those lacking confidence and those simply unacquainted with the subject area). Perhaps education could usefully spend more time on framing questions and less time on facts that may or may not provide wanted or needed answers.

The example of encountering an animal can easily be extended: for instance:

Q: **What does it eat?**
How does it find/catch/kill it?
Does it eat all parts?
Where does it eat it?
How does it digest it?
How much does it eat?
Is the food available at all times?
How does it know where to find it?

Q: **What eats it?**
How can it escape or avoid being eaten?
At what stage (egg, young, old, diseased) is it eaten?
Can it protect itself?
Why aren't they all eaten?

One way of evaluating an initial question might be how readily it generates secondary ones.

---

ing problems related to systems clearly requires a 'systems approach' (see Chapter 7) which recognizes that such things exist and have characteristic properties.

Some problems are better solved by teams and some by individuals. For example, crossword puzzles are usually solved faster by groups of people but they are best constructed by individuals.

There are also difficulties with solutions. How do you arrive at a solution? What makes us suppose that there is only one? How do you choose between possible solutions?

It is possible to explore solutions systematically and to construct a hierarchy of possible kinds of solutions.

15.5   Solution scanning

Most people when confronted with a problem of the form 'What would you do with this material (resources, waste, by-product)?', can see no way of ensuring that all possibilities are considered, partly because they can see no way in which they (or anyone) can even be aware of all the possibilities that exist. It is easy to accept the reasonableness of this position but it is unfortunate if this results in no attempt being made to tackle the difficulty.

This problem was encountered in performing a study (of this form) for a commercial undertaking and led to the idea of using a 'hierarchy of possible solutions' (Spedding and Wagner, 1978), of the kind described below.

15.6   A hierarchy of solutions

The essential idea behind this scheme is to recognize:

1. that logical distinctions can cover all initial possibilities;
2. that most situations contain constraints that operate continuously to eliminate many **classes** of possibilities;
3. that additional expertise and knowledge can always be sought once the need is classified.

Since these propositions may not be entirely self-evident, they are expanded below.

Point 1 – Imagine a question of the form 'How shall I travel from A to B?' My argument is that the answer depends heavily on being able to examine all the interesting possibilities (i.e. excluding illogical ones, such as, for example, those that involve unnecessary costs or disbenefits). Solution scanning assumes that all classes of possibilities can logically be examined, independent of detailed knowledge. Thus, a hierarchy could be constructed, gradually, as follows:

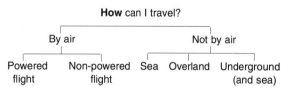

How can I travel?

By air — Not by air

Powered flight — Non-powered flight — Sea — Overland — Underground (and sea)

At each point in the hierarchy it is possible to say that no class of possible solutions has been excluded, simply by the exercise of logic.

This does not, of course, mean that all possible solutions have been **specified**. This would not be possible, even for the range of available solutions, quite apart from currently unimaginable ones.

The apparently logical bifurcation illustrated in the first step of the figure only relates to choices of one method of travel. It may be objected that combinations of different methods may be best, but these can be derived from the classification scheme by combining any of the possible solutions specified anywhere in the figure.

The idea is not to specify particular solutions but simply to provide a framework for an otherwise unaided mind to systematically examine the range of available solutions.

Point 2 – Furthermore, at any point, the customer (i.e. whoever requires the answer) can greatly simplify subsequent procedure by saying 'I do not wish to travel by air (or sea or land, etc.)'. Reasons do not have to be given, or even known, and a whole category of possibilities can be excluded, without ever having to express them in detail. In practice, therefore, a hierarchy does not necessarily get out of hand by virtue of enormous expansion.

Point 3 – The fact that possible solutions are being classified means that it is not too difficult to identify people or organizations whose help can be sought in relation to any one class. Thus, as soon as one has established a category 'travel by air' it would be possible to identify others who could be consulted to establish the possibilities within that class. However, it is not usually at this level that outside help is required. This would most commonly occur further down the hierarchy, where classes of solution might have become so specialized that only specialists would be able to elaborate further.

This is therefore an example of the value of a classification scheme in harnessing available knowledge (it is, after all, what libraries are based on).

15.7   Learning

Problem-solving is only one way of describing the contributions that any citizen may wish or ought to make to society. To be unwilling to make a contribution is to contract out of society: to be unable to make one is, in a very real sense, to be excluded from society – a fate to which the old and incapacitated are most vulnerable. (It is an interesting question as to what extent we make it possible for children to contribute.)

This raises an important distinction between what Hills (1995) has called knowledge (know-what) and technology (know-how): a variety of skills, he points out, are required in order to transform the first into the second.

Some have considered that the problem of education is how to get ideas out of experience but the basis of education is usually to learn from the experiences of others. Although it is often held that personal experience is the best way to learn, there are distinct limits to the practicability or desirability of this – death is an obvious example where it would not appear to be possible to learn about it directly. But, one purpose of education must be to save everyone having to make all the mistakes that others have already made.

If mistakes are made, however, it is important to learn from them but this does not seem to be automatic. It seems likely that one has to learn how to analyse mistakes in such a way that the right lessons are learned.

Similarly, all the learning processes, observation, questioning, formulating and testing hypotheses, all appear to require training. Of these, questioning is the vital spring. After all, you cannot actually observe everything, however small, in the natural world, any more than you can observe all the people you pass in the street.

So, observation has to be focused by a question, the most obvious being 'I wonder what that is?' 'Is it an animal or a plant?' leads to observations about movement, legs, colour, etc. (Box 15.1). If it is an animal, ask 'What does it eat and what eats it?' (Chapter 6): if it is a plant, ask 'What is it made of and where does it get it from?' (Chapter 5).

Ignorance, provided that it is identified, is the starting point for learning. (The frightening people are not those who are ignorant, but those who are unaware of it!)

What is learnt has to be built into some system of knowledge and this must be more than an unorganized collection of facts. Thus, knowledge consists of facts and the relationships between them: without this degree of organization, isolated facts are difficult or even dangerous to use.

15.8 The needs of the citizen

In Chapter 14, it was argued that the citizen needed to be in a position to understand and form views on the major issues of the day, many of which have a biological basis. So, one relevant starting point is to consider how much biology a citizen needs to know.

Furthermore, it is not possible to absorb all the relationships that occur in any one subject, much less a wide range of subjects. It is therefore necessary to concentrate on principles (see Box 15.2).

15.9 Issues of importance

It is clear that the interactions between agriculture and the food industry, its dominant role in land use and its major effect therefore on the environment of almost everyone, and its effect on the diet of the consumers, should make agriculture of concern to all citizens, whether they are directly engaged in it or not. This is as true today as when Socrates said 'No man qualifies as a statesman who is entirely ignorant of the problems of wheat'. (In fact, the reference to grain here reflected the dependence of Athens on imported cereals and the vulnerability of its citizens to interruption of this trade.) Yet this is not a general view and there are now many, in developed countries, whose knowledge or understanding of agriculture is very small indeed. Of course, we are all ignorant of most industries, but agriculture does affect some of the most important parts of the lives of all citizens – food and environment.

Some of these areas are the focus of biological specializations. Many of them are concerned with biomedical topics and the importance of human disease prevention and control hardly needs emphasizing.

Demographic problems require attention to population control and, in common with most of these vital issues, impact on the education of the citizen throughout life. This requires that the citizen should be capable of participating in the great debates of the day – a major problem, since, as pointed out in Chapter 14, no one can be well informed about all the issues on which, in a democracy, the citizen is expected to have a view. It is thus not so much a question of information but of understanding.

Box 15.2 Principles

Some time ago (Spedding, 1983), I defined a principle as 'a true statement that can be used as a basis for action' or, at the least, for further thought. Agricultural principles should therefore help in making the choices and decisions that are necessary and should provide guidance for action, leading to the achievement of agricultural objectives.

Achieving an agricultural objective involves decisions on the balance of inputs and outputs in the interests of sustained profit and requires guiding principles (of management or farm economics) to help in making these decisions. But it also requires additional knowledge, in an applicable form, about the relationships between feed intake, its quality and quantity, and milk yield, for example. The knowledge needed relates to both big and small decisions, to whole production systems and to processes within them: and for all of these, guiding principles are required.

It nearly always turns out, therefore, that principles are hierarchical, that is, they can be arranged as in figure (a). This classifies them, with major ones each giving rise to several lesser ones as the diagram is expanded. Thus, there are principles involved in the breeding of crops and animals, just as there are principles governing the feeding of animals and the fertilizing of crops.

In short, wherever one can identify an action in agriculture that needs to be carried out, or a decision that needs to be taken, it is likely that a principle will be needed to guide one in how to apply what is already known and is relevant.

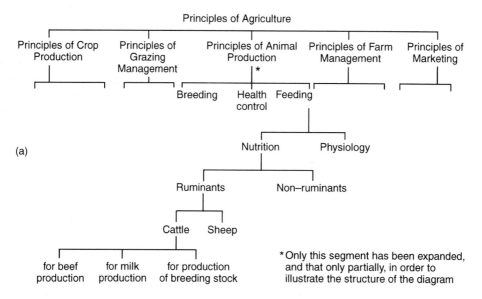

(a)

All this relates to agriculture but the same argument can be applied to virtually all subjects. Agriculture provides a good illustration because of its obvious use of many resources. All these resources represent the major elements from which agricultural systems are constructed but they do not themselves tell us much about the ways in which they interact or the ways in which they can profitably be combined. This requires guidelines or rules on how best to construct systems and these, in turn, depend upon organizing facts in such a way that they can be used for a preselected purpose. A collection of unrelated facts is not very useful and the conversion of facts into usable knowledge depends upon the formulation of specific principles.

Clearly, an overview – such as this book attempts – of agriculture and many related topics, cannot possibly be constructed of very detailed accounts of all the subject areas. Any overview has to sacrifice detail for comprehensive coverage – just as the view of the ground from an aeroplane is bound to do. This means that a broad picture has to concentrate on the essence of each constituent area and this ought to be expressed in its principles. This is very rarely done explicitly but would be an immensely rewarding task with an immensely important product.

It has been said that 'For the Greeks, the central problem of education was not the pursuit of knowledge, but rather the pursuit of significance. Learning was not a passive cognitive acquisition of data, but also involved emotional commitment, intellectual integration, value formation and expression, and practical action.' It is important to learn from one's own experience but this is a very narrow base and we also depend, as mentioned earlier, upon 'learning from experience the consequences of which we do not experience' (Senge, 1990). This includes learning from the past. As Isaac Newton (1675) observed: 'If I have seen further than others, it is because I have stood on the shoulders of giants'. Our debt to the past is largely ignored, both in terms of individuals (and not only 'giants'), but also of currently used technology, much of which is so familiar as to be invisible. So, alongside the public suspicion of science and research, there is constant use, acceptance of and, indeed, dependence on cars, television, computers, X-ray machines, dental cement, laser surgery, microwave ovens, jet engines, etc.

Perhaps the advancing edge of science would look more in proportion if there were more awareness of past and present achievement. That is not to ignore the obvious fact that knowledge and technology can be just as readily used for ill or even evil purposes. Indeed, the importance of biological knowledge and understanding relates, in very large measure, to informing debate on just such issues of public policy.

It may seem that asking for an awareness of the past is an unnecessary added burden – a kind of educational last straw. However, the purpose is actually to make it easier to understand both present and future, by seeing them as parts of a continuum in the development of knowledge and understanding, represented by what has happened up to now.

Attitudes to current and future medical advances may be different against a background of knowledge of how vaccines were first received (and developed). To see clearly how things were (and still are in many parts of the world) is to appreciate just how much change we have already accommo-

dated. This should not, of course, lead to uncritical acceptance of all that is new but it should help to encourage a more balanced appraisal.

Since knowledge advances, rather than accumulates (see Box 15.3), it follows that education has to be continuous throughout life. Indeed, it could be argued that an educated citizen is one who is both able to and does continue to learn throughout life. This suggests that there should not be too sharp a distinction between formal education (attendance at school, university, etc.) and the rest, both before and after periods of formal instruction.

Furthermore, education should be for everyone, even though capabilities and interests vary between individuals and everyone cannot study everything. However, it has often been argued that there are some components that should be common to all. Some time ago (1959), Snow (1964) described, in his Rede Lectures, what he called the 'two cultures' resulting from early specialization in either arts or science subjects. He argued that a complete education required both and suggested two criteria that could be used to assess the possession of such an education. Snow postulated that no one could claim to be educated unless, as a minimum:

1. they could describe the Second Law of Thermodynamics (this is variously stated as, for example: 'The entropy or degree of randomness of the universe tends to increase' and 'No energy conversion system is ever completely efficient');
2. they had read a work of Shakespeare.

Whatever one may think of the proposition in general, or these two criteria in particular, it is tempting to add as a third requirement that no one can be regarded as educated if they think that milk comes from a bottle. This, of course, is a rather trite way of expressing it and it is very revealing to try and express positive criteria that really fill the need.

Two important criteria suggest themselves in the following terms:

1. that one understands that photosynthesis fixes solar energy;

Box 15.3 Knowledge

The nature of scientific knowledge and how it is advanced is further explored in Chapter 16. Arguments are also deployed that research is an essential component of education because an education has to include an understanding of the ways in which new knowledge is obtained and how it can be evaluated. This requires an appreciation of how research results can be assessed, in relation to the design of experiments and the statistical analysis of results.

For example, no one who observed that a change of the diet of a dairy cow was associated with more or less milk, would jump to the conclusion that the same thing would be true for all cows under all circumstances. So all users of knowledge need to have some capacity to assess its validity and the ways in which it can be used. There is a huge literature dealing with a variety of different views about how scientific knowledge is acquired and how it grows.

It is well known that science is considered to advance by the statement of testable hypotheses and their rigorous testing. The fact that a failure of a hypothesis can be said to be more meaningful than an apparent confirmation has led to the well-known view of Sir Karl Popper (1963) that progress is actually made by the use of extremely rigorous tests that refute a hypothesis.

Every hypothesis has to have built in to it a statement of the conditions under which a proposition is held to be true. One failure, within these conditions, demonstrates the falsity of the proposition, whereas an apparent confirmation needs to be repeated almost endlessly to be sure that it is always true.

Whilst this can easily be illustrated it is often held (Kuhn, 1962) that this does not represent 'normal' science. Kuhn's concept of a 'paradigm' has dominated much discussion in this area (see Lakatos and Musgrave, 1970 and Ravetz, 1971). (A paradigm has been defined in more than 20 different ways but usually refers to a universally recognized scientific achievement that provides model problems and solutions for practising scientists, at least for a time.)

2. that one has heard of nitrogen fixation by rhizobia in the roots of legumes.

However, both of these are essentially biological, rather than agricultural, and the most useful, characteristically agricultural, criterion may be an understanding of the role of the ruminant in feeding people. Having regard to the rather sweeping nature of the proposal that without such an understanding no one is agriculturally educated and cannot claim to be a complete citizen, it is worth considering carefully whether it is reasonable and what other criteria might serve. In considering how the ordinary citizen's education could or should be influenced by agriculture, it should not be supposed that the agriculturalist's education is necessarily satisfactory.

15.10 Education of the agriculturalist

As mentioned in Chapter 7, there is a great need for those who can view agricultural systems as a whole, without being bounded by their disciplinary training. Yet most agricultural education is organized in this discipline-divided way and most graduates tend to be specialists in crop production, plant physiology, veterinary medicine, soil science, economics, animal production, parasitology or one of a host of other specialisms. Not that specialists are not needed, but their ability to apply what they know would be greatly improved if they also had some capacity to use a systems approach (see Chapter 7), both to the definition and the solutions of the problems they study.

Snow's two cultures related to different sorts of specialization but the division between specialists and generalists is also of importance. It has been said that a specialist is someone who gets to know more and more about less and less until they finish up knowing practically everything about practically nothing. The generalist is, of course, quite different and gets to know less and less about more and more, finishing up knowing practically nothing about practically everything.

Archilochus (the ancient Greek poet) said 'the fox knows many things but the hedgehog knows one big thing', pointing to the difference between very clever people and those with vision – one way of describing specialists and generalists. Of course, both are needed.

Naturally, changes occur continually in the form and content of agricultural education. Universities such as Reading have long included some systems thinking in the degree of agriculture. The Open University in the UK has applied it to whole courses and Hawkesbury College (now part of the University of Western Sydney), NSW, Australia, based a whole degree on it. Postgraduate research on a systems basis is now carried out at many universities: taught postgraduate courses have been provided, for example, in Thailand. Thus, the systems approach is slowly spreading, entirely based on the simple recognition that farms and agricultural enterprises are systems and best understood by thinking of them in this way.

Although there are many different kinds of agriculturalist (teachers, advisers and extension workers, research workers, farmers and so on), all would benefit, in my view, by awareness of a systems approach and the use of the systems techniques that are appropriate to their work. This is mainly in order to cope with complexity without being overwhelmed. For the systems we wish to improve really are complex and cannot sensibly be oversimplified. As Poul Anderson put it: 'I have yet to encounter any system, however complicated, which, when looked at in the proper way, did not become still more complicated'.

Even so, for different purposes (such as writing a book of this kind), there has to be some way of focusing on essentials and the only way to establish essentiality is with reference to the purpose. (This is obvious rather than profound. For example, the essential for a knife used for the purpose of cutting meat is its sharpness: for cutting bread it may be its serrations.)

For extension work ('extension' is a wider concept than 'advice', including extending research into practice by development, involvement and participation of farmers), there has to be practical knowledge (to establish credibility) as well as skills in communication and in involving other people.

However, as Corcoran and Dent (1994) have pointed out, success also depends upon the level of educational experience in the farm families to be influenced.

The following quotation from Bunting (1979) is still entirely valid:

> The knowledge system should involve the producers – not merely as the targets of advisory exhortation, as pupils at farming training centers, or as the passive victims of development done to them by remote government from afar: they have much to tell us about soils, weather, crops, animals, diseases and pests, as well as about their own purposes and difficulties . . . How many of us, who are so wise in international gatherings about what other people should do, could emulate them in winning subsistence, survival, dignity and fortitude in the face of calamity from the meagre resources of traditional rural society in tropical environments?

Gartner (1990), too, emphasized the human, non-science dimension of agriculture and his experience with villagers in north-east Thailand. He describes solemn discussion of ideas for agricultural development 'but unless the implementation of the idea has an element of fun in it, very little will be done'. He argues that establishing 'something in common' is essential to the process of communication in extension education and speculates that perhaps professors and peasants have in common this sense of serious fun. But then, he had a special sort of education! (The relationship between education and research will be touched on in the next chapter.)

However, in the wider view that I am advocating of agriculture and its context, there are some areas of special educational need. Education for the food industry is an important example. Youngs (1989) considered that it should include industrial experience and commercial awareness. This simply recognizes that practical applications in agriculture and the food industry take place within an economic framework and have commercial objectives.

Campbell-Platt (1989) further recognized that

the food industry includes food distribution and retailing and Malik and West (1989) considered the special needs of catering staff, who increasingly have social services responsibilities (e.g. community feeding of vulnerable groups, such as the elderly).

It may not be feasible to include agricultural education in schools but food production, diet and health can be introduced (Turner, 1989) and the art of asking questions is as relevant here as anywhere.

15.11 Education for the citizen

This brings us back to the citizen's education because this is where it starts (apart from the home). As an example of a special need for one section of the community, it would be hard to find a more important one than education about infant feeding (Clarke, 1989). There are many more such examples, of course, and the problems of including all of them in the education of a citizen are daunting. However, we are talking about a lifetime's experience of a very broad pattern of education.

The relevance of agriculture, especially in the wide sense used in this book rests on: (1) the importance of agriculturally-related issues; (2) the vital importance of multi-disciplinary problem-solving (which is a characteristic of agriculture); and (3) the fact that, by definition, it is a human activity that involves science, animals, plants, economics, law and, above all, people.

Just a final word about 'success' in education, because this is fundamental to the educational process. Education, perhaps like life, is not – or should not be – about success, although 'achievement' is important. Individuals have different abilities, capacities and potentials, whether we are talking about physical, mental or, probably, spiritual dimensions. No one would set a standard for high-jumping success, for the obvious reasons that people vary in height, weight, length of legs and so on. In fact, high-jumping is a useful analogy: if you have not failed, your target was too low! This has some general truth for individuals and societies – if we set our aims low enough, we shall all succeed; if we set our aims high, we shall all fail. Which is the better path?

Questions

1. How can education throughout life be funded?
2. How would you assess an educated person?
3. If a broad education is desirable, what subjects should be included?
4. Is there a value in knowing a little about a lot of subjects? Does this apply to languages?
5. Should failure be avoided?

Chapter sixteen

The need for research

A peasant will stand for a long time on a hillside with his mouth open, before a roast duck flies in.

Chinese proverb

On none of the subjects dealt with in this book do we have enough information. None of the major issues discussed is adequately understood and given the sheer magnitude of them this is not surprising. So, there is a continual and urgent need for new knowledge and better understanding: how is this to be obtained?

As mentioned in the previous chapter (and see Bunting, 1979), we should not ignore the knowledge accumulated by people (such as farmers) in the course of their working lives. Such knowledge may not be organized or recorded but this does not diminish its value, only its accessibility.

Research, therefore, is not the only way of increasing knowledge and the value of experience is that it has often been culled from much longer periods than are usually available for research. None the less, research is intended to increase knowledge and understanding by systematic and accepted procedures, so that we do not finish up merely with an additional range of opinions. One obvious function of research is to advance the subject (whether broadly or narrowly defined). But this should not be merely by an extension at the boundaries of existing knowledge. The idea that there is a central core of knowledge that can be

regarded as known for certain for a given subject, that there are fluid areas that have not yet crystallized out and other areas of great ignorance, is an unhelpful oversimplification. It suggests that research is confined to the fluid areas and not concerned with the central core; furthermore, it suggests that it can be considered separately and in isolation. One reason for this chapter is to argue that research is an extricable part of a developing subject and should not be so separated.

16.1 Research as an integral part of the subject

The fact is that subjects do not simply have central cores of certainty and peripheral uncertainties. The centre itself is in a state of continuous change, but in any case is not a clearly worked-out, unambiguous, well-organized body of knowledge. The nature of a subject is really much more hazy and the subject content consists of facts and relationships, some of them hypothetical. (The nature of scientific knowledge and how it is increased represent a very large area of discussion: the following references will serve as a useful point of entry: Popper, 1959; Kuhn, 1962; Waddington, 1968–70; Koestler and

Smythies, 1969; Lakatos and Musgrave, 1970; Ravetz, 1971. The result is that research and re-thinking permeate all parts of a subject and the centre is as much the subject of continual reclarification and restatement as are any of the more peripheral areas.)

It follows quite naturally from this that there cannot be a sharp separation between the subject as taught and that part of it that is the concern of research workers. Any such separation leads to an overdogmatic view of teaching, the material of which becomes arid and lifeless, and it also tends to lead to a restriction of research activity to the less important parts of the subject. If this kind of situation develops, it is bound to be followed at some time by a major upheaval, during which the subject appears to undergo a violent change brought about by a revolution external to it.

It is this essential integrity of a subject that is the basis of arguments that teaching, research and advice (or extension) should not be separated. Agriculture has been treated in a variety of different ways in different parts of the world and it is inevitable that some people should specialize in one branch or another. The problem is largely an or-ganizational one of maintaining contact between them, but the importance of this will only be clear if the essential integrity of a subject is recognized.

Applied research is a good example of an area in which this is not always recognized and, of course, it is always possible to conduct some useful research of a trial and error nature that has little to do with the development of subjects: it is more concerned with the development of practical expertise. How-ever, the ability of research workers to solve practi-cal problems depends ultimately on two quite different contributions. The first is an adequate un-derstanding of all the processes involved in the prac-tical situation, at an appropriate level of detail. The second is a clear statement of the problem in terms that make it susceptible to research. Each of these contributions might seem to come naturally from different kinds of people and yet each requires some familiarity and understanding of the other area. Some of the implications of these propositions will

be discussed in a later section; the relevant point here is that all activities within a subject are linked and there are risks in cutting them off from each other. If they are cut off, applied research becomes superficial and more basic research may become totally irrelevant to practical needs.

Another aspect of this relates to the contribu-tions that one subject may make to another. For example, agriculture is sometimes thought of as a practical subject to which sciences such as botany, zoology and biochemistry may make important contributions. However, for the reasons mentioned earlier, research in these subjects will be primarily concerned with the development of those same subjects: it will be largely a chance bonus if results of such research have direct importance for agri-culture. The main research that will benefit agri-culture is agricultural research and this requires that the scientific basis of agriculture be recognized as a subject. Incidentally, it is not implied here that agricultural research is not done, or should be done, by, for example, zoologists and botanists, or that developments in, for example, molecular biology may not be of the greatest significance for agricul-ture. But agriculture and agricultural systems are more than just scientific subjects; they involve peo-ple, management, economics, processing and mar-keting. However, they do have a scientific basis, largely biological, that is different from botany, zoology and all the other disciplines involved. It is this subject that needs to be developed and within which research is needed.

It is not suggested that it does not exist anywhere or that no such research is being carried out, but this way of looking at it is certainly not universally accepted or even consciously recognized even by those who tend to operate in the manner advocated. It is suggested that conscious recognition would allow the subject to advance and develop more rapidly and would encourage the necessary integra-tion of different contributory specialisms.

Thus, the biology of agricultural systems has a research content that, like that of any other prac-tical subject, needs a continuous range of activity from the most applied to the most fundamental.

16.2 Kinds of research

Opinions differ greatly as to the necessity of classifying research into different categories and, having done so, as to what they should be called. The idea that there is only 'good' and 'bad' research illustrates the impatience with which some view the whole discussion, but, in fact, also suggests that research can be unambiguously put into one or other group. The great range of kind and quality in research makes this unlikely and, indeed, it is still necessary to state for what (or for whom) the research is 'good' or 'bad'.

One thing is clear: whatever the merits of classifying research and whether research can be best described in one way or another, there are certainly different reasons for undertaking it. This suggests that it might be better to base research categories on these different reasons, without any implications as to the quality or nature of the research involved in each category, the kind or quantity of the resources required to sustain it or the quality of the people who engage in it.

There are, of course, many reasons why someone engages in a particular piece of research: some of these reasons are quite personal, of the kind that influence people in deciding whether or not to engage in any other activity. The reasons that concern us here are those adduced for undertaking research at all, irrespective of personalities and excluding reasons of commercial or government prestige (the reasons in these cases are for spending resources in that particular way, rather than reasons for the research). These reasons are more general than the objectives of particular pieces of research but both need to be stated very clearly. Although this need is self-evident in applied research, it is surprisingly hard to specify. This is chiefly for two reasons. First, it forces a clarity that may at times be uncomfortable and exposes the fact that there are generally several reasons and objectives and no wish to choose between them. Secondly, it forces a recognition of the high cost of obtaining even very limited amounts of information.

16.2.1 Applied research

The broad reason for undertaking applied research is to benefit practice. But practice itself may have many objectives and there are innumerable different ways of helping to achieve each one. So it is essential to be sure at the outset of applied agricultural research whether the object is to benefit the producer financially in the long or short term, or the food consumer, or the government in terms of improved balance of payments or independence of imports, or to create employment opportunities or to improve the living standards of the people within a region.

Since economic objectives are often considered to be the most important, it may be worth while pursuing this aspect a little further. If it were possible, by research, to put one farmer in a position to make much more profit from one particular enterprise, several processes would immediately begin. One would be that other farmers would adopt the same methods and erode the first individual's competitive position. This might be followed by a reduction in the price of their product and thus of the profitability of their enterprise. If profits remained high, other farmers would start such enterprises and still further spread the total gain, erode competitiveness and increase demand for the resources used in this particular enterprise. The last effect would spread the benefit to yet others but also reduce profitability. Since total production would be increased by most of these changes, the price would ultimately fall and, if it did not, surpluses would result.

Thus, *merely* making a greater profit might be of little benefit, except to a minority in the short term. On the other hand, if research results in the more efficient use of resources, including money, the benefit should be lasting and widespread. Results will clearly not be adopted if they lead to loss of profit and will be adopted faster if profit is substantially increased, but it is doubtful whether a marked and sustained increase in individual profits is a realistic objective of a particular piece of research: it is, of course, a normal commercial objective of a continuous research activity.

Increased efficiency in the use of resources, to produce what people want and are prepared to pay for, is therefore a key objective. Thus, economic assessment is one of the indispensable tools of applied research. Since this proposition is so frequently misunderstood, it must be emphasized that it is one tool among many, that there is no particular merit in not using it and that what is done with the results is entirely up to the research workers involved. Of course, there are dangers if anyone places undue weight on an economic assessment, as there are for any other single assessment judged in isolation. But this should not be allowed to distort the role of so valuable a technique. After all, the economic outcome is of some consequence and sensitivity tests that tell us which factors affect it, and by how much, are of great value.

16.2.2 Basic research

It is not intended to suggest that these research categories can or should be sharply distinguished from each other. Starting with any practical problem, it is possible to ask questions directly concerned with it and then, successively, to ask questions relating to levels of understanding far removed from the original problem. In many research activities, including agricultural research, it is highly desirable that much of the basic work is related, by clearly defined steps, to the applied research from which it arose and through which it will itself benefit practice. Even when all these steps and connections are clearly established, however, there comes a point at which it is unhelpful to describe the research as applied.

In addition, research is carried out in order to understand the ways in which agricultural animals and plants function and to explore the role of components and processes. Such information is often described as 'basic' to our understanding of applied problems and provides essential support for activity at a more practical level.

The danger is that since understanding seems so obviously desirable, it may be thought that it will automatically benefit the applied effort. It may do

so eventually, of course, but there can be no guarantee of this. After all, you, the reader, are trying to understand me at this moment, but from a particular point of view and for a particular purpose. You neither wish nor need to understand me as a whole person, if, indeed, this were possible, nor even as, for example, a gardener. It would distort my present purpose if I tried to convey a wider or additional understanding at the same time and it would interfere with your reception if you tried simultaneously to enlarge the kind of understanding that you achieve.

The fact is that all understanding must have a purpose. This is not surprising: since all understanding can be stated in model form and it has already been concluded that multi-purpose models have obvious drawbacks. It should perhaps be emphasized, at this point, that the number of purposes that a model can serve depends a good deal on the way they are phrased. In some senses, all models can serve several purposes, but there is a limit and no model can serve all the possible, or even probable, purposes for which a system may wish to be understood. 'Total' models are never achieved and would be totally unusable if they were.

Thus, basic research is simply thought of as characteristically not directly related to practice but as aiming at achieving an understanding of, for example, biological processes from an agricultural point of view. The definition of inputs and outputs is therefore based on relevance to agriculture and, further, to the particular agricultural purposes selected. To be practically effective, it seems likely that basic research ought to be related to, or at least aware of, the nature of the agricultural system (or subsystem) it is intended to understand. This relates to the 'systems approach' considered in Chapter 7 and to 'operational research' (see Box 16.1).

16.2.3 Fundamental research

At one end of the range of basic research the purpose is chiefly in support of applied research and there is no sharp separation. At the other end, the role is similar to that of fundamental research

Box 16.1   Operational research

The definition of operational research adopted by the UK Operational Research Society (see Duckworth, Gear and Lockett, 1977) was:

> . . . the application of the methods of science to complex problems arising in the direction and management of large systems of men, machines, materials and money in industry, government, and defence. The distinctive approach is to develop a scientific model of the system, incorporating measurements and factors such as chance and risk, with which to predict and compare the outcomes of alternative decisions, strategies and controls. The purpose is to help management determine its policy and actions scientifically.

There is no difficulty in stretching the term to fit application to agriculture (Spedding, 1980) but the stress on the word 'scientific' seems a little odd when the objectives include the management of people and money.

A systems approach (see Chapter 7) seems to embrace all that is intended in the above definition.

and, again, no sharp boundary lines can be drawn.

Fundamental research implies, to most people, very abstruse thinking about very detailed levels of understanding. It is more commonly visualized as ultramathematical or ultramicroscopic than dealing with, for example, buffaloes. But, of course, applied, basic and fundamental are relative terms and what is a very detailed and fundamental issue to a farmer or a scientist concerned with whole cows may be relatively applied to a cytologist or a biochemist.

Thus, one of the greatest weaknesses of these terms is that they mean quite different things in different subjects. However, this could be a strength if it firmly demonstrated that you cannot label people in this way, because the same person doing the same things could be labelled with each of them in turn, within different subjects. Within subjects like the biology of agricultural systems, applied research is chiefly concerned with the operation of biological systems within agriculture, and basic research includes the biology of component animals, plants and processes.

What then is the place, nature and role of fundamental research? It must be to speculate freely, uninhibited by economic or practical considerations, or even by current biological possibilities (hence it is sometimes called 'speculative' research). It harnesses unfettered curiosity and has greater freedom to consider propositions that may appear nonsensical. This is its chief merit, for clearly the most far-reaching propositions must all seem very strange and probably ridiculous when they are first proposed. This is not quite true, in the sense that some simple propositions of great importance seem quite obvious once they are pointed out but may remain hidden for long periods otherwise. This latter notion lies behind Szent-Gyorgyi's description of research as 'looking at what everybody else has looked at but seeing what no-one else has seen'.

The scientific character of the activity rests on this combination of curiosity, speculation and rigorous testing (usually by experiment) of the resulting hypotheses. Since experiments tend to occur at stages in the speculative process, they may well be designed primarily to check that the speculation so far is not false. If it cannot be disproved, there are grounds for proceeding with further speculation, in whatever direction curiosity leads.

All this having been appreciated, it is usual to say that this is fine but few people are good at it and, in any case, we cannot afford to employ a lot of people and facilities just to satisfy curiosity. This ignores the practical benefit because it is very difficult to be sure that there will be any. Over the whole of such an effort, it would be expected that there will be large potential practical benefits, though these may not be exploited at all, or, at any rate, by those who funded the fundamental research.

So, in fact, we cannot be sure what is the best approach and it seems probable that a balance between them would be most rewarding. Since agricultural systems contain non-scientific components and unquantifiable relationships, it often appears that science can only have a limited role in their improvement. However, orthodox science subjects also contain relationships that can only be quantified in a statistical sense, so that prediction is concerned with an estimate of probability.

The important questions, really, are whether the knowledge required for the improvement of agricultural systems is scientific or not and whether such knowledge can be applied within a practical context.

16.3   The knowledge required

In practical farming, which is, of course, only one aspect of agriculture, important questions are often unanswerable with any degree of confidence or precision. For example, will it be more profitable to plant beans or barley on a particular field in a particular year? No answer can be given unless a great deal of additional information is provided, including such data as future costs and prices, future weather, disease incidence, demand for the product and even the occurrence of accidents (to the farmer or to their equipment). Furthermore, such information has to be predicted well before sowing, or even before decisions are taken about cultivation or the ordering of seed and fertilizer.

But no one and no method can answer such questions by the use of reason or logic. Science can only answer questions that are posed in a particular way and this must include a statement of all the values to be attached to all the relevant variables. Clearly, agriculturally important questions can be framed in this way and are no different from those in any other field in which values have to be postulated in the question, whether they can be predicted for some future date or not. However, it is not obvious over what range of subjects these scientific questions can be posed. For example, a biological or chemical question would be assumed by everyone to be acceptable but an economic question might easily be considered inappropriate.

It is difficult to defend this kind of distinction. There seems to be no fundamental difference between a 'biological' question concerned with the use

an animal makes of its feed and an 'economic' question that relates a quantity of output to a quantity of input, however these quantities are expressed. Monetary expressions are no more arbitrary, necessarily, as a means of weighting things according to some scale of values, than expressions of quantities on a dry matter (or protein or energy) basis because we consider that this expresses the 'value' of the product more accurately than, say, the weight of wet roots or the liveweight of an animal.

If the questions that science can deal with cannot be distinguished in this way, then how can they be distinguished? The only satisfactory answer seems to be the rather circular one that science can deal with questions that are susceptible to the scientific method.

16.4   Science and the scientific method

Science is usually thought of as a body of knowledge, relating to certain subjects in particular and to knowledge of a peculiarly reliable kind. Thus we talk of scientific knowledge, tending to mean 'known for certain', and of scientific subjects, meaning those with a basis of provable facts, usually expressed quantitatively. No one argues that this is the only kind of knowledge, or the only knowledge worth having, but its distinguishing features are highly prized and jealously guarded.

Closer examination reveals that the 'certainties' are not quite what they may have seemed and that scientists may regard them as anything but the final truth. Indeed, scientists probably doubt whether anyone possesses the final truth about anything, except in matters of logic (two + two **must** equal four, for example), or could ever know whether they did or not. I say 'probably' because generalizations about groups of people are rather rash: 'scientists' are no more alike or uniform in their views than are farmers, students or readers.

A scientific proposition or 'fact' is characterized not by being correct but by being testable or provable. If a proposition cannot be tested it cannot be called scientific. If it can be tested, it generally is, unless this demands immensely expensive equipment (as with some physics) or impracticable conditions (as could occur in astronomy) or intolerable activities (such as experimentation on human beings) or impracticable periods of time (as in some evolutionary propositions). It is generally regarded as important that tests can be made by many different people and it is accepted that only one 'disproof' of a proposition is needed to render it false. Until proved false a proposition may continue to be regarded as true. Scientific facts are therefore those propositions that have successfully resisted all attempts to prove them false: but a new attempt could succeed at any time and the 'facts' would then undergo revision to restore the position, that they could not be shown (or have not been shown) to be false. Scientists thus live with a healthy scepticism about their knowledge and recognize that it is almost certain to change with time. In fact, they are mostly engaged in trying to change it.

It may seem curious, therefore, that the common view of scientific knowledge is so different from that held by the scientist. The absolute certainty often read into science simply does not exist in the minds of its practitioners. What does exist, however, is the rigorous insistence that a 'fact' is not scientifically acceptable if it has not been exposed to this procedure of formulation in a testable proposition, followed by systematic attempts to falsify it.

A proposition that is testable but has not yet been sufficiently tested is called an 'hypothesis'. After a good deal of testing it may be termed a 'theory' and after a very great deal of testing, provided that it is of sufficient importance, it may be dignified by the term 'law'.

It is this insistence that a statement cannot be regarded as embodying any scientific facts or knowledge unless it has been through this scientific process that tends to imply undue certainty in those propositions that are accepted.

Since scientific truth and knowledge depend entirely on this testing process, and since one clear falsification is deemed sufficient, it is not surprising that the testing procedures are themselves the sub-ject of detailed and meticulous scrutiny. They usually involve both logic and experimentation or observation, and they are subject to rules of procedure, carefully agreed conventions and terminology and a recognition that others must be able to repeat any test that is carried out. The whole enterprise is based on immense trust in the honesty of other scientists, combined with this capacity to check by repetition. The importance of this latter capacity cannot be overemphasized. Facts are not acceptable if the observation leading to them cannot be repeated, given adequate conditions. Sometimes, of course, the conditions are not repeated and cannot be controlled. In general, however, if it cannot be repeated, it probably is not so. This casts a special light, incidentally, on talk of 'duplication', as if that was always undesirable, unnecessary and wasteful.

Science, then, strongly implies not only the body of scientific knowledge but also the way in which this is acquired, since this is what renders it scientific.

16.5 The scientific method

This term has been used for a long time to describe the characteristic way in which scientists acquire knowledge. In thinking about it, the meaning that scientists attach to knowledge is very important. At the beginning of this chapter, attention was drawn to the oversimplification that each subject consists of a 'core' of knowledge that can be regarded as known for certain, surrounded by peripheral areas with which research is mainly concerned. Now, this might be so if research consisted simply of the addition of new information to some great existing heap – and there is sometimes a danger of this happening. But unorganized heaps of information do not constitute knowledge about anything very important. They represent isolated bits of knowledge and these are often important, but they relate to rather unimportant parts of the natural world. This problem is easily visualized in terms of the construction of a building. All the bricks, mortar, pipes, wires and planks needed to construct a building could be placed in a large heap and a great deal could be known about the weight, size, strength and composition of each unit, but there is no way that this would represent a building. The latter depends upon relating all these units to one another and this depends upon further knowledge about the possible relationships between them and how they interact with each other.

This, like all analogies, has its limitations. Incidentally, some people regard analogies as dangerous: but so is a knife, or any other useful tool. Everything depends upon the educated and responsible use of tools. Analogies are intended to provide an insight that is difficult to acquire by direct study. So it is possible to help someone who has never seen a kangaroo to understand how it jumps by saying that it is a little like a frog. (This example is chosen partly because it is a bad analogy, but it does point to what a good analogy should do.)

All this has a considerable bearing on the way in which scientists acquire knowledge and the role of particular activities, such as experimentation. The first thing to note is the magnitude of the intellectual activity that is required to comprehend whole spheres of knowledge sufficiently to reshape them. The second is that the idea of doing so (never mind how or in what direction) simply does not occur to everyone. Indeed, it occurs to very few and it is only the intellectual giants who can stride, fearlessly and with a unique combination of humility and confidence, across such large areas of human knowledge.

This discussion has seemed necessary in order to counterbalance the frequently encountered view that the scientific method is preoccupied with such things as experimentation. The latter is concerned with how new information is gained in a rather practical sense, but the process of assimilating such new information and creating knowledge requires a great deal more than experimentation.

16.6 Experimentation

This is certainly a characteristic feature of the scientific method but is best preceded and followed by much thought.

A great deal of discussion is often centred on the choice of experimental method when, in fact, this should largely follow from sufficiently well thought out objectives. However, in case this sounds too simple or too easy a solution, it is as well to admit to the extreme difficulty that often surrounds this first task of stating the objectives of an experiment.

Consider an everyday example. We wish to know how to grow more food. Superficially, this sounds a very clear and desirable objective, so why not do an experiment to find out? The number of possible experiments that would have some bearing on our objective is, of course, legion, and we could argue for a long time about which was the right one to do. However, it does not take much thought to expose the inadequacies of the original statement.

For instance, we already know how to grow more food – we can keep twice as many cows or plant wheat on three times as much land. So we have a further question: 'More food than what, when or for whom?'. That tells us immediately that the statement was useless, simply because we cannot tell from it what constitutes 'more' or how we should know when we had succeeded in producing it. But more critical is that no one ever means 'more', in the sense of producing it, without also meaning more per unit of something – a hectare, a person, an hour or some other resource.

So suppose we agree that we mean more food (we might even specify tomatoes or dietary energy or milk protein) per hectare of land (per year). Surely this is clear enough? Yes, it is, but it is probably not what we mean. It implies that we wish to know how to produce more food per hectare without regard to the cost of doing so, for instance. Now, we have to be very careful, for the next steps are crucial. Of course we do not mean 'regardless of cost', but what is a tolerable cost? Or do we mean at the lowest possible cost, because, if so, it may not even be possible. It usually turns out that we wish to impose a whole series of conditions: we do not wish to heat the soil, irrigate with nutrient solution each day, enclose the hectare under glass, employ enough people to remove every weed and pest by hand, harvest every single potato produced however hard we have to search, and so on. All of these conditions really have to find expression in a statement of objectives that is fit to base experimentation on and which has any chance of successful achievement.

Very often, a scientist finishes up rather disappointed at the enormous experimental effort that is still required to answer what finally becomes a very limited question, with nothing like the obvious importance that they originally attached to it. These are all commonplace problems of the working scientist engaged in experimentation and they have to do with the kind of information produced by the experiment.

In addition, and quite fundamental to the idea and purpose of scientific experimentation, we need to know whether the results are true or how we should interpret them. After all, you could experimentally throw a brick at a tree and immediately be struck by lightning: but what is the result of the experiment?

Several important principles have been evolved to guide the experimenter in interpreting the results of their experiments. First, the experiment is usually designed to test an hypothesis, so the experimenter is saying in advance what the experiment is about and what will count as a result and this determines what is measured. Measurement is a characteristic feature, partly because it is helpful to quantify effects but chiefly because it increases **objectivity**. This is extremely important: we are not very interested in whether I thought one thing was larger than another, we need to know that it was so by some objective test that does not depend on my judgement, eyesight or opinion. Next, we need to know whether it was larger because of the treatment being tested. Suppose we grow more potatoes where we have applied more fertilizer: how do we establish a connection between the two?

One common device is to set up a 'control', in which everything is the same except for the fertilizer treatment: if less fertilizer results in less yield, it looks as though there is a positive relationship. But it is very difficult, in fact, to make everything else

the same. Perhaps the soil in the control plot has more nitrogen in it, purely by chance. The answer is usually to replicate treatments, so that we do not depend upon one plot of soil, and, by statistical techniques, to estimate the effect of chance, to try and gauge the probability that any result could have occurred by chance.

Thus the methods of experimentation, including design and interpretation, are essential parts of the process by which confidence can be gained that the data collected can bear the meaning that we attribute to them.

However, where natural variation is high – and this is certainly so within agriculture – this confidence may not be very great and it is necessary to repeat experiments in different years, on different soil types, at different altitudes and so on. This is partly, of course, because we cannot control weather and seasons, and, in any case, we may wish to know the answer for a range of weather or even climatic conditions.

'The answer' sounds a perfectly reasonable way of describing the results of an experiment that posed a question, but it is necessary to recognize different sorts of answers.

If we measure potato yields in response to amounts of fertilizer applied, we envisage the answer as a response curve showing the quantitative relationship between yield and fertilizer. In spite of experimental design and statistical treatment, however, this result might still be due to chance or might only occur on the sites on which the experiment was carried out. We are, of course, rarely interested in such a result: we usually hope to generalize from these particular results to the whole of England, or to all clay soils or to areas with a given rainfall. After all, if the result of an experiment on a cow only applies to that cow, we have not made a very valuable advance.

It is when we wish to generalize widely that the value of trying to falsify hypotheses becomes evident. The argument here is that we can never be sure that an hypothesis is generally true, because the next test may falsify it. Do we then have to go on and on, and for how long? At least we would have more confidence if we designed the experiment to try and prove the hypothesis wrong: if it still survived intact it would suggest that the generalization must apply to all less stringent tests.

So, a most important experimental technique is to design a situation in which a proposition can be shown to be false, because this only has to be done once to be completely certain. Such a falsification does not, of course, demonstrate that there are no circumstances in which the proposition would be true, only that the generalization cannot be true.

One of the problems with agricultural research is that, ultimately, it is particular solutions to problems that are required: but they have to relate to a particular farmer's field and not just to the particular experimenter's field. This is different, however, from expecting useful generalizations that will apply to all fields. Only very simple propositions, such as that more fertilizer (up to some level) will produce more potatoes, are likely to be generally true. The question then is how to find out exactly how much fertilizer a particular farmer should use on a particular field in a particular year.

A great deal can obviously be learned by observation, generally repeated, of associations between events. This is often the way in which scientific hypotheses are formed and the experimental tests subsequently imposed still depend on observation of their results.

What experiments do is to control the framework within which observations are made and they do this to varying degrees. In general, the practical difficulty is that the more controlled the situation, the less relevant it is and the more relevant, the less the degree of control.

The systems approach discussed in Chapter 7 is intended to provide a relevant framework within which experiments can be carried out. Incidentally, these can be conducted on models, over a wider range of conditions and over longer periods, than would be possible in real life. This often raises two different concerns: (1) whether a systems approach is then in conflict with science and (2) whether 'theory' can be trusted.

Point 1. It may then be asked, what is the difference between a systems approach and a scientific approach? Are they in opposition to one another?

The main difference is really in their immediate and direct relevance to real-world problems. Science, in dealing with controlled experiments is generally deriving results from very artificial situations, about which it is relatively easy to state hypotheses. But real-world situations are very complex and, in addition to using science to determine constituent relationships, a systems approach also concerns itself with very complex hypotheses, often only statable as computer models. This capacity to deal with very complex hypotheses, or hypotheses about very complex systems, characterizes a systems approach and, in a sense, represents an extension of the scientific method in order to achieve relevance.

One important general point remains to be made. If the hypothesis to be tested scientifically is to be useful in agricultural practice, it must relate to a recognizable agricultural system or to a sufficiently independent part (or subsystem) of one. Furthermore, this relationship must be explicit in the statement of the hypothesis. Now this is more difficult than would at first appear.

For example, try to formulate an hypothesis about a bicycle without actually mentioning the word 'bicycle'. An hypothesis might concern the proposition that whenever the pedals were depressed the rear wheel revolved (or any other proposition you may like to imagine), but it is meaningless without relating it to a bicycle and, even then, including the word only solves the problem if everyone already knows, or can find out, what a bicycle is. (We will ignore, for simplicity in this argument, that there are many sorts of bicycles, having chosen a proposition that probably applies to all of them.) But suppose we are dealing with a system that is not readily denoted by a word (milk production, for example: from what, where, how managed, with what inputs, for what purpose, etc.?), then we are in the position of having to describe it in our stated hypothesis. This is difficult enough for a bicycle – just imagine the problems with agriculture! Clearly, the first thing is to aim at a description of the essentials of the system (this is often called a 'model'): the last thing we want is a description that is cluttered up with trivial detail.

Point 2. The farmer is inclined to ask of a theoretical proposition: 'Yes, but will it apply to my cows (sheep etc.)?'. Research cannot always be conducted on the actual farmer's cows (which farmer?), although it is hoped that the results will apply to them. Similarly, although the research has to be carried out on somebody's actual cows, it is certainly hoped that the results will not only apply to them. In short, the information required relates to theoretical rather than actual cows and, because it does so, it should apply sufficiently well to both sets of cows and to many more of the same kind. Theory is a powerful tool and needs recognition: subjects advance by development of theory, rather than by the accumulation of lore relating to particular experiences. As Brezhnev is reported to have said: 'There is nothing so practical as a good theory!'

The problems of experimentation should not be underestimated. Stating a problem is extremely difficult: even more so in terms that permit experimentation. The obvious experiment is frequently quite useless: as is obvious from everyday experience. For example, if we removed one leg from a four-legged table and it remained standing, we would not argue that the leg was not essential: for two main reasons. First, we are conscious that we could demonstrate, successively, that none of the legs was essential, but only meaning, of course, that any one of them could be dispensed with at any one time. Secondly, three legs are insufficient in many of the circumstances for which tables are used. So the answer is only that in given circumstances are three legs enough: in some special circumstances, even two legs are adequate.

16.7 The nature of research

Research is done by people and not everyone is any good at it. It requires curiosity and imagination combined with acceptance of rigorous procedures – of experimentation, recording and reporting.

The best scientists thrive on freedom and generally wish to spend the maximum amount of time on their science, as opposed to fund-raising, writing proposals and reports. However, scientists, like all other categories, vary greatly and generalizations about them must be treated with caution. Thus, for every original mind, full of imaginative leaps, there may well be a need for an army of relative plodders, because there is often a need for much tedious work, requiring great attention to detail and yielding very little in the short term.

The issue of scientific freedom will be returned to later in this chapter but one of the reasons that it presents some difficulty is, of course, that scientists have a cost. This is not merely the salary and, increasingly, substantial overheads, but, especially in some subject areas, the cost of facilities (e.g. equipment, laboratories, workshops, farms) may be very high. There has, therefore, to be some way of dealing with accountability, whatever the source of funding. The problem is to reconcile this with the nature of research: it is well illustrated by the confusion over 'duplication'.

16.7.1 Duplication

A common attitude to research is that duplication is wasteful and must be avoided. This is in sharp contrast to commercial activity, where the existence of two or more organizations doing the same job is seen as essential to allow competition!

Why scientific competition should not also be encouraged is a mystery, especially where many of the great discoveries have been associated with research teams in vigorous competition with each other. But the denigration of duplication in research is even more serious than this suggests; it is an attack on one of the basic principles of science.

Whatever a scientist discovers, if no one else can repeat their findings, they are not accepted. They may be believed but they are not accepted as new scientific knowledge unless they have been repeated independently – in other words, duplicated.

This is why scientific papers are commonly written like recipes: they have to record exactly what was done, under what conditions and using what methods, in order that someone else can repeat the experiment. If they then repeat exactly what was done but do not obtain the same results, it is not believed. Now, in order to repeat someone else's work, it is necessary that suitably qualified scientists, in the same field, have access to the same kind of equipment and other facilities. This must mean duplication of facilities, people and research activities: how else would it be possible? Yet, 'rationalization' usually takes the entirely opposite view.

Another important part of the nature of research is innovation.

16.7.2 Innovation

This is another concept that everyone appears to favour, perhaps because they define it differently. To a scientist, innovative thinking generally implies the generation of new ideas, new relationships, new ways of looking at the world. To the industrialist, it usually means the process that results in a new product reaching the market.

To many of us, the second depends on the first but, of course, new ideas do not have to come from scientists. However, in some areas (e.g. pharmaceuticals) this is more likely to be the case. In farming, it is often otherwise and many practical advances may occur long before anyone knows why or how they work.

Industrial innovation may be limited by lack of access to risk capital or the difficulty of justifying investment. Sometimes, it is possible to work on the assumption that, although only a small proportion of ideas may prove to be winners, the only way to find out is to examine a very large number (of ideas, substances, etc.). But some developments are very expensive and some probability of success is required at an early stage.

Whatever decisions are taken about such matters, advances in human welfare very often require innovative (or creative) minds and the question is 'how are they to be found and encouraged?' Since there is no certain way of spotting winners here

either, there remains an as yet unsolved problem. What is quite clear is that an excess of bureaucracy, accounting and accountability may squeeze out innovative thinking entirely.

There is some tendency currently to judge people and proposals by peer review, by established track records and by conformity to current fashions. But, as Ford (1994, 1995) has pointed out (Table 16.1), some of the most far reaching ideas were generated by people with no relevant track record, few or no relevant qualifications and against the judgement of their peers. In fact, the historical record suggests that, quite often, advances are made by those who have changed their field of interest or endeavour and thus come with no intellectual baggage, or preconceived notions of the truth or the right ways of finding it. Such people, of course, would be unlikely to have relevant qualifications or experience. This changing from one field to another may often occur somewhat later in life than that at which potential would usually be assessed.

What to do about all this is a different matter but it needs thinking about. One possibility would be to encourage those women who take a substantial period away from their science, to rear a family, to deliberately change their field of research.

16.8 Research policy

The meaning of policy (and related terms) is considered in Box 16.2: it is suggested that there must be at least two elements: (1) objectives and (2) a timescale for achieving them.

Simply to state the objectives as 'advancing knowledge' is not enough to justify funding and, indeed, recognizes no priorities amongst the infinite number of ways in which it could be done. Accountability therefore demands objectives that can satisfy certain criteria.

Currently, in the UK, these criteria have been established (White Paper on Science and Technology *Realising our Potential*, HMSO, 1993) as (1) the creation of wealth and (2) the improvement of the quality of life. (In practice, Government Departments use a whole range of criteria for funding

TABLE 16.1 Famous innovators and their backgrounds (from Ford, 1994, 1995)

Innovation	Innovator	Background	Training
Electricity	Michael Faraday (1791–1867)	Bookbinder's apprentice; Humphrey Davy's assistant at age 21	None
Relativity	Albert Einstein (1879–1955)	Clerk at Zurich Patent Office; 'My pen is my laboratory'	None
Uranus	William Herschel (1738–1822)	Musician; oboist, Bath church organist	None
Plate tectonics	Alfred Wegener (1880–1930)	Climatology; amateur philosopher	None
Plastic (bakelite)	Leo Baekeland (1863–1944)	Belgian chemist, later USA	None
The cell	Robert Hooke (1635–1703)	Philosopher's assistant (Robert Boyle)	None
Microbiology	Antony Leeuwenhoek (1632–1723)	Draper; town official	None
Vaccination	Edward Jenner (1749–1823)	Surgeon and naturalist	None
Cell nucleus	Robert Brown (1773–1858)	Surgeon's assistant	None
Germ theory	Louis Pasteur (1822–1895)	Chemistry	None
Bacteriophage	Félix d'Hérelle (1873–1949)	Outlaw	None
Genetics	Gregor Mendel (1822–1884)	Monk of Brno, Moravia	None
Antibiotics	Alexander Fleming (1881–1955)	Farmer's son, trained bacteriologist	None
DNA helix	Francis Crick (b. 1916) and James Watson (b. 1928)	Biologists working after hours	Graduates
St Paul's	Christopher Wren (1632–1723)	Professor of Astronomy	None
Geological maps	William Smith (1769–1839)	Canal navigator	None
Computer	Charles Babbage (1792–1871)	Gambling mathematician	None
Pneumatic tyre	John Boyd Dunlop (1840–1921)	Veterinary surgeon	None
Automatic dialling	Almon B. Strowger (c. 1889)	Undertaker	None
Aircraft	Wilbur (1867–1912) and Orville Wright (1871–1948)	Bicycle makers	None
Tape-recorder	Valdemar Poulsen (1869–1942)	Telephone company troubleshooter	None
Colour photos	Leopold Mannes (c. 1923–1935) and Leo Godowsky	Musicians, concert performers	None
Ballpoint pen	Ladislao Biro (1900–1985)	Sculptor	None
Photocopying	Chester Carlson (1906–1968)	Patents lawyer; amateur experimenter	None
Vinyl discs	Peter Goldmark (?)	Record executive; home experimenter	None
Minicomputer	Clive Sinclair (b. 1940)	Technical journalist, no degree	None
Float glass	Alastair Pilkington (?)	Glass company director; machine broke	None
Hovercraft	Christopher Cockerell (b. 1910)	Marconi radio researcher	None
Spinhaler	Roger Altounyan (?)	Bootleg researcher	None

research, such as those given in Table 16.2). In fact, of course, research alone cannot easily accomplish either of these two main objectives but may provide the knowledge needed to do so.

Oddly enough, this situation is changing, partly because knowledge may be patentable and therefore represent wealth, and partly because the commercialization of research activities means that knowledge that was hitherto made freely available to the world, is now bought and sold (with or without an advisory package on how to use it) (see Boxes 16.3 and 16.4).

Wealth creation, as a concept, sounds eminently desirable but it generates a whole series of questions

TABLE 16.2 Criteria used by Government Departments to judge the funding of research (in no particular order)

Potential for new advances in scientific knowledge
Potential for new advances in production for the market
The need for improving customer safety/environmental protection
Potential for expanding the market
Recent trends in production and sales (output value)
Cost/effectiveness of carrying out the research
Contribution of the producers in the sector to research funding
Perceived value to the industry
Statutory and legal requirements

Box 16.2 Policy, strategy and tactics

These terms are used so frequently, and often so loosely, that it is difficult to know what they actually mean on any one occasion. The following are definitions that I have found useful and simply indicate the meanings attached to the words in this book.

Policy

A policy is a combination of objectives with a strategy for achieving them. A single objective (such as increasing the prosperity of the country) may be satisfactory as a general aim but is insufficiently precise to be called a policy or even the objective of one (in the sense used here).

Objectives would normally have to be multiple, related and compatible with each other. They have to be achievable and they cannot be simultaneously achieved or sensibly sought if they are not compatible.

Since the definition includes the word strategy, this, too, has to be defined.

Strategy

This represents the way(s) in which policy objectives are planned to be achieved. There are three main elements:

1. a detailed programme (of actions, decisions, etc.);
2. a timetable for implementing the programme (and all parts of it), often shown as a chart;
3. an analysis of the resources required and available to implement the programme to the agreed timetable.

An overall strategy implies general plans that are intended to be adhered to.

Tactics

It has to be recognized, however, that life is characterized by major uncertainties and there have to be ways of adjusting policies and strategies in the light of unforeseen (and unforeseeable) events. These include the activities of others, whose policies and strategies may be in direct conflict with one's own.

Tactics are the short-term adjustments made in response to relatively minor changes in the circumstances under which a strategy is being pursued.

Box 16.3 Intellectual property rights

Such 'rights' aim to promote technical innovation and creativity by recognizing 'ownership' of innovative works, so that unauthorized use and exploitation by others are prevented.

Recombinant DNA technology is the area of greatest current difficulty and has given rise to considerable discussion about the patenting of genes, for example (see Box 16.4).

The need is because academic scientists (as opposed to commercial scientists) are not in the business of keeping their findings secret. Publication is part of the business of a scientist but, where ideas have commercial potential, some degree of protection is now seen as necessary.

Patent law, copyright and Design Right are available to protect intellectual property (Hird and Peeters, 1991) but the present situation is far from satisfactory. Total freedom of information (i.e. without protection) may hold back advances, since the, often heavy, investment in research cannot be recouped. On the other hand, protective behaviour may discourage collaboration in the areas of science (possibly 90%) that are non-patentable.

such as: 'what is wealth?' (surely there is a difference between creating wealth and making money?) and 'how can it be measured?'. But the most difficult question is 'how can one possibly tell, in advance, what research will lead to knowledge that, when applied, will generate wealth?'. The more fundamental the research, the less predictable both the potential and the time that might be taken to achieve it. It therefore cannot even be costed, so how can any attempt be made to assess costs and benefits?

As Williams (1995) put it, describing the development of nuclear magnetic resonance (NMR), leading to the magnetic resonance imager (MRI) and the use of superconducting magnets: 'What would "Foresight" have made of superconductivity in the 1960s? . . . How could it have been predicted that basic research at Nottingham, Aberdeen and Oxford would not only lead to a revolution in medical diagnostics but would also produce billions of pounds worth of business?'

Box 16.4 The use of patents

A patent is a form of contract between an inventor and society, in which 'certain limited monopoly rights are granted in return for the publication of information specifying the nature of the invention' (HMSO, 1995).

To be granted a patent, a process or product has to be

novel;
non-obvious (involving an inventive step);
capable of industrial application.

The new patent is then valid for 20 years, during which time no one else can exploit the invention.

New, genetically-manipulated plants and cells, DNA sequences not found in nature and genes in identified and isolated forms, can all be patented (Science Museum, 1994). It is this 'ownership' of living matter that raises the sharpest moral concerns but there are others, concerned with 'who actually benefits' and 'who should benefit'.

Patents can be challenged and overturned but this can be very expensive: however, defending a patent is also not without cost.

Similarly for 'quality of life'. Whose quality of life? Judged by whom? For how long? Are the qualities essentials or luxuries? After all, providing water to an African village and giving a cigar to a rich man may both be said to improve the quality of life.

These difficulties are real but are not rehearsed in order to decry perfectly proper concerns about accountability (someone has to pay for research and those in receipt of funds have a duty to account for their use), priorities amongst topics and choices between scientists. Not every topic (or even priority) or every scientist (or even the best) can necessarily be supported.

Broad objectives are easy to criticize and purposive research is successfully carried out. Cures for diseases are sought and found, so research can certainly be directed to an end, even though there can be no guarantee of success. The danger is that highly original research, the generation as well as

the answering of questions and totally unforeseeable ideas and developments, may be squeezed out by an attempt to force all research into the same mould. Policy also has to take into account the need to provide facilities, qualified and trained scientists and a career structure that will retain them.

Research cannot be turned on and off like a tap. At moments of acute and urgent need for research, such as the sudden and unexpected outbreak of a dangerous disease for which no treatment or cure exists, the necessary scientists and facilities may not be there. Trained and experienced scientists cannot be conjured up at short notice and some problems, e.g. caused by viruses and even less understood agents, require quite elaborate facilities involving a high degree of security. So, the most applied research needs, that everyone has to accept must be met, may not be evident today and removing research capacity now may leave us all very vulnerable tomorrow.

The future is characterized by uncertainty and research capacity is some insurance that we may be able to respond to it.

16.9 Research funding

The following section is based on the conclusions of the Institute of Biology Natural Resources Policy Group (Institute of Biology, 1991).

Since publicly-funded R&D will generally be justified as being in the public interest (i.e. contributing to public 'well-being'), this would be a sensible way to categorize it. 'Public interest' is preferred to 'public good' for two main reasons:

1. 'good' implies value judgements, including of a moral kind, that will be made differently by different people;
2. there is the risk of confusion, since 'public goods' carries other meanings (for example, to economists).

'Public good' R&D has been defined in the UK as 'R&D which fulfils a specific policy need or addresses a problem such as reduction of agrochemical usage, animal health or food safety.'

The responsibility for assessing which R&D justifies the expenditure of public funds must rest with those disbursing the funds, but wider and more open consultation would lead to better and more generally acceptable decisions. Clearly 'the public' is entitled to express its concerns but appropriately formulated information is needed in order to overcome undue influence by sensational sections of the media. There is a case for involving responsible representatives of different sectors of the public in the formulation of the questions that require answering, in order to establish trust and confidence that concerns are not being brushed aside. This problem does not yet seem to have been satisfactorily resolved (but see Box 14.3). Nor, indeed, has the problem of clarifying the issues that need to be confronted in deciding how 'public' and 'private' interests in research can be better defined.

The fact is that very few R&D proposals or projects fall wholly and exclusively within any of the categories used. Just because there is eventually a group of proposals judged to justify funding in the public interest does not mean that such a category can be used to assess which proposals should fall within it.

'In the public interest' is one argument, amongst many others (including 'near-market') that can be deployed about any one proposal. Whether to support it with public funds is a decision that has to be based on the relative strengths of all the arguments put and it is important that all the relevant arguments are put, rather than overstressing the one that is considered crucial to get the proposal into the desired category.

The central difficulty is the unpredictability of the outcome of research or its area of practical application. Only in the case of relatively applied research is it possible to specify benefits that are likely to accrue in the short term. Some would argue that research with a highly predictable outcome is really 'development'.

Much fundamental research is often undertaken with a practical benefit in mind (e.g. cancer research) but with little possibility of predicting when it will be achieved. In some cases, such research is supported by both public and private funds with the scientists involved left to determine their own research programmes.

Thus, it is not only Government that is able and willing to fund fundamental research and no one can stop commercial companies from engaging in such research. So it is not appropriate to use the distinction between 'fundamental' and 'applied' research as a criterion for public funding. Indeed, it is already the case that industry provides considerable resources for research within publicly-funded Institutes. In this connection, the importance of freedom to publish (even if this has to be constrained by some delay) must be recognized. Such freedom is important to scientists, to the scientific community and to the general public. It is clear that the freedom not to publish is also implied but, in relation to 'public good' research, it can be argued that there is an obligation to publish and to do so comprehensively and promptly. However, it is not only in the private sector that the results may be kept confidential and not available to the scientific community (defence is the best example of this).

It may still be argued that Government has no business funding applied work where the benefits appear to accrue directly to one section of society or one group of people. However, even this is not as simple as it may seem: research undertaken for the benefit of those who suffer from a rare disease could not possibly be funded by those who suffer from it, or even by the somewhat larger number who could be said to benefit. Indeed, it is possible to argue that most research that appears to benefit primarily one particular sector in the first instance will, by definition, ultimately benefit society as a whole. This is certainly true of any research that creates new products, or lowers costs of production, or raises the efficiency or reliability of production processes.

The difficulty of 'picking winners' has led the UK Government to institute a 'Technology Foresight' programme (see Box 16.5). The fact is that society wishes research in many fields to be carried out for the benefit of society in general, and recognizes that some results will arrive more quickly than others and will not benefit everyone equally.

The same argument applies in many other directions. The heavy use of agrochemicals in agriculture is also a matter of concern to society as a whole, but the perceived benefits of R&D aimed at reduced usage may appear to be associated directly with: (1) those who may gain commercially (e.g. if farmers are able to lower costs); (2) those concerned with wildlife (where the benefits are extremely hard to quantify, especially in monetary terms); (3) consumers (in terms of reduced risks – real or perceived); and (4) farm employees involved in field operations. However, these are only the immediate or first-round beneficiaries; other gains are generated and gradually diffuse widely through the system. (There will also be some losers.)

It can be argued that those who benefit commercially should fund R&D. This comes close to the UK Government's use of a 'near-market' category of R&D, which, if it is wanted badly enough, will be funded by those who are going to benefit commercially. If such R&D were entirely left to be funded through the market, however, the following consequences would need to be recognized.

1. Small sectors of activity might not be supported by R&D at all.
2. Areas that could not generate large profits would be neglected. (Even if the activity was extensive, it might be difficult to achieve a profit or, indeed, any return at all to those who fund the research.) Alternatively, even if the expectation of profit were sufficiently great in a mathematical sense, the very high risk associated with this might be too much for individual commercial organizations to accept.

Box 16.5 Technology foresight

This UK White Paper on Science, Engineering and Technology, 'Realising our Potential' (HMSO, 1993), led to the Technology Foresight Programme.

Such techniques are not new and have been used in Japan for decades: in the 1980s Martin and Irvine (1989) proposed such a scheme for the UK (it was regarded at the time as too 'interventionist'). Since 1993 the Office of Science and Technology has carried out an exercise which brings together scientists and industry, in order to attempt a systematic identification and promotion of areas of research that are scientifically strong and commercially desirable (likely to yield the greatest economic and social benefits).

Foresight is intended not to be 'picking winners', forecasting or predicting the future, but is described as a systematic way of assessing those market opportunities and technical developments likely to have a major influence in wealth creation and the quality of life. It depends upon joint discussion between those who know about what is scientifically possible ('science-push') and those who know about gaps in the market-place for products or services ('market-pull').

There was also a 'pre-foresight' period of preparatory work, including some form of the 'Delphi' technique, which asks a wide range of people what sort of advances are feasible and on what time-scale, and their assessment of their relative importance. This was intended to result in a consensus view (or not, as the case may be), which scientists, industry and others can take into account in their own thinking.

Not everyone believes that Technology Foresight will be beneficial and some are concerned that it may result in damaging the necessary freedom of basic scientists. These dangers have also been highlighted by supporters of Technology Foresight such as Williams (1995), who stressed 'the necessity of a vigorous community of world class scientists, competitively funded and focusing above all on scientific excellence'.

It is important to recognize:

1. that prediction of future needs and possibilities is rarely possible;
2. that 'Foresight' is primarily applied to technologies currently foreseeable, so that industry can better judge where best to invest the very large sums that are required for development;
3. that neither future technologies nor scientific developments can be foreseen;
4. that some new scientific research may arise when scientists are made aware of industrial problems.

3. Some important areas might not be served at all, because the information generated might be counter-productive to the commercial interests of those most involved in the commercial activity.

4. Results of some R&D might be revealed selectively: only those results favourable to commercial interests might be published and those unfavourable might be suppressed.

5. The information publicly available might not always, therefore, be regarded as credible or reliable – and certainly it could not be regarded as independent. In some cases, this would not be of serious concern and, in the long term, would affect commercial reputation and success.

6. There might be very little contribution to the 'community of science' and to its publicly-available, peer-reviewed literature. This is important because ultimately most scientific advance has to be based on the accumulated knowledge produced within this framework.

7. Attention would not be drawn to side-effects, unforeseen circumstances, effects on animal welfare and the environment, associated with the products sought or the production processes used.

Some of these raise ethical issues. First, it is necessary to be aware that virtually all areas may have ethical implications. All research that is relevant to feeding people may be said to have ethical dimensions: indeed, anyone concerned about this has to consider not merely the benefits of such research (see Box 16.6) but whether progress is possible without it. On the other hand, it is often argued that research is neutral and that it only provides tools that can be used for good or ill.

There is no doubt, however, that some issues are seen as raising very acute ethical issues (Spedding, 1995; Heap, 1995). In some cases, this is because an ethically important area requires research: examples include animal welfare and environmental protection. In others, the reverse is the case and it is believed that research should not be undertaken,

---

> **Box 16.6 The Benefits of research in food production**
>
> In a splendid recent book, Tribe (1994) summarized the economic returns to international agricultural research. Not all the effects are measurable but, even for those that are, the overwhelming majority of all published figures fell between 30 and 100%.
>
> Cost analyses have shown that, for each dollar invested in agricultural research, the average return is usually between $10 and $15. Numerous studies have shown (Schuh, 1988) 'rates of return that range between 25 and 100% per year in perpetuity – even studies that cover all of the agricultural research in a country, successful as well as unsuccessful ventures'. Tribe argues powerfully for increased donor support for the CGIAR centres (see Box 10.10).

---

either because the methods used are offensive or because the objectives are judged to be undesirable (or even evil).

16.10 Research methods

There is already some distrust of methods that involve interference with animals, especially those that may cause suffering, but such procedures are covered by legislative and other control mechanisms.

There is also suspicion of genetic manipulation although, curiously, there seems to have been little concern about traditional breeding, which has often produced gross deformations of the species involved.

This kind of ethical concern does not apply in the same way for plant species, although one basic cause is the same, namely the view that it is wrong to 'interfere with nature' in 'unnatural' ways. This proposition needs to be considered separately from worries that there will be ill consequences of such interference.

In the case of crop species, these worries relate to loss of control, if genes are able to wander about between species in an unintended or unexpected manner. The general concern is that scientists would be putting at risk aspects of our total environment without any real knowledge of the possible risks involved.

16.11 Objectives

Some concern about objectives may be described as ethical but idiosyncratic. For example, the idea that R&D was designed to increase productivity is simply pandering to the needs of greedy producers may be understandable but is a quite unrealistic concern within the current framework of society.

Other concerns overlap those about methods; these include notions that 'unnatural' animals, crops and, indeed, systems should not be developed. These have nothing to do with the 'neutrality of tools' argument: they embody the view that such tools should not be produced.

Given the benefits to humanity (in medicine, for example), it is difficult to sustain such a general proposition, although some religions do appear to take the argument to its logical conclusion. Probably the core ethical issue here is whether an argument based on the concept of 'naturalness' (e.g. what 'God' intended) is sustainable.

16.12 Naturalness

It is hard to find a meaningful definition of 'natural', although it is often thought to be self-evident that something is 'unnatural'. But where the terms are useful (such as in human behaviour – natural versus contrived, violence in a violent person), naturalness does not necessarily imply goodness or desirability. Indeed, many laudable aims involve the control of natural behaviour, whether of people or, for example, lions in relation to people.

Certainly, the use of the term 'natural' to defend something regarded as desirable (and vice versa) cannot be generally right. Large parts of disease control, surgery, food production, home and personal hygiene are unnatural and what might be regarded as a proper ecological balance of organ-

isms in organic farming would be inappropriate in a hospital. There seem to be no advantages and much confusion attached to the use of the term 'natural' in ethical debate.

16.13 Bioethics and research policy

It has to be accepted that society will determine what research methods are acceptable and on what material, especially if this consists of living and sentient animals.

It may seem more likely that ethical considerations will apply to the objectives of applied rather than basic research, but it may be more helpful to distinguish between the immediate objectives of research and the reasons for carrying it out.

Given acceptable methods and material, it is hard to see why there should be ethical objections to the acquisition of knowledge: but suspicion may attach to the reasons for wanting it and what it is going to be used for. This, however, is part of the 'tools' argument: is it right to learn how to make a knife, a gun or an aeroplane? Even words can be misused.

Questions

1. If duplication is inefficient, how can competition be encouraged?: is it inefficient to have several runners in a race?
2. How would you measure the performance of a scientist seeking a cure for cancer?
3. Why should you believe what a scientist says?
4. Could scientific research involve more people as amateurs, pursuing it as a hobby?
5. If research requires specialists, how can people who have no relevant background move into a new area and achieve spectacular advances?
6. How can we distinguish a valuable original thinker from a crank?
7. Does 2 + 2 = 4 for rabbits?
8. Is it natural to have bred dogs, poultry and goldfish so grossly different from their ancestors?

Chapter seventeen

The future

We cannot predict the future,
we can do better than that,
we can invent it.

Anon.

17.1 Thinking about the future

This is a vital activity, in which we all engage, and is quite different from prediction: our inability to predict the future does not relieve us of the responsibility for thinking about it. In most of our activities, whether personal, social or economic, we recognize the need to think about the future and, indeed, to engage in planning for it, even though we also recognize that major uncertainties exist, often related to events beyond our control. Prediction, however, is unhelpful. Even the forecasting of scenarios, because each one combines a particular set of components, is more akin to prediction than to broadly-based thinking. Predictions may actually constrain thinking about the future: however, they may also stimulate the thinking of others.

Thinking about the future may be best based on a study of the main (or even possible) determinants of change, in order to see how they may move. In this way, probable changes may be identified, but the rate at which they may occur appears to be quite unpredictable.

The reasons for thinking about the future of food and agriculture rest on the contributions that they make to people and their environment, but the agricultural industry is just as vulnerable to uncertainty as any other. Following Chernobyl, the changes in Eastern Europe and in the Middle East, no one needs persuading about the impact of un-

foreseen (and possibly unforeseeable) events on all our lives.

However, the functions of thinking about the future include improving our ability to foresee possibilities and contingency planning about how we might improve our capacity to respond and adapt to, or even to exploit, such possibilities. But there are major difficulties about actually doing all this. Most people are interested parties in relation to the very aspects of the future that concern them most. If you are part of a meat processing company, for example, you may find it hard to make an objective assessment of the future development of vegetarianism, even though it may be very much in your interests to do so.

These sorts of difficulty are fairly obvious and the dangers of self-fulfilling prophecies are clear. Furthermore, just because there are many circumstances quite outside our control, does not mean that we have no influence, that our actions can make no difference. There is no reason to suppose that the future is predetermined and that we somehow have to try and guess what it contains. Part of our thinking about the future needs to focus on the probable consequences of a range of decisions or actions that we might take.

In agriculture, for instance, whatever view you take about planning and its effectiveness, the fact is that policy decisions are taken and implemented that may change (indeed, are intended to change) the economic framework of the industry and thus the direction in which it moves. At the very least, therefore, in devising policy, its likely impact has to be foreseen, just as those affected by it need to consider the consequences of a range of responses on their part. It would be no bad thing if thought were also given to how the impact could be monitored, so that policy could be changed if we got it wrong or if changed circumstances rendered it less appropriate.

It is therefore a responsibility of individuals and organizations involved to engage in forward thinking and planning, but, as mentioned earlier, it is hard to be objective. It is also hard to assemble the relevant information and thinking about the future is extremely difficult: indeed, few people seem to have considered how it should be done, what we have learnt from trying to do it, what methods are available and who should really be doing it.

Until recently, in the UK, it has been very difficult to obtain funds to support uninhibited thinking about all this. This probably reflects a mixture of disbelief in the usefulness or importance of thinking about the future, genuine misunderstanding about what is involved and what is possible, confusion (with prediction, for example) about what is meant and a reluctance to spend money on thinking. The last point affects many kinds of research, leading to overemphasis on visible activity (experiments, reports, etc.) often with inadequate time for thought before or afterwards.

17.2 Reliance on trends

One of the easiest ways of thinking about the future is to suppose that current trends will continue. Of course, sometimes they do – at least for substantial periods. A good example in the UK has been the decrease in numbers of farmers and those employed on farms. On the other hand, no linear trend, up or down, can continue indefinitely: it is not possible. (A horizontal line is, of course, possible on a continuing basis.) So the more interesting questions concern when and in what way the trend line will change and, most important, for what reasons.

Trends are useful bases for discussion and may be good predictors in the short term, but their interpretation is vastly improved by identifying the factors that are responsible for them continuing and that might bring about change. This leads to the more general proposition that, although it is difficult to predict change, it is often possible to identify the determinants of change.

Before considering this it may be helpful to give one or two examples of misleading trends and assumptions. One of the best illustrations concerns oil, not an agricultural commodity but one on which modern agriculture depends. Past predictions of oil price and oil consumption, based on trends, have been quite wrong, usually because of the occurrence of unforeseen (and quite often unforeseeable) events. The price of land has also been subject to unpredictable fluctuations.

Another trap for the unwary is the assumption that a trend in one area will naturally be mirrored in another. Twenty years ago, it was commonly assumed, though not by everyone, that the kind of intensification that had taken place in the UK with poultry, pigs and, to a lesser extent, with cattle would automatically occur with sheep. Few would now predict any such thing.

17.3 The determinants of change

Any current situation can be thought about in this way, simply by asking what kinds of things would change it. No great skill or intellectual power is required. As the old proverb has it: 'The thing to do is to pick up the stick; you will soon discover which end you have got hold of!' This is an important point, since few of us quite know where to start. It really does not matter greatly. Once started, the process leads on to the identification of other factors and then to a family of factors that affect them, eventually generating a network of 'determinants of change' which can be arranged hierarchically. Of course, experience and practice do lead to improved skills in arriving at more systematic arrangements but, essentially, it is a self-generating process.

Actually, we all engage in it to some extent. For example, if we wish to get from A to B, we can immediately classify the possible ways (air, sea, land) and then elaborate each in more detail (as discussed in Chapter 16). The advantage is that we are not confronted by thousands of irrelevant possibilities (e.g. by camel?) because we can easily eliminate whole classes (e.g. animal transport) if we wish, just as anyone afraid of air travel can eliminate that whole class. This example illustrates the fact that there may be techniques that can be employed to assist our thinking and it is helpful to know whether they have been found to be useful or not.

Having identified a possible determinant of change, such as vegetarianism as a factor that would influence meat consumption, it is possible to ask what is likely to affect the numbers of vegetarians and how rapidly things might change.

17.4   Rates of change

One of the lessons most of us have learnt in doing this is that the rate at which things may happen is one of the hardest questions to answer. Something that appears to be logically inevitable may take a remarkably long time to change.

Equally, there are plenty of past examples (e.g. the balance of cattle breeds in the United Kingdom or New Zealand) where change occurred much faster than expected – even by those who expected any change at all.

17.5   Dangers

Amongst the dangers one has to beware of are:

desire for certainty;
unwise generalization;
accepting received wisdom;
untested assumptions;
use of logic on inadequate information;
dismissal of possibilities that are currently judged unlikely;
dismissal of ideas as impossible.

There have been several illustrations in recent years of dismissal of possibilities currently judged unlikely in the UK, including the farming of deer, goats (especially Angora goats), snails and llamas. All of these were frequently dismissed as laughable until suddenly a fashionable bandwagon developed and prices of breeding stock soared. The net result has been the very reverse of orderly development.

There is also a general danger of believing that, because decisions have to be taken, this has any bearing on whether they are right. Readiness to recognize that a decision was wrong depends a good deal on recognizing that possibility from the outset.

17.6   Recognition of important principles

It is vital that important principles should always be borne in mind. Thinking about the future requires a rare combination of boldness (not arrogance) and humility, recognizing that one may be wrong, that the future will contain uncertainties, that logical outcomes do not indicate rate of occurrence and that we mostly start from a position of considerable ignorance of relevant facts.

17.7   Aims

Realistic aims of those engaged in this activity are:

the projection of possibilities – that one should be aware of;
the description of scenarios – amongst which one can choose (although this has its dangers);
the identification of the determinants of change.

These are modest starting points but they can lead to systematic examination of future possibilities and an assessment of their likely consequences.

17.8   Determinants of agricultural change

Agriculture uses solar radiation, 'support' energy (mainly fossil fuels in developed countries but human and animal power in developing countries), water and, increasingly, agrochemicals (as pesticides, weedicides and drugs).

Most food is processed, if only by cooking, and, in developed countries, the food industry uses more 'support' energy than is used in production.

It must be clear that there is no single 'agriculture' that may change and few conclusions are possible that could have any general validity, but it is worth examining those that are currently fashionable.

17.9 Current global propositions

1. Limitations on land area, against projected huge population increases, suggest that more food will have to be produced per ha;
2. the fertilizer input required may result in controls designed to avoid nitrate pollution of water supplies;
3. the pesticides etc. associated with intensive farming may have an undesirable impact on the environment and may be serious sources of pollution;
4. non-renewable sources of energy will have to be used to a greatly reduced extent;
5. for these last three reasons, what are often called 'sustainable' systems may have to replace intensive, high input/high output systems.

Chapter 2 considered how the first proposition could then be dealt with. Of course, as pointed out in Chapter 8, if a cheap, safe and plentiful source of energy were devised/discovered all this would change (and who can predict the chances of this?). Energy can solve many problems including (by desalination) water supplies.

Against this background, the shape of agriculture in the future will be determined by the following:

Climate   Whatever effects of global warming (see Chapter 10) may actually occur.
Resources Land
            Solar radiation

Water supply ⎫
Support energy supply ⎬ see Chapter 8
Available plants ⎫ may be changed
Available animals ⎬ by biotechnology
Manpower and skills – see Chapter 15
Knowledge – see Chapter 16
Economic demand

17.10 Economic demand for products

As emphasized in Chapters 2 and 3, if human need is not translated into economic demand for products, these products will not be produced. If, on the other hand, the demand is sufficiently powerful, the products certainly will be produced (unless some other constraint applies). Need will reflect population changes. Demand will be affected by the degree of affluence and consumer preferences, mainly related to environmental impact (Chapter 10), animal welfare (Chapter 12) and concerns about human health (Chapter 13). Some of these will operate through legislation but some (such as vegetarianism) would simply change the kinds of product demanded.

The economic framework within which agriculture will operate is extremely hard to foresee: it will be fashioned increasingly by global trade agreements.

The socio-political framework will also be important and is probably even harder to foresee. It is hard to isolate the future pattern of development for any one country within Europe. Most projections have been based on the fact that current overproduction has to be contained: the total cost is simply too high. This means that the agricultural industry has to be reduced in size or develop new markets. Reduction in size has to be designed to reduce output (largely of food), either by a considerable reduction in the area farmed (North, 1990) or by a reduction in intensity, as advocated by those favouring the 'organic' option (Hodges and Scofield, 1988; Holden, 1989; Lampkin, 1990). Future options considered by others are given in Box 17.1.

Box 17.1 Projected scenarios

Scenario	Reference
Major contraction of agricultural area	North (1990)
Expansion of 'conservation' area	HMSO (1990)
Rural policy to sustain rural populations	Neville-Rolfe (1990)
General extensification of farming	Taylor and Dixon (1990)
Increase in organic farming	Hodges and Scofield (1988)
	Holden (1989)
	Lampkin (1990)
Major development of biofuel production	Carruthers and Jones (1983)
	Rexen and Munck (1984)
Larger-scale production of raw	Rexen and Munck (1984)
materials for industry	Barnoud and Rinaudo (1986)

Of course, there is no reason to suppose that the whole of agriculture has to conform to one or other of these scenarios and, indeed, Whitby (1990) argues in favour of multiple land use. It is quite possible to imagine highly intensive production on the more fertile soils, in those areas where inputs are not constrained by environmental impact restrictions, and extensive farming on cheaper or less fertile land or in areas that are classified as environmentally sensitive.

Indeed, as is currently happening with set-aside and partial organic conversion, it is possible to have mixtures of high-and low-intensity farming within one farm. One could go further and see the incorporation of unploughed, unsprayed headlands as representing a mixture of intensity within one field.

It may be a mistake therefore to see options and scenarios as mutually exclusive; it is more a question of adjusting the nature of the farming to the prevailing pressures and constraints operating in an area. This is, in any event, a more realistic approach than to suppose that particular systems or patterns can, will or should apply across the wide range of conditions to be found within Europe.

Out of all this, there appear to be three major possibilities that have not been dealt with in detail in any of the foregoing chapters and that are susceptible to further analysis. They are:

lower-input farming;

vegetarianism;
biotechnology.

17.11 Lower-input farming

The dangers of a two-tier structure of land use have been highlighted by Oliver-Bellasis (1991), in terms of the best land being used for low-cost production with damage to ecosystems and the poorer land not generating enough money for good stewardship. Some would prefer to see a general extensification (Taylor and Dixon, 1990), though not necessarily to the point of organic farming.

'Extensive' and 'intensive' are terms used in a variety of ways but there is a strong public theme that inputs such as agrochemicals are not needed, since we have an embarrassment of surpluses.

Less intensive methods of farming, such as organic and lower-input systems, claim to be better for the environment, partly because that is their intention. But these claims need to be rigorously tested and this is difficult to do. It is certainly true that organic and lower-input systems use less agrochemicals (by definition) but this may not guarantee all such systems are better for the environment. There should be less risk of chemical residues in foodstuffs but naturally-occurring substances may be equally toxic and the risk from residues is generally extremely small.

Grazing animal systems may not differ greatly in terms of usage of insecticides and weedicides but may differ a good deal in fertilizer input and the routine use of animal medicines. These may influence wild fauna and flora but hedgerows and untreated boundaries can be maintained.

Probably, most people, including producers, would favour minimal use of inputs such as sprays and fertilizers, provided that food production and profit were adequate. It may be that we can learn to be as productive and profitable at much lower levels of inputs but currently that is not usually the case. Thus, two important questions are commonly posed about such systems: (1) can the world be fed in this way? and (2) does the price of food have to rise?

Point 1 The first question is currently (largely) irrelevant. As discussed in Chapter 3, the world is not adequately fed now and hungry people are so because they are poor. Part of the inadequacy of developing country farming, however, is related to the fact that poor farmers cannot afford inputs and soil fertility steadily diminishes, especially if dung has to be used for fuel (a process that might be regarded as 'sustainable' on the basis that the fuel is renewable – but only if you have animals and forage to feed them on!). At the same time, developed countries are overproducing, in the sense that they cannot sell all they produce and are taking action to reduce the quantities produced. Such control is inevitable, since the costs of storage and disposal are so high.

These matters are complex but the crucial issue here is whether the world's food needs could be met by lower-input farming, supposing that these needs could be translated into economic demand. It can certainly be argued that the current inequalities across the world are 'unsustainable', in the sense that they are intolerable (but see Chapter 11), both in terms of acceptability and the fact that they will not be tolerated but will lead to unacceptable degrees of social unrest. If so, then the food needs of the world will have to be met and this will require greatly increased production, mainly in those countries where it is needed. Such increases will proba-

bly require inputs but there is no reason to suppose that this would have to be at levels currently regarded as 'intensive.'

Indeed, it is possible that systems mainly based on recycling could be based on fertility levels greatly raised by 'one-off' injections of plant nutrients. Current 'organic' thinking seems constrained by notions of fertility maintenance, or a relatively slow build-up.

By comparison with the current situation in many developing countries, inputs will certainly be needed, not only to improve soil fertility but to control weeds, pests and diseases, both during production and post-harvest.

Point 2 In countries that have adapted to high-input, high-cost systems, including high-priced land, lowering inputs generally leads to lowered profit and the need for premium prices. Thus, most organic systems in the UK are judged to be less profitable than conventional farming (though the latter is also held to be not very profitable even with present levels of subsidy). Initially, therefore, food prices would need to be higher for lower-input systems but this might not be so over the longer term.

Indeed, the starting point has to be the proposition that crops need protection. Agriculture is always to an extent unnatural and thus operates in a hostile environment. It is carried out for specific purposes and normally involves a concentration on a very limited number of species, animal and crop, in order to produce what is required. Natural forces resist this concentration and agricultural crops usually need protection from the unwanted species. Not that the wanted species are confined to those directly producing agricultural products; but there is a whole range of species that compete with the agricultural ones for resources and, indeed, for life itself.

In primitive agriculture, and still in many of the developing countries, protection is needed from large animal species that would otherwise consume or damage crops. Fencing often has to be high and strong to keep out such animals, most notably with, for example, elephants.

In developed countries, most of these large competitors have been eliminated or confined to non-agricultural areas and fences are primarily to keep domesticated animals in.

But pests, weeds and diseases cannot be physically excluded in most agricultural systems and they result, worldwide, in losses on an enormous scale, during production, in storage and during transport. These losses are extremely serious and it may often be the case that, where production is inadequate for people's needs, the priority should be to reduce losses rather than to try and produce more. The question remains as to how this is to be done and, clearly, there is no general answer that will apply to all crops and all environments.

However, the arguments tend to rest on different general approaches or philosophies, broadly related to the use of inputs. Some take the view that inputs should be minimized, or at least restricted to 'natural' (as opposed to 'artificial') elements, substances and species. Others see no reason to place restrictions on the use of inputs other than those imposed by legislation or 'good practice'.

Curiously, both extremes are most strongly represented in affluent countries where, on the one hand, inputs can be afforded and, on the other hand, the higher-priced food generally associated with minimum inputs can also be afforded.

In developing countries, there is generally a great need to increase productivity and a lack of money to pay for inputs: at the same time, low-input systems are associated with extremely low outputs. It is worth dwelling on this link and asking whether it is inevitable.

17.12 Low-input/low-output systems

In any production system, there is a direct relationship between **essential** inputs and outputs up to a certain point (beyond which further inputs are no longer essential). Obviously, if excessive inputs are provided, they can be reduced without affecting output. Equally, if inputs are very much lower than what is needed, output will rise if they are applied. This is true for providing nutrients for crops and for feeding animals. However, the inputs have to be **balanced**, in the sense of the desired proportions

between one element and another, and it is now seen clearly that they have to be related to external effects, i.e. outside the system, in terms of pollution and other undesirable consequences.

But there is another argument concerning the effects of inputs on the system itself, viewed holistically. The application of agrochemicals is an important example. The use of heavy applications can result in systems that depend upon them, because they have reduced their 'immunity' or 'resistance' to pest organisms. When inputs are reduced in such 'dependent' systems, output falls, but this does not show that 'independent' systems cannot exist. Skilled management may be able to encourage predators and parasites of pest organisms to the point where no agrochemicals are needed and where their use only interferes with the system created. However, as may be seen in both developing and developed countries, this may not always be possible and may rarely be easy.

The links between inputs and outputs are, of course, different for nutrient inputs (feed and fertilizer) and for crop protection chemicals. In the first case, nutrients are needed but the supply may come from another (biological) source – as with clover nitrogen. In the second case, the chemicals may not be needed but an alternative method of pest control may be. This will not be so where it is possible to achieve systems in such biological balance that no species ever constitutes a pest. We may not know much about this yet but the possibility exists. Where it does, it is possible to aim at the 'zero option' of using no applied (or manufactured) chemicals at all. The potential for biological control of pests and weeds is discussed in Box 17.2.

17.13 The 'zero option'

There is no need to make assumptions about the ultimate productivity of this option, or whether it could ever feed the world, or what price its products would have to be. In any case, all these issues may be greatly affected by other factors. It is an option worth considering because it can make contributions to environmental control, consumer

Box 17.2   Biological control

Biological control may be simply defined as the use of biological agents (usually living organisms) to control others that are regarded as undesirable (e.g. pests, parasites, carriers of disease).

In the sense that all organisms are eaten by (usually a host of) others, biological control goes on all the time. As a human activity, however, it is aimed at controlling, or reducing, the numbers of plants and animals that we regarded as undesirable. When we engage in it, it is important to remember that our target is selected because, for some reason, the normal controls are inadequate – from our point of view. For, although the natural balance is achieved over time, there are peaks and troughs for most populations.

The key feature is that, particularly if the control organism feeds only on the target species, its numbers can only increase when the target is plentiful. Thus, for example, the numbers of ladybirds and their larvae increase after their food supply (e.g. aphids) has built up large populations. This implies, as van Emden (1987) has pointed out, that complete control of a pest or a weed is unlikely ever to be achieved by biological control alone.

Integrated pest management (IPM) is based on this need to combine biological and chemical techniques in order to achieve the necessary control. Conway and Pretty (1991) give the following examples.

1. Rice in Orissa, India
   For gall midge, brown planthoppers and stemborers;
   IPM resulted in halving the insecticide applications.
2. Cotton in Texas, USA
   For boll weevil, pink bollworm and tobacco budworm;
   IPM reduced production costs and increased yields.
3. Apples in Nova Scotia, Canada
   For codling moth, European red mite and apple maggot;
   90% of apple growers have been using the IPM programme continuously since the 1950s.

Biological control is often defined in terms of the use of 'natural enemies' but the concept can be rather wider than this. It was first practised against insect pests but it has now been used against other arthropods (e.g. mites), snails and slugs, rabbits and weeds.

There is now an enormous literature on biological control (e.g. Debach and Rosen, 1991) and the successes and failures are well documented.

Successes

The importation of species alone, has resulted in permanent (but not necessarily complete) control of 164 species of pest insects. For example:

1. control of the Citriculus mealy bug in Israel (1939–1940) by parasites (e.g. *Clausenia purpurea*);
2. control of the Klamath weed in California (1944–1946); by beetles (e.g. *Chrysolina quadrigemina*);
3. control of the Fruit fly in Hawaii (1947–1951) by parasites (e.g. *Biosteres* spp.);
4. control of moth pests by *Trichogramma* spp., an egg parasitoid, now used worldwide on 15m ha.

Failures

These are of two kinds:

1. unsuccessful introductions;
2. introductions that have got out of hand, destroyed the target species and then gone on to other (unintended) food sources.

Point 1   Conway and Pretty (1991) list the following:
   (a) introduction of over 50 spp. of natural enemies failed to control Bermuda cedar scales (1946–1951);
   (b) the nematode *Remanomermis calcivorax* failed to control mosquito larvae;
   (c) nuclear polyhedrosis virus failed to control *Heliothis* spp. (though not for technical reasons).

Point 2   The most spectacular example is the introduction of cane toads into Australia to control sugar cane beetles: it did so but multiplied enormously to become a major pest, even in urban areas.

confidence, cost reduction and our understanding of agricultural biology.

To be in favour of exploring such an option need imply nothing derogatory about other options or those who operate them and need not imply that it would be the most desirable option in all circumstances. But it is legitimate to ask why one should aim at 'zero' use of non-biological or artificial substances, having regard to the ways in which we run the rest of our lives. Toothpaste can hardly be said to be natural and fleas most certainly are, but we choose the less natural repeatedly.

Farming is itself 'unnatural' except as an activity of humans. The rabbit is not interested in theoretical notions of ecological balance (and getting eaten) but, like us, does have an interest in survival. Wild species, however, have to tolerate population peaks and crashes, and the suffering that goes with them that may be natural but would not appeal to us. So, working with nature does not seem to me to exclude the use of substances that are manufactured (any more than metal implements are excluded), even though we may have very good reasons – sometimes based on knowledge and sometimes (just as soundly) based on ignorance – for minimizing the amounts used.

The main arguments for the 'zero option' then have to relate to principle (that there is some powerful – perhaps moral – reason for using none at all of the substances in question) or practicality (that only 'zero' can be policed, guaranteed or relied upon to be low enough – in the absence of hard data on acceptable levels of all the substances involved).

These practical issues can be of very great importance in attaching a meaning to any low-input system, simply because of the difficulties of establishing credibility for the products so produced.

17.14 Lower-input systems

Curiously enough, neither those who favour organic farming nor those who argue that no unnecessary restrictions should be imposed are usually in favour of intermediate levels of inputs. The first

fear confusion in the mind of the purchaser of products and the fact that false claims could easily be made. The second see no reason to restrict the producer from using any substance that, in the amounts used, is neither dangerous nor damaging to the environment. There are also concerns that ill-founded worries might prevent the use of substances that may increase productivity, profit and food safety.

Theoretically, the very characteristics of complex ecosystems render them capable of absorbing shocks and inputs: if the disturbance is too great for adaptation, a new ecosystem emerges that is adapted. This occurs even after volcanic eruptions, for example. However, the fact that there are obviously natural disasters, poisons and suffering should not be used to attack the concept of 'what is natural' where 'appropriateness' is really implied.

It is often argued that 'organic' management is much more difficult than 'conventional' management but it could be that the intermediate lower-input systems actually require even more knowledge and skill in order to integrate inputs and the harnessing of biological organisms successfully.

It would be possible to set standards for these intermediate levels of inputs, although they might have to include different levels for different crops, but policing would present major problems. The development in the UK of 'conservation grade' products is an example of this.

It can certainly be argued that lower-input systems would increase safety margins, increase public confidence, decrease risks and might even increase profits. It can also be argued that, for environmental reasons, conventional agriculture should move in this direction and that such a change would affect much greater volumes of food and areas of land than would ever be involved in organic farming.

17.15 Organic farming

In this subject, perhaps more than any other, it is wise to start with some definitions.

17.15.1 Definitions

Probably the most comprehensive of the many definitions available is that of the United States Department of Agriculture (USDA) (USDA 1980):

> Organic farming is a production system which avoids or largely excludes the use of synthetically-compounded fertilizers, pesticides, growth regulators, and livestock feed additives. To the maximum extent feasible, organic farming systems rely upon crop rotations, crop residues, animal manures, legumes, green manures, off-farm organic wastes, mechanical cultivation, mineral-bearing rocks, and aspects of biological pest control to maintain soil productivity and tilth, to supply plant nutrients and to control insects, weeds and other pests.

This definition, however, is factual rather than philosophical and thus does not convey the positive attitudes of the organic movement.

Lampkin (1990) argued that a short, sharp, clear definition of organic farming could not be framed because of confusion about nomenclature and because conceptual understanding is as important as specific practical techniques.

The principles and practices that lie behind organic farming have been concisely expressed in the standards document of the International Federation of Organic Agriculture Movements (IFOAM) as:

> to produce food of high nutritional quality in sufficient quantity;
> to work with natural systems rather than seeking to dominate them;
> to encourage and enhance biological cycles within the farming system, involving micro-organisms, soil flora and fauna, plants and animals;
> to maintain and increase the long-term fertility of soils;
> to use as far as possible renewable resources in locally organized agricultural systems;
> to work as much as possible within a closed system with regard to organic matter and nutrient elements;
> to give all livestock conditions of life that allow them to perform all aspects of their innate behaviour;
> to avoid all forms of pollution that may result from agricultural techniques;
> to maintain the genetic diversity of the agricultural system and its surroundings, including the protection of plant and wildlife habitats;
> to allow agricultural producers an adequate return and satisfaction from their work including a safe working environment;
> to consider the wider social and ecological impact of the farming system.

In the minds of the majority of the public, organic farming probably means the absence of agrochemicals on crops and little thought is usually given to what it means for animal production or any process beyond the farm gate.

To the organic farmer (and grower), it means much more than this: it represents a holistic approach to the whole farm, that works with nature, strives for ecological balance, and represents a better way to treat soil, crops and animals. It involves no use of 'artificial' aids (except machinery and fossil fuel), especially in the form of synthetic chemicals, including medicines, drugs, plant nutrients, hormones, growth promoters, pesticides and medicines.

Control of pests, weeds and diseases thus depends upon the use of naturally-occurring substances, management (such as crop and animal rotations) and operations such as flame-gun control of weeds.

This total avoidance of 'non-natural' substances has led to concerns that, for example, animal welfare may be neglected. In theory, this could not be further from the truth.

17.16 Animal welfare

It is one of the principles of organic farming that animal suffering must be avoided where possible and minimized at all times. Thus, it is laid down in all well-developed Organic Standards that no animal should be allowed to suffer if this is avoidable and that every animal must be treated appropriately if it becomes injured or diseased. This includes the whole range of conventional veterinary treatments. Whether the animal can thereafter retain its organic status is another matter: this may have to lapse until some agreed 'withdrawal' period has passed or the animal may have to be moved elsewhere for a time. However, the arguments are rarely about curative treatment but about prevention.

Organic producers recognize the need to take preventive actions, which is why rotations are encouraged and stocking rates restricted, but they wish to avoid unnecessary dependence upon synthetic chemicals as preventatives, particularly where they are used simply on an insurance basis.

However, where it can be demonstrated that, without such medicaments, disease will certainly occur, and where there is no other effective preventive action, these treatments are generally allowed. Thus, in the United Kingdom Register of Organic Food Standards (UKROFS) (see Box 17.3), both vaccination and appropriate anthelmintic drugs are permitted.

17.17 Ecological balance

The real problem is that full organic farming is not just the absence of chemical use and takes time (often up to five years) to establish. Thus, simply stopping certain conventional practices does not provide the appropriate environment for organic practices immediately. Applying no more manufactured fertilizers may leave soil with few soil organisms; applying no more pesticides may leave few natural predators of pests. So 'natural' controls may not operate right away and the continued use of conventional methods may delay the development of a full organic system.

This raises the question as to whether ecological types of farming necessarily require the zero inputs insisted upon by organic standards. It also has to be remembered that the term 'zero inputs' only applies

to 'artificial' or 'manufactured' products. Water will commonly be used, both for drinking and irrigation from outside the farm and animal manures may also be imported. The latter can contain heavy metals, such as cadmium, zinc, copper and lead, and these may accumulate in soils (see Chapter 10).

As to the future for organic production, this probably depends on whether the currently high premiums will continue to be needed. If organic farming did not actually require such premiums for economic viability, but could compete on price without them, then the potential market might well be vastly higher. After all, if it costs no more, carries none of the possible or alleged risk of residues etc., and involves 'kinder' methods of animal production, many people might prefer organically-produced food.

The matter is not, of course, as simple as that because there may be other features of such foods that make them less attractive. More blemishes, less uniformity, greater incidence of pests and pest damage, and poorer keeping quality represent possible drawbacks. That is not to say that these cannot be overcome but these factors could certainly influence demand.

The development of organic farming is much more advanced in some countries than in others (see Table 17.1). Within Germany, for example, the numbers of farms and the areas farmed (over 250 000 ha in 1993) organically are much higher than in most other countries, but France, Denmark and Italy all show strong movements in this direction.

TABLE 17.1 Numbers of organic farms in Europe in 1994 (Vogtmann, 1993)

Country	No.
Luxembourg	12
Portugal	120
Greece	150
Belgium	150
Ireland	250
Netherlands	467
Denmark	669
Great Britain	700
Spain	758
Italy	3000
France	3650
Germany	4950

Box 17.3   The United Kingdom Register of Organic Food Standards (UKROFS)

In 1991, the EU Council of Ministers adopted Regulation 2092/91 which established a European scheme for organic farming, aimed at imposing common standards, with legal force, throughout the EU. The regulation came into force in July 1991 and has been in operation since January 1993, obliging every producer/processor producing or importing organic foodstuffs to register with a designated regulatory body in his/her country.

UKROFS, which was actually established in 1987, in anticipation of the need, was designated as the UK control authority (in 1991) responsible for

setting organic standards;
establishing a register (at a fee) and a certification scheme;
approving applications and monitoring standards;
inspecting those registered.

All the main UK organic sector bodies are registered with UKROFS, they are:

The Soil Association;
The Scottish Organic Producers Association;
Organic Farmers and Growers;
Bio-Dynamic Agricultural Association (Demeter);
Organic Food Federation.

Setting the organic standards took full account of the organic philosophy on which these bodies are based but it also incorporated what processors needed and what consumers wished to mean by an 'organic' label.

All the foregoing considerations played their part in determining that the National Standards for organically-produced foods, set by UKROFS, reflected the holistic approach of the organic producer. Only those chemically-extracted or synthesized substances are allowed that are part of a legal requirement, are the only means of curing or preventing animal suffering or can be demonstrated to be essential in some fashion. Currently, no agrochemicals are included in these categories.

The standards relate to production, including processing, and thus relate to the materials used, the way products are produced, treated, transported, packaged and sold. No claims are made about the products themselves except that they have been produced according to these (policeable) standards.

The UKROFS Board is both independent and neutral. There is a market for organically-produced foods and it is essential that producers and consumers alike can be sure that the label on them carries a credible guarantee. Furthermore, the standards are published so that anyone can find out, in detail, what they are.

It is illegal to use the word 'organic' (or its agreed equivalent in other European countries) unless the food has been produced according to the standards laid down. In 1995, the numbers of those registered in the UK were producers 784, processors 280 and importers 30: the total area under production was 46 068 ha.

Vogtmann (1995) argues that there is an increasing recognition that the conventional system of food production is flawed and that organic farming is 'socially essential'. It is also claimed that organic production would only result (in Germany) in an increase in food prices for individual households of 2%.

17.18   Vegetarianism

Vegetarianism differs from **organics** in that, amongst other things, it focuses primarily on the consumer rather than the producer. There are probably few consumers who eat only organically-produced foods and there are probably extremely few producers who produce only food for vegetarians. However, vegetarians are not easy to define either (see Box 17.4 for details of vegetarianism in the UK). Essentially, they are people who eat no meat, but this sometimes includes fish and eggs and sometimes does not. Those who consume no animal products of any kind (including milk) are termed **vegans**. Their reasons for being vegetarians also vary and may include beliefs that animal products are either unnatural and/or unhealthy components of the human diet, that animal production is currently (and maybe inevitably) cruel, that killing animals for food is wrong and that the exploitation of animals at all is wrong.

Those who disagree point out that, in developing countries, crop production depends upon animal power, that milk and egg production also always involve the slaughter of animals at some stage, that even pets eventually have to be put down, unless they die in accidents or of disease (or starvation), and that some peoples (e.g. the Innuit) depend entirely on animal products for their food.

The agricultural and food industry already recognizes that people are entitled to their own views about what they wish to eat and are not obliged to defend their preferences to anyone else, unless what they do offends others. Of course, vegetarianism does raise moral issues but no one objects to people wishing to eat only vegetarian foods. It is therefore the business of the agricultural and food industries to produce them, to the extent that this can be done profitably. This should raise relatively few problems since crop products are inherently more cheaply produced than are animal products.

What is of interest is the size of the market and this depends upon demand. The latter is influenced by the number of people who are vegetarians but also by any change in the proportions of crop production in the diets of those who are not vegetarians.

The fact is that there is a significant market for vegetarian products. But there is no difficulty in meeting demand, unless (1) foods are required in particular forms or (2) there are also associated requirements as to how these foods (or their raw materials) are produced.

There is little doubt that a vegetarian diet can be nutritionally satisfactory, with some reservations about the supply of riboflavin, methionine and tryptophan (Guggenhein, Weiss and Fostick, 1962), and even vegans appeared healthy when their diets contained added vitamin B_{12} (Kurtha and Ellis, 1970).

Box 17.4   Vegetarianism in the UK

Since vegetarianism is practised to varying degrees by different people, statistics are difficult to assemble. One way of judging the trend, however, is to look at sales of vegetarian food. These are thought to have risen by about 26% between 1988 and 1992, reaching £11.1bn but it is hard to be sure that even specialized vegetarian foods are only consumed by vegetarians.

The number of vegetarians, according to the Vegetarian Society statistical data have risen from 0.2% in c. 1946 to 2% in 1980 and 4.3% by 1993, probably mostly among the under-25s and especially in the age group 11–18 years old. Averages are not necessarily very meaningful and it is thought that the 1993 figure is made up of c. 5.4% of the female population and 3.2% men. The motivation seems to be a mixture of health and ethical concerns.

There are many others (perhaps a further 6.5% of the population) who avoid red meat: by contrast, there are probably less than 0.25m vegans.

Many countries are now encouraging greater consumption of fruit and vegetables and a reduced consumption of meat. It is difficult to generalize about vegetarian diets but the following are examples.

Food	g head^{-1}
Bread	235
Pulses (in Israel)	39
Nuts and seeds	103
Oils and fats	34
Fruits and vegetables	1718
Sugar and sweets	31

The difference between vegetarian and vegan diets is illustrated below.

	Vegetarian (%)	Vegan (%)
Milk	15	–
Cheese	2	–
Eggs	2	–
Cereals	15	16
Potatoes	6	9
Pulses and vegetables	20	24
Fruit and nuts	16	27
Sugar and beverages	11	10
Butter or margarine	11	13

In terms of supplying the nutrients required for a vegetarian diet, there is no great problem since fewer resources are required to feed people on crop products and the latter are generally cheaper to produce, even for protein. Calculations have been made of the resources required to produce a largely plant-based diet (Thompson and Braun, 1978) but the impact depends a great deal on whether conditions are also laid down for methods of production.

17.19 Implications for agriculture

The implications of any major increases in either organic farming or vegetarianism are substantial but almost opposite in their effect. Organic farming involves extensification and thus the use of more land in production but less input, especially of agrochemicals. Vegetarianisms, by contrast, would involve much less land in food production because a vegetarian diet requires a smaller area to feed people than is needed for conventional agriculture designed for conventional diets (Table 17.2, Figure 17.1)

The land area required for feeding the UK population on organically-produced foods depends upon a number of assumptions about the nature of

TABLE 17.2 Land area required to feed one person on vegetarian, vegan and conventional diets in the UK. (source: CSO, 1988; FAO, 1985; MAFF, 1988, 1989; Spedding, Walsingham and Hoxey, 1981)

Diet	Area (ha)
Conventional diet[a]	0.32
Vegetarian diet[b]	0.14
Vegan diet[c]	0.07

[a] Calculated from present UK land use for agriculture:

Land use for agriculture, 1987	18 652 000 ha
Less land used for woodland, etc.	559 000 ha
i.e.	18 093 000 ha
Divided by UK pop. of 56 763 000	0.32 ha

[b] Calculated to provide at least 68.7 g protein and 9.9 MJ (2370 kcal) per day, i.e. the intake of these requirements by the UK population, 1987. (Diet including eggs, milk and milk products, peas and beans, potatoes, sugar and wheat).
[c] Calculated as for [b] but excluding eggs, milk and milk products.

the diet and, above all, the average level of production per hectare that would be achieved.

Some would argue that it would not be possible for the whole of agriculture to farm in this way, because the inputs of animal manure, for example, would not be available in the necessary quantities (Jollans, 1985). This is because it is impractical to return all of the nutrients which leave the farms for the towns and cities. However, the biggest losses are of nitrogen and there are biological ways of fixing atmospheric nitrogen in the soil.

17.20 Implications for the agricultural input industries

Certainly organic farming, and possibly vegetarianism, have considerable implications for the manufacturers and suppliers of inputs to farming. By far the largest input supplier is the fertilizer industry (see Table 17.3) but agrochemicals and pharmaceuticals are also very big sectors, in both volume and monetary terms.

All would be affected significantly by large changes in methods of production associated with the production of organically-grown or even vegetarian foods.

17.21 Implications for the food industry

If there should be a significant move towards vegetarianism, there would be several very important consequences to the food industry. The most obvious are: (1) the impact on meat producers, processors, retailers and distributors of a reduced demand for meat; and (2) the impact on all branches of the

TABLE 17.3 Magnitude of the main input industries (source: Bennett, pers. comm., 1989)

UK inputs 1988	£m
Fertilizers and lime	693
Pesticides	386
Animal health products, 1982	> 100

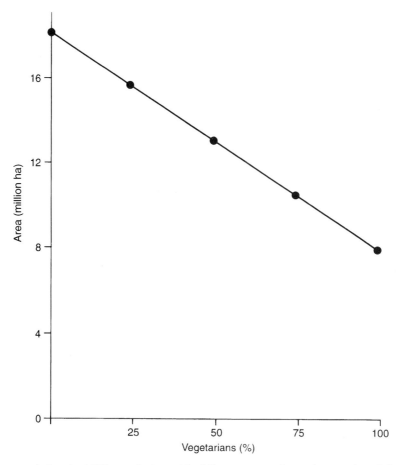

FIGURE 17.1 Area needed to feed UK population with different proportions of vegetarians (after Spedding, 1989).

food industry of an increased demand for vegetarian foods, involving different kinds and different formulations.

The first is rather obvious but would also affect importers, transport, docks and a host of other industries. Consider one only: the leather industry. Currently, leather is an important product and most of it comes from animals slaughtered for meat.

The second is also fairly obvious but it, too, has some less immediately evident aspects. A significant proportion of current vegetarian diets is imported, so the balance of imports and exports could be affected. But many people who became vegetarians might do so for reasons which had nothing to do with not liking the taste of meat. So the implications for meat substitutes could be substantial.

Indeed, perhaps the biggest industrial impact of vegetarianism would be on the potential for further development of meat substitutes. If the market for such products became larger, it might accelerate the development of attractive and cheap products that tasted like meat but had none of the associations with meat production that some people may wish to avoid. Such a development could then pose a competitive threat to meat and meat products, even amongst consumers who hold no strong views about the ways in which food should be produced.

17.22 Organically-grown foods

The implications for the food industry relate primarily to labelling and identification. Consumers will need to be confident that they can rely upon the fact that, when they see the statement 'organic' on a product, it conforms to their expectations. But labelling also presents problems, especially on certain products and commodities, and the use of labels also has to be policed.

Whereas vegetarians only purchase vegetarian foods, it must be expected that a wide range of consumers will purchase organically-grown produce and few may confine themselves to such foods. This has implications for marketing and leads to a mixture of products (organic and others) at retail points, thus exaggerating the identification problems.

17.23 Conclusions regarding organics and vegetarianism

It is clear that major growth in either organic farming or vegetarianism could have major effects on all parts of the human food chain. Some of the large-scale effects are fairly easy to predict; others are more difficult, even though they could be large. Some problems are actually worse while such sectors are minor. For example, the problem of abattoir provision is relatively straightforward if all the raw materials are of the same kind, whichever they are, and not too difficult if each sector is substantial. But the scattered requirement for a minor sector may pose much greater difficulties.

One advantage (in this context) of organic farming is that it requires at least a two-year transition period and this provides an early-warning system of what may be needed. Apart from this, however, it will be difficult to predict changes in demand and the safest, as well as the most satisfactory, procedure is to ensure choice. If the consumer is offered credible choice of different products, complaints are minimized and demand can be assessed by direct measurement.

There is little point in trying to guess what changes may occur and everything to be said for allowing consumers to make up their own minds. That still leaves major areas of education and persuasion available to influence how choice will be exercised.

17.24 Biotechnology (see Box 17.5)

Biotechnology was defined by the House of Lords Select Committee on Science and Technology (1993) as 'the use of biological processes, like fermentation or selective breeding, for making useful products'. In its modern form it is based on genetic modification (or genetic engineering) using the techniques of molecular biology to move genes from one organism to another in a controlled and highly specific way (Box 17.6). Applications include 'gene therapy' to treat hitherto incurable conditions.

Except for pathogens, the Select Committee Report saw no great cause for concern, even in relation to the release of genetically-modified organisms (GMOs) 'except where bacterial or virus vectors, live vaccines or modification of the genome of animals are involved'.

Recent developments were considered to fall into six categories:

1. recombinant DNA (rDNA) – cutting and splicing of DNA;
2. fused cell technique – creating a new cell from two others;
3. micro-injection – injection of DNA into a cell or its nucleus;
4. 'biolistics' – shooting DNA into a cell from a particle gun;
5. viruses as vectors – to carry DNA into a cell;
6. fermentation and cell culture.

The potential applications in agriculture include plant and animal breeding (by transgenics), diagnosis of disease and the development of pharmaceuticals to control animal health.

These potential benefits are enormous but, in fact, our lives are already full of examples of successful biotechnology (e.g. insulin, vaccines, interferon, antibiotics). There has been little appli-

Box 17.5 Biotechnology

In the UK, the Biotechnology and Biological Sciences Research Council (BBSRC) was formed in April, 1994. It defines biotechnology as 'the application of biological systems and biological processes to the manufacture of new products'. For the future, this means, for example:

improving cereals by direct gene transfer (e.g. a transformed cereal called Tritordeum);
disease-resistant crops;
bioremediation – environmental clean-up using hyper-accumulators; plants that mop up soil contaminants such as heavy metals.

But the definition could be used to cover all future developments in agriculture that aimed at new products.

Other definitions are even simpler, citing the use of living organisms to improve plants and animals or provide goods and services. Some of them include such ancient processes as cheese-, wine-, and bread-making and the selective breeding of plants and animals. This group of processes is often referred to as **traditional biotechnology.**

By contrast, **modern biotechnology** refers to techniques of genetic manipulation and recombinant DNA (see Box 17.6). It is, of course, the modern version that is thought to raise moral and ethical concerns (see Straughan, 1992). Modern biotechnology includes, in addition to genetic manipulation:

diagnostic procedures for the detection of contamination by diseases;
tissue culture: the rapid multiplication of plants from small pieces of tissue – indeed the regeneration of whole plants from single cells;
molecular vaccines for the control of animal diseases;
embryo multiplication to speed up animal breeding programmes;
rapid food safety contamination tests;
the use of enzymes in food processing;
biopreservation without the use of synthetic chemical preservatives;
production of meat substitutes from fungal protein;
micro-algal culture for food from seaweeds;
non-food crop production;
treatment of wastes from agriculture and the food industry.

However, one must also recognize the equally enormous potential outside agriculture, especially in medicine. Here, possibilities exist for:

new and better drugs:
limitless quantities of blood and bone marrow cells;
human vaccines (e.g. for the hepatitis B virus).

Many of these offer enormous potential benefits for the future but they also raise public concerns (see Straughan, 1989; Martin and Tait, 1992), perhaps none more so than genetic manipulation (see Box 17.6).

cation yet to food, however, and there are public concerns in this area (see Chapter 13) and more general fears about the 'escape' of modified organisms with unknown impact on the environment (see Chapter 10).

In the UK, all aspects are subject to rigorous controls and, if anything, the Select Committee considered that they could and should be relaxed. In fact, the UK Biotechnology and Biological Sciences Research Council (BBSRC) aims to harness biotechnology to clean up industrial wastes and pollution (known as bioremediation), to develop drugs to combat disease, to produce new non-food crops for industrial use and even to produce animal vaccines from plants.

Interestingly enough, although biotechnology is often viewed as being at odds with organic farming, there are possibilities of genetically manipulating crop plants in ways that would allow a reduction in inputs of agrochemicals. Indeed, Miflin (1992)

Box 17.6 Genetic manipulation

Genetic manipulation (also called genetic engineering) means transferring genetic material from one species to another: the products are known as 'transgenic' plants or animals.

The techniques involved include recombinant DNA (deoxyribonucleic acid) – the molecular basis of heredity. Each molecule contains two complementary strands arranged in a double spiral, called the double helix. Pieces of DNA are transferred from one species to another (in which they do not naturally occur), thus incorporating new genetic instructions. The production of human insulin from modified bacteria (instead of having to extract it from the pancreas of cows and pigs) was an early success (1979).

The significant part of the DNA molecule is the gene (a protein) and recombinant DNA is based on the use of enzymes to cut specific parts out of the DNA of one species and insert them into others. The actual transfer may require a vector (usually plasmids in bacteria or viruses, because of their ability to invade specific hosts and multiply there). All this requires not only the sort of techniques mentioned, but also a map showing where all the genes are (and knowledge of what they do). Such maps may show the sequence of genes in a chromosome or the whole genome, showing the relative positions of thousands of genes.

Attempts are being made to construct a human genome map (24 different chromosomes and 3bn nucleatide building blocks that make up the genome) and progress appears to be accelerating in an astonishing way (see, for example, Coghlan, 1993).

has argued that 'genetic alteration of plants through plant breeding using both conventional and rDNA methods is an important method for developing sustainable crop production systems.'

Genetic engineering is also reported to make possible the use of cultured human cells for a pharmaceutical test that can reduce the need for animal experimentation.

Although the foregoing has drawn its examples from the UK, there is a great deal of research being carried out in most developed countries and the rate of change in this field is set to accelerate. But it should not be thought that the benefits will be confined to such countries. Indeed, there are massive gains to be made by the application of biotechnology in developing countries where some of the most intractable problems are to be found.

Of course, as with all scientific advances, there is also potential for misuse, in, for example, biological warfare, and there are real ethical concerns about the use of modern biotechnology (see Box 17.7).

17.25 Possible future scenarios

Since all future possibilities could occur in countless combinations, the value of postulating specific scenarios is very limited. However, there are clear indications that the whole industry should use more solar radiation and less support energy: certainly energy should not be wasted. Recycling of resources using solar radiation would be of benefit.

Recognizing that only technology which is economic will be applied, it is still wise to ensure that new technology is not dismissed because of either premature economic assessment or excessive public concern.

A change in the balance of production from food to non-food products could have beneficial effects in those countries currently producing food surpluses by:

1. preserving agricultural production capacity that may be needed for food production at some time in the future;
2. reducing overproduction of food by diverting resources to non-food production;
3. contributing directly to energy supplies from renewable resources;
4. increasing the total market for agricultural products by diversification;
5. offering less intensive animal production by combination in agroforestry systems.

Whatever system are devised, they will have to be economically and energetically efficient, probably

Box 17.7 Ethical concerns about biotechnology

It is increasingly being recognized that scientists cannot just stand aside from the ethical concerns about their work and the way it is used. Heap (1995) has recently reminded us that Aristotle, more than 2000 years ago, wrote 'that an ability to think and act with a rationality not linked to an understanding of values could seriously impair the social and physical wellbeing of individuals, communities and societies'.

Public concern has focused especially on the 'emerging technologies' which the Banner Committee (1995) was set up to examine, in terms of their ethical implications. In their report, the Committee identified three principles as a framework within which the present and future uses of animals should be assessed.

1. harms of a certain degree and kind ought under no circumstances be inflicted on an animal;
2. any harm to an animal, even if not absolutely impermissible, none the less requires justification and must be outweighed by the good which is realistically sought in so treating it;
3. any harm which is justified by the second principle ought, however, to be minimized as far as is reasonably possible.

Heap (1995) has pointed out that the ethical implications of gene transfer, examined by the Polkinghorne Committee (1993), are actually much wider than those raised for animals.

Recently, Postgate (1995) has argued that we must debate now the issues raised by the fact that, eventually, 'we shall learn the genetic bases, such as they may be, of social inadequacy, criminality, and other behavioural aberrations – even perhaps, of aspects of intelligence and creativity'.

The word 'eugenics' (which actually means good for the genome) raises the spectre of the breeding of 'better' or 'purer' human beings, which might lead to the exclusion of possibilities of preventing hereditary diseases such as cystic fibrosis.

It is clearly important that there should be widespread and informed public debate about how to use wisely the scientific understanding which will continue to grow, probably at an accelerating pace.

It is also important to be aware of the existing regulations governing the use on Biotechnology, including the EU Directives of GMOs (see POST, 1994).

---

within a lower-cost framework, and they will have to satisfy the public concerns referred to earlier.

Finally, it is important, whilst being aware of tremendous uncertainties and the need to be able to respond appropriately, also to see the future as presenting opportunities. The most important reason for thinking about the future is in order to change it – for the better.

Awareness of current and potential problems should not obscure the need for vision. We should be thinking internationally about how the world could be transformed for the benefit of people. Grand schemes would become attainable if conflict could be contained and sources of cheap, safe energy discovered or devised. There is no shortage of such schemes, such as:

the elimination of major diseases;
the provision of plentiful, safe supplies of water (perhaps by desalination);

the recapture of deserts;
the restoration of forests.

But, as the whole theme of this book has stressed, it is important to recognize that such schemes cannot be successfully implemented by regarding them purely as agricultural/medical/scientific enterprises. If the schemes do not embrace a wider socio-economic dimension they will probably fail.

This implies that the schemes should not be seen as ends in themselves but means to much larger ends. After all, none of them are ends in themselves: they are intended to benefit people and it is essential, at the outset, to have a clear idea of who is going to benefit and in what ways.

It is important to avoid corruption of good, but perhaps unstated, intentions, resulting in exploitation by the greedy rather than the needy.

It is important to avoid grandiose plans that are unrealistic about their aims and their chances of success. It is important to involve, from the beginning, the intended beneficiaries and to start modestly, with pilot schemes properly planned and evaluated.

Much of this is what is intended by those who advocate 'sustainable development'. As discussed in Chapter 11, I believe the word 'sustainable' is actually inappropriate and unhelpful – not because it may not be a necessary criterion but because it is a wholly insufficient one.

Nevertheless, a major reason why the concept is so widely embraced is that people are searching for a foundation (and a name) for the kind of development which can be sustained in resource-use terms, does not pollute the environment, makes economic sense, does not lead to greedy exploitation and benefits the poor, the needy and the disadvantaged – in short, the majority.

To be realistic, however, all this has to be achieved in the real world, as it is now, and in a political context. Conway and Barbier (1990) identified a series of five criteria that would have to be met:

1. 'political will' – Governments perceiving that it is economically essential;
2. 'economic analysis' – proper evaluation of the options;
3. 'appropriate incentives' – the need for economic incentives for and the involvement of small farmers and pastoralists;
4. 'institutional flexibility' – the need for an appropriate institutional framework;
5. 'complementary infrastructure' – the need for the development of a complementary infrastructure to reinforce the agricultural/medical/scientific development policies and programmes.

These criteria would be hard to satisfy but realism is an essential concomitant of vision.

Vision (as Proverbs 29, 18 puts it: 'without which the people perish') is too often dismissed by those whose minds can neither generate nor encompass it, as 'pie in the sky', utopian or idealistic. Lack of realism about the difficulties and time-scale for achieving it make such dismissal easier.

But it is the clarity of long-term goals that inspires people to extraordinary effort and sacrifice. Of course, they cannot be achieved overnight and the cost (in all senses) is usually vast: any more than one can leap to the top of a mountain – these things are done in small steps and the first cannot be taken without a sense of direction and impulsion that can only come from a 'vision' of the peak. That is the essence of 'vision': the objective may not actually be visible and may even not be imaginable to the majority.

The point of trying to understand the issues discussed in this book would be largely lost if it did not help to create some worthwhile vision of the future. As President Kennedy once put it: 'Others look at the world as it is and ask "Why?". I look at the world as it ought to be and ask "Why not?" ' Ashdown (1989) attributed an earlier version to Aeschylus: 'Some people see things that are and ask themselves "Why?". I dream things that never have been and ask myself "Why not"?'

Questions

1. Do you accept that you cannot predict the future?
2. Do you expect the sun to rise tomorrow?, and on 1 June 2040?
3. Do you know where you will be next Wednesday?
4. Which are more predictable, events that can be controlled/influenced by humans or those that cannot?
5. Why can't the vast amounts of rubbish in developed countries be used to put organic matter into deserts?

Proverbs

Proverbs often enshrine some profound insight, particularly into human behaviour, based on vast human experience, and expressed succinctly and often humorously. Sometimes, however, some thought is required to penetrate the use of analogy, parable and indirect articulation of truth: indeed, this is part of their purpose.

In this book, the proverb (sometimes rather widely interpreted) at the head of each chapter has been chosen for its insight into one of the central themes of the chapter, but, of course, it may have little relevance to the rest of the text.

In the unlikely event that its relevance is not immediately obvious, it is hoped that the following explanatory notes will be helpful.

Chapter 1

The human mind, like the parachute, is best kept open when in use.

This quotation has all the best qualities of a good proverb (humour, contrast, surprise, elegant phrasing and succinctly expressed wisdom) and could well become one.

Chapter 2

Never stand between a hippopotamus and the water.

The simple picture of the danger involved in getting in the way of a lumbering (on land) hippo is inherently amusing, but the subtlety lies in the likelihood that the animal means no harm but can see no reason why it should not go directly to its goal (safe refuge and preferred milieu).

This suggests a value in simply recognizing realities, of the sheer power and inevitability of some of the forces we confront. Such forces abound in the world and confront anyone trying to work within it. It is not a question of wanting to stop the action, merely of recognizing the consequences of not being alert to it.

Chapter 3

Whether elephants make love or war, the grass suffers.

Rather like the previous proverb, there is humour in the faintly ridiculous pictures conjured up.

The most pressing food needs are those of relatively powerless 'little' people (the grass), but there are big players out there (Governments and their departments and agencies, international aid organizations and commercial companies) which, whatever their intentions, may inadvertently crush those whose powerlessness prevents them even from getting out of the way.

Chapter 4

Wisdom is rarer than emeralds, and yet it is found among the women who gather at the grindstones.

This is the last sentence from a longer paragraph, as follows:

Be not arrogant because of your knowledge. Take counsel with the ignorant as well as with the wise. For the limits of knowledge in any field have never been set and no one has ever reached them.

The whole message warns those who wish to help (anyone, with anything) that those who need help do not lack knowledge or experience and helping requires humility.

Chapter 5

> When a neighbour is in your fruit-garden, inattention is the truest politeness.

This proverb illustrates the relatively rare occasions when paying no attention to your crops may be wise. But, of course, it is not really advocating ignorance, only inattention – or the appearance of ignorance. It is a recognition of the fact that one still has to live with one's neighbours even though their behaviour may fall below the highest standards.

The relevance to this chapter (as to Chapter 10), is that many of the plants we depend on have come from other parts of the world, which have benefited little from their exploitation. The process is still going on and causes resentment in poor countries whose plants are being 'mined'.

Chapter 6

> You will walk for a long time behind a wild duck before you pick up an ostrich feather.

Clearly, walking behind a wild duck with any expectations is faintly amusing, but with such a ridiculous ambition is ludicrous. So, the folly of false and exaggerated expectations or the hope of illogical or unnatural outcomes is made more vivid and entertaining.

However, other interpretations are also possible, related to the fact that one ought to be able to anticipate what will be found by following known creatures (people, organizations, groups) – because certain attributes are in their nature. Thus, what someone will say (or write) may be anticipated from knowing their background/interests/political affiliation, etc.

The implications are not always obvious for animals. For example (see Chapter 12), the fact that a suffering cow or antelope will try to avoid revealing its suffering to the watching predator means that we cannot always judge welfare by observing behaviour.

Chapter 7

> The camel driver has his plans – and the camel has his!

Systems approaches tend to imply that all the relevant information can be put into a model or the outcome predicted. But we all know that, however well planned, systems are vulnerable to uncertain and unpredictable reactions – often by other human beings rather than camels – that may render the whole exercise futile.

This is so easy to visualize in the case of the camel and the person on top who is nominally in charge.

Chapter 8

> If you do not change direction, you will finish up where you are going.

At its simplest, this simply insists that the end of the road is often quite foreseeable and, if you do nothing differently, that is where you will arrive. Thus, in the use of non-renewable resources, it is inevitable that they will eventually run out (it is extremely difficult to predict from its direction how long a road may be!).

More generally, it can focus attention on whether the final goal is what is actually wanted (and by whom?). In many of these agricultural matters, it is clear that many people are not happy with the way we appear to be going.

Chapter 9

> Foolish is the man who tries to pick up two water melons with one hand.

If efficiency is sought by reducing the number of hands, the result may be total disaster – well illustrated by two water melons (who knows where they might finish up!).

Yet again, there is humour in the mildly ridiculous attempt, so obviously idiotic that one cannot really imagine it happening. Even if, with big hands and small melons, it could be done, greater efficiency would still be bought at the risk of loss of effectiveness.

Chapter 10

> The higher the baboon climbs, the more he shows his less attractive features (the last three words are commonly substituted by 'bottom').

This proverb is a reference to the fact that people in high positions may exhibit their worst features, not only passively but also actively – because they can get away with doing so.

But it also relates to the dangers of a lofty view of life, remote from the detailed realities confronting those below. This is evident in some of the pronouncements of those who pontificate about what the world should do about the environment. Having less energy, doing without the car, avoiding pollution are all familiar examples of matters that are easy to pontificate about from a position of lofty indifference or ignorance.

Chapter 11

> He who speaks the truth should have one foot in the stirrup.

Most people immediately identify with the dangers of being truthful and the advantages of a rapid departure afterwards.

There is a more general point that perhaps only those who are able to avoid the adverse consequences of speaking the truth may actually do so and, very often, this is a very limited number of people (old, retired, with no need to protect their backs and no future to protect).

However, there is also a more positive interpretation (less amusing but nobler), that 'one foot in the stirrup' can equally indicate willingness and, indeed, preparedness to fight in defence of the truth

and one's right to speak it. [In Latin: *Verax quia conscendere equum paratus.*]

The relevance in this chapter relates, of course, to the challenge it represents to so many fashionable claims to worth, in which it is assumed that inserting the word 'sustainable' confers ultimate respectability.

Chapter 12

> It is false economy to burn down your house in order to inconvenience your mother-in-law.

This is usually regarded as the funniest of the proverbs quoted in this book, partly due, no doubt, to the inestimable humorous value added by the magic words 'mother-in-law'. It is used in this chapter to emphasize the fact that overreaction, as in so many confrontations between welfarists and livestock producers, is counter-productive. Burning down your house is clearly an overreaction to almost anything your mother-in-law may have done; it is also obviously extremely damaging to the aggrieved person (of whatever sex).

In the context of animal welfare, confrontation benefits no one (although it may make some feel better), including the animals, and does not move the debate forward at all. Unfortunately, it may prove necessary in order to persuade some to take the matter seriously but, even then, the nature and scale of the confrontation needs to be carefully thought out.

Avoiding confrontation usually depends upon having some respect for the opposing view. As Reinhold Niebuhr pointed out, there is virtue in 'residual awareness of the possibility of error in the truth in which we believe, and of the possibility of truth in the error against which we contend'.

Chapter 13

> Into the closed mouth, the fly does not enter.

This is usually interpreted as pointing out the advantages of keeping quiet and is paralleled by 'Whilst the sheep bleats it loses its mouthful' (Flemish proverb). Thus, there are negative benefits of keeping silent and positive benefits of not wasting time complaining or even talking at all. Behind both are notions that what goes in at the mouth matters, an obvious truth for the connections between diet and health.

The relevance to this chapter is somewhat tenuous but perhaps real concern about what is or is not consumed could be regarded as more important than fashionable talk about what is good for you.

Chapter 14

> He who takes the middle of the road is likely to get crushed by two rickshaws.

The picture of being crushed by rickshaws coming in opposite directions is sharper than the idea of simply being at risk from both sides. Those who try to take moderate, middle-of-the-road positions frequently do get attacked from both sides and compromise is sometimes seen as weaker than firmly taking sides.

It clearly depends on the issue: moral questions may demand that sides be taken; more complex questions, especially if knowledge is less than complete, may benefit from a balanced view. The citizen is confronted by both situations and needs to be aware of the risks.

Chapter 15

> Even the cleverest monkey may drop a coconut.

It is not enough to be clever and education should aim at more than this. But, in fact, such accidents can happen to anyone and we must expect that everything may not always go as planned.

The combination of 'clever' and 'monkey' is a common reinforcing device, and losing a coconut is presumably of considerable consequence to a monkey. It is also likely to cause it embarrassment and it is hard to think of anything else that would!

The importance of asking questions made the following Chinese proverb a strong contender for this chapter:

He who asks a question may look foolish for a short time; he who does not, may look foolish for the rest of his life.

Chapter 16

> A peasant will stand for a long time on a hillside with his mouth open, before a roast duck flies in.

In other words, things, however desirable, do not happen by themselves, and that standing about in hope and expectation is simply futile.

If we are to shape our future and retain control, as far as this is possible, we need to ensure that relevant research is carried out. Without soundly-based research we shall not sufficiently understand the world we are trying to improve.

Chapter 17

> We cannot predict the future, we can do better than that, we can invent it.

This quotation from a Nobel prize winner, whose name now escapes me, is not a proverb (yet) but uses a ringing phrase to emphasize that, in spite of uncertainties and imponderables, we should aim to shape our own destiny. Of course, without an agreed vision of what we are aiming at, we do not know what we are trying to achieve. Invention may use completely new ideas and techniques but the purpose has to be clear.

Proverbial postscript

A common feature of the best proverbs is humour: this is not surprising if you agree with the following quote:

> Common sense and a sense of humour are the same thing moving at different speeds. A sense of humour is just common sense, dancing.

Also, much can be forgiven if the tone is humorous. As Montaigne said: 'No-one is exempt from talking nonsense; the misfortune is to do it solemnly.'

It is also the case that humour can make the truth more palatable and more likely to be listened to: it also seems to be memorable – because people actually wish to remember it.

Occasionally a proverb appears so opaque that it is impossible to tell whether it is humour or not. This last one (a Bajan proverb) defeats me still:

Calabash button suit crocus bag trousers!

Glossary

Additives Substances added to feedingstuffs (for preservation or flavour).

Aerobic digestion Decomposition of organic matter by micro-organisms in the presence of oxygen.

Agricultural system A system with an agricultural output and containing all the major components.

Agro-forestry Cultivated mixtures of trees, crops and/or animals.

Ammonification The formation of ammonium ions from nitrogenous compounds.

Anaerobic digestion The breaking down of organic matter in oxygen-free conditions.

Anaerobic glycolysis Biological liberation of energy in tissues without requiring oxygen.

Annuals Plants that flower and complete their life cycle in the same year in which they are raised from seed.

Anthelmintics Drugs used to remove parasitic worms (helminths) from their hosts.

Anthesis The time of flower opening.

Apical meristem Growing point at tip of root and stem in vascular plants (see also meristem).

Aquaculture The cultivation of aquatic organisms for the production of human food (mainly, e.g. fish farming).

Ark A movable, often triangular, shelter for pigs or poultry.

Artificial insemination (AI) The collection of sperm and its use in impregnating females.

Available visible radiation Wavelengths between 0.4 and 0.7 μm also referred to as 'usable' radiation: equal to *c*. 45% of direct solar radiation or *c*. 50% if the different quality of diffuse component is included.

Bagasse Fibrous residue remaining after crushing sugar cane and removing the juice.

Battery cages Cages, usually metal, in which poultry are housed.

Biennials Plants that only flower in the year following that in which they are sown.

Biochemical Oxygen Demand (BOD) The amount of oxygen required by micro-organisms, usually in polluted water, slurry or industrial effluent, for oxidation processes. Generally measured as milligrams of O_2 taken up by 1 l of the sample when incubated at a standard temperature (20°C) in five days.

Biofuel Fuels derived from biological materials, including crops (especially trees) and animal wastes.

Biological control (Biocontrol) The control of one organism by deliberate use of another.

Biological efficiency The efficiency of organisms or biological systems.

Biological system A system consisting essentially of biological processes.

Biomass The total weight of living material, of all forms.

Biotechnology This has been defined in many ways. BBSRC defines it as 'the application of biological systems and biological processes to the manufacture of new products'.

Boundary The conceptual limits of a system, penetrated by outputs and inputs but not by feedback loops.

Bovine spongiform encephalopathy (BSE) A disease of cattle (so-called 'mad cow disease') caused by an agent which is neither a bacterium nor a virus. Thought to be a form of 'scrapie': both diseases have a very long incubation period.

Brassicas Plants of the group of Brassica crops (e.g. turnip).

Break crops Crops grown between periods of continuous cultivation of a main crop.

Breeding season Period when animals are sexually active.

Broiler Young chicken grown for meat (usually slaughtered at 42–45 days).

Broken-mouthed ewes Ewes having some teeth missing.

Browsing A method of feeding by herbivores, in which the leaves and peripheral shoots are removed from trees and shrubs.

Calving interval The interval between one calving and the next.

Carnivores Animals that feed on other animals or on material of animal origin.

Carrying capacity The number of animals that an area of land can support (feed).

Cash flow Movement of funds through the business.

Catchcrop Crop utilizing land between two longer-term crops.

Chemotherapy The treatment of disease by substances that have a specific antagonistic effect on the organism causing the disease.

Chloroplasts Discrete photosynthetic organelles within plant cells, containing chlorophyll.

Cloning Producing a stock of individuals all derived asexually from one sexually produced.

Closed system A system which does not exchange matter with the surroundings; it may, however, exchange energy with the surroundings.

Coccidiosis An intestinal disease of livestock caused by microscopic protozoa.

Component An identifiable unit within a system: it may be capable of independent physical existence or be an entirely conceptual entity.

Conservation Protection and preservation, in relation to: (1) soil; (2) herbage; (3) species; or (4) the environment.

Controlled atmosphere (CA) Regulation of levels of oxygen and carbon dioxide as well as temperature to improve storage of fruit and vegetables.

Coprophagy Consumption of own faeces (normal in the rabbit) (see also Refection).

Cost/Benefit analysis An assessment of the costs

and benefits associated with a process, action, enterprise or system.

Cows Adult female cattle which have had one or more calves.

Critical temperature (lower) The environmental temperature below which the metabolic rate must rise if the animals' deep-body temperature is to be maintained.

Crude fibre (CF) A constituent of animal feedstuffs comprising mainly of cellulose, lignin and related compounds.

Crude protein (CP) An approximate assessment of the protein content of animal feed; usually calculated as 6.25 x N(%).

Culling Removal of animals (culls) from a breeding population, generally on account of some physical or performance deficiency.

'Dedicated' fuel cropping Use of resources solely towards the production of a crop destined for use as a fuel or fuel feedstock.

Deep litter A system of keeping housed poultry on litter (*c*. 15 cm deep).

Depot fat Fat reserves localized within the animal body and used in times of poor nutrition.

Desalination The removal of salts from water (generally sea water).

Desiccant A chemical causing drying.

Deterministic A deterministic situation is one in which given inputs lead to predictable outputs.

Detrivores Animals that feed on dead plant or animal material.

Developing countries Both low-income and middle-income economies. Income is measured in terms of GNP per capita.

Development (biological) Sequential organizational changes in an organism of a qualitative kind, often associated with growth.

Digestibility Proportion of feed digested by animals, expressed as a ratio, a percentage or a coefficient.

Digestible energy (DE) That part of the feed energy that is available to the animal after digestion.

Dry (cows, sheep) Not producing milk.

Ear emergence Main heading date (e.g. for a sward, the date at which 50% of the inflorescences have emerged).

Economies of scale Unit cost reductions which result from increasing total output. For many industries increases in total output can lead to a reduction in the average cost of producing each unit, up to a point: such economies evolve from the more efficient utilization of resources.

Ecosystems Systems which include both living and non-living substances interacting to produce an exchange of materials between the living and non-living units.

Ecto-parasite A parasite which lives on the outside of its host.

Efficiency A ratio of output (or performance or success) to the input(s) (or costs) involved, over a specified time and in a specified context.

Endo-parasite A parasite which lives within the body of the host.

Energetic efficiency An efficiency ratio where both numerator and denominator are expressed in terms of energy.

Energy flow The rate of energy transfer between elements of an ecological system.

Entropy The degree of randomness or disorder of a system.

Eutrophication The accumulation of excessive concentrations of plant nutrients (especially nitrogen phosphate) in water courses.

Evapo–transpiration Loss of water by evaporation and transpiration from the above-ground parts of the plant, dominated by meteorological conditions, especially when water supply is not limiting.

Extensive systems Systems which use a large amount of land per unit of stock or output (see also Intensive systems).

Fallow(ing) Resting land from deliberate cropping: not necessarily without cultivation or grazing, but without sowing.

Farmer's lung Disease of the lung caused by micro-organisms in mouldy hay.

Farrowing The act of parturition in the sow.

Feather-pecking (or-picking) An unfortunate habit of hens, usually when close-confined, of pecking at other birds' feathers, often resulting in considerable damage.

Feedback (loop) The use of information produced at one stage in a series of operations as input at another, usually previous, stage.

Feedlot An area of land, used to accommodate animals (very commonly beef cattle) at a very high density, not contributing at all to the production of animal feed (all of which has therefore to be brought to the animals from outside the feedlot).

Field capacity The state of saturated soil when all the soil moisture that is able to freely drain away has done so.

First Law of Thermodynamics In all processes, the total energy of the universe remains constant.

Flow diagram The diagrammatic representation, usually with conventional symbols, of the structure of a system in terms of physical and information flows between compartments.

Fodder Generally refers to dried feeds such as hay.

Food, feed A useful but arbitrary distinction is often made between 'food', consumed by humans, and 'feed', consumed by or fed to animals.

Food chain Includes all those businesses involved in the transformation of raw materials into food. It can be short, for example, where eggs are bought direct from the farmer, or highly complex for processed products which move from the farm, for example the transport/slaughterhouse/transport/processor/cannery/transport/retailer/consumer chain which beef destined for a tin of casserole may follow.

Forage/Forage crops Leafy crops that are grazed.

Forbs Broad-leaved herbaceous plants.

Fossil fuels Biological materials which have been subjected to long-term geological effects, e.g. oil, coal, natural gas and peat.

Free-range A system of poultry keeping in which hens are allowed to range over a relatively large area.

Genetic engineering The science of modifying the genetic constitution of plants and animals directly.

Genome The full set of chromosomes of an individual.

Geothermal energy Energy contained in the earth's heat (as steam, water or hot rock).

Graminoids Grasses and grass-like plants.

Grazing A method of feeding by herbivores, characterized by repeated removal of only a part (generally the leaf) of the plant (most commonly herbage, such as grasses and clovers).

'Greenhouse effect' Global warming due to build-up of atmospheric carbon dioxide.

Gross domestic product (GDP) The total final output of goods and services produced within a country in a year by residents and non-residents, regardless of allocation of domestic and foreign claims.

Gross national product (GNP) The total value of goods and services produced in an economy over a particular period of time, usually one year. GNP is made up of consumer and government purchases, private domestic and foreign investments in the country and the total value of exports.

Growth Increase in size.

Growth-promoters Substances given to farm animals to promote growth.

Growth regulator A natural or chemical substance that regulates the enlargement, division or activation of plant cells.

Herbivores Animals that feed on plant material.

Hermaphrodite Bisexual. In a flowering plant, having both stamens and carpels in the same flower: in an animal, producing both male and females gametes.

Hibernation Dormancy during winter: metabolism is greatly slowed, and in mammals temperature drops close to that of surroundings.

Hierarchy A structural relationship in which each unit consists of two or more sub-units, the latter being similarly subdivided.

Homeotherms Warm-blooded animals whose body temperature is maintained above that of usual surroundings.

Hormone A secretion from special glands within the animal's body that effects various body functions.

Hybrid The first generation offspring of a cross between two individuals differing in one or more genes.

Hydrological cycle The cycle of water evaporated from water surfaces and plants and precipitated as rain.

Information flow Where a component influences another component without the physical transfer of material.

Infrastructure The availability of roads, power supplies, education and health facilities, for example, which all industries share and for which they do not pay directly.

Integrated control Integrated use of both biological and, for example, chemical methods of controlling pests or weeds.

Intensive systems Systems in which cropping is frequent and yields are high per hectare; or where stock numbers are high per unit area (see also Extensive systems).

Intercropping or Mixed cropping The growing of more than one species on the same piece of land at the same time.

Invertebrates Animals without backbones.

Leaching Removal of nutrient materials in solution

from the soil (usually in gravitational water).

Leaf area index (L or LAI) The area of green leaf per unit area of ground.

Legumes Plants of the family Leguminosae (e.g. peas).

Lux(LX) Unit of illumination

$$1 \text{ lx} = 0.0001 \text{ phot}$$
$$= 0.1 \text{ milliphot}$$
$$= 0.09 \text{ foot-candle.}$$

Maintenance requirement The food required by an animal to maintain its body weight or to prevent catabolism of its own tissues.

Mathematical model A model using algebraic expressions to represent relationships within the system.

Meristem Localized region of active cell division in plants (see also Apical meristem).

Metabolic age Age related to the proportion of mature weight achieved: calculated as age multiplied by mature weight raised to the 0.27th power.

Metabolic energy (ME) The energy of feed less the energy of faeces, urine and methane.

Metabolic size or weight The size of an animal to which its metabolic rate is proportional. Mean standard metabolic size of mammals is often expressed as (body weight)$^{0.75}$.

Methaemoglobinaemia A medical condition (also called 'blue baby' syndrome) where nitrite combines with the haemoglobin of blood to form methaemoglobin, which cannot carry oxygen.

Microchip (micro-electronics) A large number of electronic circuits in miniature form.

Micro-nutrients or Trace elements Nutrients which are required in very small amounts.

Mixed grazing More than one type of animal grazing the same area at the same time.

Model A simplified representation of a system (which may be expressed in word, diagrammatic or mathematical terms).

Monocropping or Monoculture The growing of the same, single crop species continuously on an area of land.

Monogastric, Non-Ruminant or Simple-stomached animals Animals having only one stomach (e.g. pig, poultry, humans).

Neonatal losses Deaths at or shortly after birth.

Net assimilation (rate) Mean rate of dry matter production per unit of leaf area.

Nitrogen fixation Conversion of atmospheric nitrogen to plant compounds by micro-organisms (in soil, root nodules).

Nomadism Continual movement, of humans and animals, with no fixed settlement, generally in search of food or water.

Nutrient-film-technique (NFT) A system of growing crops in which a very shallow stream of water containing all the dissolved nutrients required for growth is recirculated past the exposed roots of crop plants in a water-tight gulley.

Omnivores Animals that feed on material of both plant and animal origin.

Open system A system which exchanges matter with the surroundings; it may also exchange energy with the surroundings.

Organic farming Commonly thought of as without

the use of manufactured chemicals, but now conforming to very detailed production standards.

Ozone layer A layer of ozone found in the stratosphere, where it absorbs harmful solar ultraviolet radiation.

Parasitoid An organism in which a free-living female lays her eggs in or on a host, which subsequently dies as a result.

Perennials Plants that continue their growth from year to year.

Permanent pasture An established plant community in which the dominant species are perennial grasses, there are few or no shrubs and trees are absent.

pH Chemical measure of acidity (< pH 7) and alkalinity (> pH 7).

Pheromone A chemical substance produced by one individual which affects the behaviour or physiology of another.

Photosynthesis The process by which carbohydrates are manufactured by the chloroplasts of plants from CO_2 and water by means of the energy of sunlight. Net photosynthesis is equal to gross photosynthesis less the sum of respiration during the day and night.

Plantation crops Subtropical and tropical perennial crops grown on plantations or large estates.

Plastids Plant cells containing chlorophyll, involved in photosynthesis.

Poaching Damage to herbage and soil caused by excessive treading by animals in wet weather.

Poikilotherms Cold-blooded animals whose body temperature varies, to a large extent, depending on the environment.

Primary production Production by plants: 'primary' in being the first use of solar radiation, the main energy source for biological processes.

Productivity A measure of efficiency, relating output of product to the use of a resource (including time).

Pulses or Grain legumes Leguminous plants or their seeds, chiefly those plants with large seeds used for food.

Pyrolysis The destructive distillation of organic material in the absence of oxygen to yield a variety of energy-rich products.

Refection The consumption of own faeces (see also Coprophagy).

Relative humidity (RH) Water vapour in air compared with the amount of water vapour held at the same temperature when saturated.

Respiration The oxidative breakdown and release of energy from fuel molecules by reaction with oxygen in aerobic cells.

Rhizome An elongated underground stem, usually horizontal, capable of producing new shoots and roots at the node.

Riboflavin Vitamin B_2.

Robotic milking Automated system that applies the teats of the milking machine to the cow and removes them without manual operation. (A 'robot' is a mechanical device capable of performing tasks with a high degree of 'mental' and physical agility.)

Rotation The growing of a repeated sequence of different crops.

Rotational grazing The practice of imposing a regular sequence of grazing and rest upon a series of grazing areas.

Rumen A large additional reservoir in which food is subjected to microbial digestion before passing to the true stomach.

Ruminant Animals possessing a complicated stomach of four parts, rumen, reticulum, omasum, abomasum, e.g. cows, sheep, deer, etc.

Scrapie Long-established disease of sheep that causes intense irritation: cause unknown but thought to be neither a bacterium nor a virus.

Secondary production Production by herbivorous animals.

Second Law of Thermodynamics The entropy or degree of randomness of the universe tends to increase.

Senescence or Leaf senescence Process in which leaves age and die, usually involving chlorophyll degradation.

Sericulture The keeping of silk moths and their larvae for the production of silk.

Set-stocking Grazing system in which stock remain in one field or paddock for a prolonged period.

Simple-stomached animals Those without major adaptation to components of their diet (especially fibre), such as pigs and humans.

Simulation The use of a model, normally mathematical, to mimic or imitate the behaviour of a system which changes through time.

Slurry Agriculturally, usually refers to a mixture of faeces, urine and water, sometimes containing bedding materials (such as straw or sawdust).

Snood A fleshy protuberance on the top of a turkey's head; it is readily damaged when animals fight and may become a focus of infection.

Soil water deficit (SWD)/Soil moisture deficit (SMD) The amount of rain or irrigation water required to restore field capacity.

Somatotrophin Growth hormone secreted by anterior pituitary gland.

Steers Castrated male cattle.

Stochastic A stochastic situation is one in which a given input leads to a number of possible outputs, each with a probability of occurrence.

Stocking density The number of animals per unit area of land at a point in time.

Stocking rate (SR) Number of animals per unit area of land over a given period.

Stolon (runner) A creeping stem above the soil surface (roots usually form at the nodes).

Stomata Controllable openings (pores) in the leaf (usually on the underside) through which gaseous exchanges take place.

Store cattle (or sheep) Animals which have been grown slowly so that their skeletal development has not been impaired, but muscular tissue is slightly below the animal's potential and fatty tissue is undeveloped. Such animals are purchased to be fattened at pasture or in yards. The fattening process involves some growth of bone, more of muscle and rapid deposition of fat.

Stratification (of sheep breeds) Spatial distribution of the different breeding components involved in sheep production. Objects include the crossing of breeds developed for different attributes and the exploitation of different environments.

Stubble That part of a crop left above ground after harvesting.

Subcutaneous (fat) Fat layer beneath the skin.

Subsystem Often used to describe any part of a system: used in this book in a more restricted sense, to describe a part of a system contributing to the same output as the system itself.

Suckler herd Beef cattle where the dam suckles its own calf (single suckling), another calf as well (double suckling) or several (multiple suckling).

Support energy All forms of energy used other than direct solar radiation (including the 'fossil' fuels).

System A number of components linked together for some common purpose or function.

Systems analysis The study of how component parts of a system interact and contribute to that system.

Terms of trade Measure of the relative movement of export prices against that of import prices. They indicate whether a country is heavily reliant upon the import of goods or is able to generate income via exports.

Tiller An aerial shoot of a grass plant, arising from a leaf axil, normally at the base of an older tiller.

Total digestible nutrients (TDN) A summation of the digestible crude protein, crude fibre, oil and nitrogen-free extract in a feedstuff.

Transhumance Situation in which farmers with a permanent place of residence send their herds, tended by herders, for long periods of time to distant grazing areas.

Translocation Movement of substances within a plant, e.g. between root, leaf and stem.

Transpiration (also called evapo–transpiration) The evaporation of moisture through the leaves.

Ultra-high temperature (UHT) Method of pasteurizing milk using high temperature (270°F) for not less than one second.

Validation The process of assessing the accuracy for a given purpose of a simulation model by comparing the model's predictions with independent results.

Variable A quantity able to assume different numerical values.

Vertebrates Animals with backbones.

Volumes per million (vpm) i.e. parts per million (ppm) by volume.

Voluntary intake The amount of feed consumed by an animal unrestricted by the quantity available. Maximum voluntary intake is used to describe the maximum quantity (usually per day) that a given animal can consume when neither quantity, quality nor time are limiting.

Water activity (a_w) A measure of the 'availability' of the water present to micro-organisms; used in food technology.

Weaning Removal of young mammals from their source of milk.

Weaning (early) Removal of young mammals from the milk source before the normal time.

Yield Quantity harvested: necessarily related to a specified crop or animal, or to an area, and to a period of time. It can also be used in the same way for non-biological processes, but the word 'harvested' might not be used ('obtained' would be more usual).

Zero-grazing Where grass and other forage is cut and carried to the animal.

Zoonoses Diseases transferable between humans and animals.

Bibliography

Chapter 1

Blundell, T. (1995) Turning green into gold. Opinion. *The Times Higher*, **July 28**, 11.

Spedding, C. R. W. (1979) *An Introduction to Agricultural Systems*, 1st edn, Elsevier Applied Science, London.

Spedding, C. R. W. (1988) *An Introduction to Agricultural Systems*, 2nd edn, Elsevier Applied Science, London.

Spedding, C. R. W. (ed.) (1989) *The Human Food Chain*, Elsevier Science Publishers, London.

Spedding, C. R. W. (1994a) Farming systems research/extension in the European context, in *Rural and Farming Systems Analysis*, (eds J. B. Dent, and M. J. McGregor), CAB International, 46–52.

Spedding C. R. W. (1994b) *Biology and the Citizen*, Presidential Address given at 45th Annual General Meeting of the Institute of Biology. *Biologist*, **41**(3), 119–21.

Chapter 2

Body, R. (1982) *Agriculture: the Triumph and the Shame*, Temple Smith, London.

Bryne, P. J. and Ravenscroft, N. (1990) The role of diversification in restructuring farms and rural estates. *Journal of the Royal Agricultural Society of England*, **151**, 51–65.

Coupland, R. T. (1974) Grasslands, in *Encyclopaedia Britannica*, **8**, pp. 280–6.

Coupland, R. T. (1979) Conclusion, in *Grassland Ecosystems of the World*, (ed. R. T. Coupland), International Biological Programme, **18**, Cambridge University Press, Cambridge.

Craig, G. M. (1991) *The Agriculture of the Sudan*, Oxford University Press, Oxford.

Craig, G. M. (1993) *The Agriculture of Egypt*, Oxford University Press, Oxford.

Davenport, M. (1994) Changes in the pattern of world trade in agricultural products, in *Priorities for a New Century – Agriculture, Food and the Rural Policies in the European Union*, (eds B. J. Marshall and F. A. Miller), CAS Paper 31, April 1995, Centre for Agricultural Strategy, University of Reading, pp. 68–81.

FFB (1992) *Fruit Sector Group of Food from Britain Report*, Food from Britain.

Glantz, M. H. (ed.) (1985) *Drought and Hunger in Africa*, Cambridge University Press, Cambridge.

Harvey, D. R. (1991) Economic Factors influencing the Agricultural Environment, Pt 1. *The Agricultural Engineer*, **46**(3), 71–6.

Kowal, J. M. and Kassam, A. H. (1978) Rangeland livestock and forest resources, in *Agricultural Ecology of Savanna*, Clarendon Press, Oxford.

Marks, H. F. (1992) *Food – its product, marketing and consumption*. Centre of Management in Agriculture and Farm Management Unit, University of Reading.

Marsh, J. S. (1992) The food and agricultural industry, in *Fream's Principles of Food and Agriculture*, (ed. C. R. W. Spedding), Blackwell Scientific Publications, London, pp. 1–14.

Nursten, H. E. (1992) The principles of processing of products, in *Fream's Principles of Food and Agriculture*, (ed. C. R. W. Spedding), Blackwell Scientific Publications, London, pp. 228–45.

Pendleton, D. F., Van Dyne, G. M. and Whitehouse, M. T. (1979) *Prediction of Grazingland Productivity under Climatic Variations*, Interim progress report to the National Science Foundation, Washington, DC.

Pickering, D. C. (1992) World agriculture, in *Fream's Principles of Food and Agriculture*, (ed. C. R. W. Spedding), Blackwell Scientific Publications, London, pp. 45–93.

Pratt, D. J. and Gwynne, M. D. (eds) (1977) *Rangeland Management and Ecology in East Africa*, Hodder and Stoughton, London.

Rosen, A. (1991) *The Reform of the Common Agricultural Policy*. Farming News, London.

Spedding, C. R. W. (ed.) (1981) *Vegetable Productivity*, Macmillan, London.

Spedding, C. R. W. (1988) *An Introduction to Agricultural Systems*, 2nd edn, Elsevier Applied Science Publishers, London.

Van Dyne, G. M., Smith, F. M., Czaplewski, R. L. and Woodmansee, R. G. (1978) Analyses and syntheses of grassland ecosystem dynamics, in *Glimpses of Ecology*, (eds J. S. Singh and B. Gopal), International Science Publishers, Jaipur, India.

Xu, G. and Peel, L. J. (1991) *The Agriculture of China*, Oxford University Press, Oxford.

Chapter 3

Anderson, J. (1994) Food and agriculture: a global perspective, in *Priorities for a new century – agriculture, food and rural policies in the European Union*. CAS Paper 31, April 1995. Centre for Agricultural Strategy, University of Reading, 21–33.

Biggs, P. M. (1985) Infectious animal disease and its control, in *Technology in the 1990s: Agriculture and Food*, (eds K. Blaxter and L. Fowden), Proceedings of a Royal Society Discussion Meeting held 17–18 October, 1984, 113–28.

Brown, L. R. *et al.* (1990) *State of the World*, Unwin Paperbacks, London.

Bunting, H. (1992) Feeding the world in the future, in *Fream's Principles of Food and Agriculture*, (ed. C. R. W. Spedding), 256–90.

Buringh, P. (1985) The land resource for agriculture, in *Technology in the 1990s: Agriculture and Food*, (eds K. Blaxter and L. Fowden), Proceedings of a Royal Society Discussion Meeting held 17–18 October, 1984, 5–14.

CAS (1979) *National Food Policy in the UK*, CAS Report 5, Centre for Agricultural Strategy, University of Reading.

Chambers, R. (1983) *Rural Development: Putting the Last First*, Longman, Harlow.

Conway, G. R. and Barbier, E. B. (1990) *After the Green Revolution*, Earthscan Publications, London.

Conway, G. R. and Pretty, J. N. (1991) *Unwelcome Harvest*, Earthscan Publications, London.

Cramer, H. H. (1967) *Plant Protection and World Crop Production*, Farbenfabriken Bayer AG-Leverkusen.

Duckham, A. N., Jones, J. G. W. and Roberts, E. H. (eds) (1976) *Food Production and Consumption*, North Holland.

FAO (1980a) *World Meat Situation and Outlook*. CCP: ME80/MISC. Food and Agricultural Organisation, Rome.

FAO (1980b) *Production Year Book*. Vol 33. Food and Agricultural Organisation, Rome.

FAO/AEZ (Higgins, G. M., Kassam, A. H., Naiken, L., Fischer G. and Shah, M. M.) (1982) *Potential Population Supporting Capacities of Land in the Developing World*. Technical Report of Project FPA/INT/513, Land Resources for the Populations of the Future. Food and Agriculture Organisations of the United Nations, United Nations Fund for Population Activities, and International Institute for Applied Systems Analysis. Food and Agriculture Organization of the United Nations, Rome.

FAO (1991) *FAO Food Outlook*, FAO, Rome, Oct. 1991.

Harrison, P. (1987) *The Greening of Africa*, Paladin Grafton Books, London.

Mellor, J. W., Delgado, C. W. and Blackie, M. J. (eds) (1987) *Accelerating Food Production in Sub-Saharan Africa*, Johns Hopkins University Press.

Oerke, E-C., Dehne, H-W., Schönbeck, F. and Weber, A. (1994) *Crop Production and Crop Protection*, Elsevier, London.

Pinstrup-Andersen, P. and Hazell, P. B. R. (1987) The impact of the green revolution and prospects for the future, in *Food Policy: Integrating supply, distribution and consumption*, (eds J. Price Gittinger, J. Leslie and C. Hoisington), Johns Hopkins University Press for the World Bank, 107.

Reddy, K. V. S. and Walker, P. T. (1990) A review of the yield losses in graminaceous crops caused by *Chilo* spp. *Insect Sci. Applic*, **11**(4/5), 563–9.

Sloyan, M. (1992) *World food resources: a grain of sense*. MLC Background Paper, Meat and Livestock Commission/Edu-Com Consultants, Milton Keynes.

Spedding, C. R. W. and Hoxey, A. M. (1975) The potential for conventional meat animals, in: *Meat*, Proceedings 21st Easter School in Agricultural Science, University of Nottingham, 483–506.

Spedding, C. R. W., Walsingham, J. M. and Hoxey, A. M. (1981) *Biological Efficiency in Agriculture*, Academic Press.

Tribe, D. (1994) *Feeding and Greening the World*. CAB, Wallingford, UK.

Walker, P. T. (1983) The assessment of crop losses in cereals. *Insect Sci. Applic*, 4(1/2), 97–104.

Walker, P. T. (1987) Losses in yield due to pests in tropical crops and their value in policy decision-making. *Insect Sci. Applic*, 8(4/5/6), 665–71.

Webb, P. and von Braun, J. (1994) *Famine and Food Security in Ethiopia: Lessons for Africa*, John Wiley and Sons, Chichester.

Chapter 4

Brandt Report (1980) *North–South: a Programme for Survival*, Pan Books, London.

Brandt Commission (1983) *Common Crisis*, Pan Books, London.

Chambers, R. (1983) *Rural Development: Putting the Last First*, Longman, Harlow.

Clayton, E. (1983) *Agriculture, Poverty and Freedom in Developing Countries*, Macmillan International College Edition, London.

Conway, G. R. and Barbier, E. B. (1990) *After the Green Revolution*, Earthscan Publications, London.

George, S. (1985) *How the Other Half Dies*. Penguin Books, Harmondsworth.

Hadjor, K. B. (ed.) (1988) *New Perspectives in North–South Dialogue*. I. B. Tauris & Co. Ltd, London.

Harrison, P. (1987) *The Greening of Africa*, Paladin Grafton Books, London.

Jackson, T. (1982) *Against the Grain: the dilemma of project food aid*. Third World Publications, Birmingham, UK.

Lele, U. (1987) Growth of foreign assistance and its impact on agriculture, in *Accelerating Food Production in Sub-Saharan Africa*, (eds J. W. Mellor, C. L. Delgado and M. J. Blackie), Johns Hopkins University Press, Baltimore and London.

Pearse, A. (1980) *Seeds of Plenty, Seeds of Want*, Clarendon Press, Oxford.

Ruthenberg, H. and Jahnke, H. E. (ed.) (1985) *Innovation Policy for Small Farmers in the Tropics*. Oxford University Press, Oxford.

Chapter 5

Agarwal, A. (1979) Blue-green algae to fertilise Indian rice paddies. *Nature*, 279(5710), 181.

CSTI (1992) *The Greenhouse Effect: fact or fiction?* CSTI Environmental Information Paper 1, 1992, CSTI, London.

Goulding, K. and Poulton, P. (1992) Unwanted nitrate. *Chemistry in Britain*, **December**, 1100–2.

Harrison, P. (1987) *The Greening of Africa*, Paladin Grafton Books, London.

Kleiner, K. (1995) Gene engineers grow better polymer plants. *New Scientist*, **1960**, 19.

Pearce, F. (1994) Counting Africa's trees for the wood . . . *New Scientist*, **11 June**, 8.

Phillips, J. D. (1992) Medicinal plants. *Biologist*, 39(5), 187–91.

Polunin, O. and Walters, M. (1985) *A Guide to the Vegetation of Britain and Europe*, Oxford University Press, Oxford.

Seager, J. (1995) *State of the Environment Atlas*. Penguin Books, Harmondsworth.

Spedding, C. R. W. (1971) *Grassland Ecology*, Oxford University Press, Oxford.

Spedding, C. R. W. and Diekmahns, E. C. (eds) (1972) *Grasses and Legumes in British Agriculture*, CAB, Farnham Royal, Bucks.

Spedding, C. R. W., Walsingham, J. M. and Hoxey, A. M. (1982) *Biological Efficiency in Agriculture*, Academic Press, London.

Stevenson, R. (1995) The second agricultural revolution. *Chemistry in Britain*, **March**, 177–8.

Wilson, E. O. (1992) *The Diversity of Life*, Penguin Books, Harmondsworth.

Chapter 6

Debach, P. and Rosen, D. (1991) *Biological Control by Natural Enemies*, 2nd edn, Cambridge University Press, Cambridge.

HMSO (1995) Report of the Committee to Consider the Ethical Implications of Emerging Technologies in the Breeding of Farm Animals. HMSO, London.

Lewis, T. and Taylor, L. R. (1967) *Introduction to Experimental Ecology*, Academic Press, London.

Lloyd, C. (1995) Fish factories to mass produce cod. *The Sunday Times*, 5 Feb. 1995.

Masefield, G. B. Wallis, M., Harrison, S. G. and Nicholson, B. E. (1969) *The Oxford Book of Food Plants*, Oxford University Press, Oxford.

Moore, P. (1995) The big and the small. *New Scientist, Inside Science*, 80, 18 March.

Rao, K. K. P. N. (1976) *Food Consumption and Planning*, Vol. 5, International Encyclopaedia of Food and Nutrition, Pergamon Press, Oxford.

Spedding, C. R. W. (1975) *The Biology of Agricultural Systems*, Academic Press, London.

Spedding, C. R. W., Walsingham, J. M. and Hoxey, A. M. (1981) *Biological Efficiency in Agriculture*, Academic Press, London.

Wilson, E. O. (1992) *The Diversity of Life*, Penguin Books, Harmondsworth.

Chapter 7

Checkland, P. (1981) *Systems Thinking, Systems Practice*, Wiley, New York.

Checkland, P. and Scholes, J. (1990) *Soft Systems Methodology in Action*, Wiley, New York.

Grigg, D. B. (1974) *The Agricultural Systems of the World*, Cambridge University Press, Cambridge.

Norman, M. J. T. (1979) *Annual Cropping Systems in the Tropics*, University Presses of Florida, FL, USA.

Ruthenberg, H. (1976) *Farming Systems in the Tropics*, 2nd edn, Oxford University Press, Oxford.

Simmonds, N. W. (1985) *Farming Systems Research: A Review*, World Bank Technical Paper No. 43, The World Bank, Washington, DC.

Spedding, C. R. W. (1971) *Grassland Ecology*, Oxford University Press, Oxford.

Spedding, C. R. W. (1973) *Grazing Systems*. Proceedings of the III World Conference on Animal Production, Melbourne. Sydney University Press, Sydney, 145–57.

Spedding, C. R. W. (1979, 1988) *An Introduction to Agricultural Systems*, 1st and 2nd edns, Elsevier Science Publishers Ltd, London.

Wiener, N. (1961) *Cybernetics*, 2nd edn, M. I. T. Press, Cambridge, MA, USA.

Chapter 8

Arad, N. and Glueckstern, P. (1981) Desalination, A review of technology and cost estimates, in *Salinity in Irrigation and Water Resources*, (ed. D. Yaron), Marcel Dekker Inc, New York, 325–61.

Barwell, I. and Ayre, M. (1982) *Harnessing of Draught Animals*, Intermediate Technology Transport, Oxon.

Carruthers, S. P., Miller, F. A. and Vaughan, C. M. A. (eds) (1994) *Crops for Industry and Energy*, CAS Report 15, Centre for Agricultural Strategy, University of Reading.

Cousins, W. J. (1975) Preliminary estimates of energy usage in energy farming of trees, in *The Potential for Energy Farming in New Zealand*, Proceedings of Symposium, DSIR Information Series No. 117, 77–82.

Gilliland, P. (ed.) (1979) *Atlas of Earth Resources*, Mitchell Beazley Publishers, London.

Goldenberg, J., Johannson, T. B., Reddy, A. K. N. and Williams, R. H. (1988) *Energy for a Sustainable World*, John Wiley, Chichester.

Hardy, R. W., Heytler, P. G. and Rainbird, R. M. (1983) Status of new nitrogen inputs for crops, in *Better Crops for Food*, CIBA Foundation Symposium 97, Pitman, London, 28–48.

Lowe, P. (1986) *Animal Powered Systems*, Vieweg, GTZ, Eschborn, Germany.

Myers, N. (1995) The world's forests. *Science*, **268**, 823–4.

NAS. (1980) *Firewood Crops: shrub and tree species for energy production*, National Academy of Science, Washington, DC.

Slesser, M. and Lewis, C. (1979) *Biological Energy Resources*, E & FN Spon, London.

Spedding, C. R. W. (ed.) (1989) *The Food Chain: forging the links in the human food chain*, Elsevier Applied Science, London.

Spedding, C. R. W. (1992) Towards Sustainable Pasture Systems. NINTII RASE Symposium *Towards Sustainable Crop Production Systems*, RASE, Monograph Series No. 11, 33–8.

Spedding, C. R. W. and Walsingham, J. M. (1975) *Energy Use in Agricultural Systems*, SPAN, 18(1), 7–9.

Spedding, C. R. W., Walsingham, J. M. and Hoxey, A. M. (1981) *Biological Efficiency in Agriculture*, Academic Press, London.

Stanhill, G. (1985) The water resource for agriculture, in *Technology in the 1990s: agriculture and food* (eds Sir Kenneth Blaxter and Sir Leslie Fowden). Proceedings of a Royal Society Discussion Meeting held 17–18 October, 1984.

Starkey, P. and Ndiamé, F. (eds) (1988) *Animal Power in Farming Systems*, Proceedings of the Second West Africa Animal Traction Networkshop, Sierra Leone, September 1986, Vieweg, GTZ, Eschborn, Germany.

Starkey, P. (1988) *Perfected yet Rejected*, Vieweg, GTZ, Eschborn, Germany.

Starkey, P. (1989) *Harnessing and Implements for Animal Traction*, Vieweg, GTZ, Eschborn, Germany.

Starkey, P. (1995) *Animal Traction in South Africa: empowering rural communities*. Development Bank of South Africa, South Africa.

Wilson, P. N. (1992) The Inputs to Agriculture, in *Fream's Principles of Food and Agriculture* (ed. C. R. W. Spedding), 17th edn, Blackwell, Oxford.

Chapter 9

Dalton, G. E. (1975) *The Study of Agricultural Systems*, Applied Science Publishers, London.

Large, R. V. (1973) Factors affecting the efficiency of protein production by populations of animals, in *The Biological Efficiency of Protein Production* (ed. J. G. W. Jones), Cambridge University Press, Cambridge, pp. 183–200.

Norman, L. and Coote, R. B. (1971) *The Farm Business*, Longman, London.

Spedding, C. R. W. (1970) (ed.) *An Introduction to Agricultural Systems*, 2nd edn, Elsevier Applied Science, London.

Spedding, C. R. W. (1973) The meaning of biological efficiency, in *The Biological Efficiency of Protein Production*, (ed. J. G. W. Jones), Cambridge University Press, Cambridge.

Spedding, C. R. W. (1975) *The Biology of Agricultural Systems*, Academic Press, London.

Spedding, C. R. W. (ed.) (1989) *The Human Food Chain*, Elsevier Science Publishers Ltd, London.

Spedding, C. R. W. and Brockington, N. R. (1976) Experimentation in agricultural systems. *Agricultural Systems*, **1**, 47–56.

Spedding, C. R. W. and Hoxey, A. M. (1975) The potential for conventional meat animals, in *Meat*, University of Nottingham Easter School, 1974.

Spedding, C. R. W. and Walsingham, J. M. (1975) The production and use of energy in agriculture. *Journal of Agricultural Economics*, **XXVII**, 19–30.

Chapter 10

Biscoe, P. V. and Dawson, R. G. (1983) Cereal production systems, in *Fream's Agriculture* (ed. C. R. W. Spedding), John Murray, London, pp. 471–93.

Carson, R. (1962) *Silent Spring*, Hamish Hamilton, London.

Cherfas, J. (1994) Article in *New Scientist*, **6 August**, 36–40.

Conway, G. R. and Barbier, E. B. (1990) *After the Green Revolution*, Earthscan Publications, London.

CSTI (1994) *Environmental Auditing*. CSTI Environment Information Paper 2, CSTI, London.

Edwards, C. A. and Lofty, J. R. (1972) *Biology of Earthworms*, Chapman & Hall, London.

Emsley, J. (1992) On being a bit less green. *New Scientist*, **17 October**, 53–4.

Fell, N. and Liss, P. (1993) Can algae cool the planet? *New Scientist*, **21 August**, 34–8.

Gotsch, N. and Rieder, P. (1995) Biodiversity, biotechnology and institutions among crops: situation and outlook. *Journal of Sustainable Agriculture*, 5(1/2), 5–40.

Hamblyn, C. J. and Dingwall, A. R. (1945) Earthworms. *New Zealand Journal of Agriculture*, 71, 55–8.

HMSO (1994) *Biodiversity: the UK Action Plan*. Summary Report. Department of the Environment, Cm 2428.

Holdgate, M. W. (1991) Defining biological diversity, in *Conserving the World's Biological Diversity; How can Britain Contribute?* Proceedings of a seminar presented by The Department of the Environment in association with The Natural History Museum, 13–14 June 1991. Department of the Environment, 8–13.

IAAWS (1992) *Welfare Guidelines for the Re-introduction of Captive Bred Mammals to the Wild*, UFAW, Potters Bar.

IOB (1992) *Conservation Objectives*. IOB Policy Studies No. 3, Institute of Biology, London.

IUDZG (1993) *The World Zoo Conservation Strategy*, Chicago Zoological Society.

Jollans, J. L. (1985) *Fertilizers in UK farming*, CAS Report No. 9, Centre for Agricultural Strategy, University of Reading.

Lawton, J. H. and May, R. M. (1995) *Extinction Rates*, Oxford University Press, Oxford.

MacKenzie, D. (1994) Battle for the world's seed banks. *New Scientist*, **2 July**, 4.

May, R. M. (1989) How many species?, in *The Fragile Environment*, (eds L. Friday and R. Laskey), Cambridge University Press, Cambridge, pp. 61–81.

McGrath, S. P., Hirsch, P. R. and Giller, K. E. (1988) Effects of heavy metals on microbial processes. Report of the Institute of Arable Crops Research, AFRC Rothamsted, 78.

Menon, S. (1994) 'Wrong paddy fields' measured for methane. *New Scientist*, **27 August**, 6.

Minnich, J. (1977) *The Earthworm Book*, Rodale Press Emmaus, PA, USA.

Myers, N. (1995) The world's forests. *Science*, **268**, 823–4.

Pearce, F. (1992) Flourishing forests mop up missing carbon. *New Scientist*, **11 July**, 10.

Pearce, D. and Moran, D. (1994) *The Economic Value of Biodiversity*, Earthscan Publications, London.

Pellew, R. (1995) Biodiversity conservation – why all the fuss? *RSA Journal*, CXLIII (5456), 53–66.

Rajan, L. (1995) Heavy metal guzzlers relish dirty waters. *New Scientist*, **27 May**, 22.

Ratcliffe, P. R. (1993) *Biodiversity: Britain's Forests*, The Forestry Authority, Edinburgh.

Royal Society (1983) *The Nitrogen Cycle of the United Kingdom*, The Royal Society, London.

Solbrig, O. T., van Emden, H. M. and van Oordt, P. G. W. J. (eds) (1992) *Biodiversity and Global Change*, Monograph No. 8, International Union of Biological Sciences, Paris.

Spedding, C. R. W. (1976) *Grassland Ecology*, 2nd edn, Oxford University Press, Oxford.

Systematics Agenda 2000 (1994) *Systematics Agenda 2000: Charting the Biosphere*. Technical report produced by Systematics Agenda 2000.

Tinker, P. B. (1985) Crop nutrients: control and efficiency of use, in *Technology in the 1990s: agriculture and food*, (eds K. Blaxter and L. Fowden), Proceedings of a Royal Society Discussion Meeting held 17–18 October, 1984.

Tribe, D. (1994) *Feeding and Greening the World*, CAB International, Wellingford, Oxon.

Watts, S. (1993) Plants suck metal from polluted land. *Independent*, **August 31**, 1993.

Whitfield, M. (1993) *UK Priorities in Marine Biodiversity*. Document prepared at a meeting held by the Marine Biological Association, 28–29 April 1993, Plymouth.

Whittemore, C. T. (1995) Animal excreta: fertilizer or pollutant? *Journal of Biological Education*, **29**(1), 46–50.

Wilson, P. N. (1973) The biological efficiency of protein production by animal production enterprises, in *The Biological Efficiency of Protein Production*, (ed. J. G. W. Jones), Cambridge University Press, Cambridge; pp. 201–10.

WMO/ICSU (1990) *Global Change*. A scientific review presented by the World Climate Research Programme.

WSPA/BF (1994) *The Zoo Inquiry*, WSPA/BF.

Young, C. P. and Gray, E. M. (1978) *Nitrate in Groundwater*. Water Research Centre, Medmenham.

Chapter 11

Brundtland, G. H. (1987) *Our Common Future*, World Commission on Environment and Development, Oxford University Press, Oxford.

Conway, G. R. and Barbier, E. B. (1990) *After the Green Revolution*, Earthscan Publications, London.

Crosson, P. (1992) *Sustainable food and fibre production*. Presented at the Ann. mtg. of the American Association for the Advancement of Science, Chicago, 9 February 1992.

Daly, H. E. (1991) *Steady State Economics*, Island Press, Washington, DC.

FAO (1989) Sustainable Agricultural Production: Implications for International Agricultural Research. Prepared by TAC for the CGIAR.

Francis, C. A., Flora, C. B. and King, L. D. (eds) (1990) *Sustainable Agriculture in Temperate Zones*, John Wiley, Chichester.

Handy, C. (1994) *The Empty Raincoat*, Hutchinson, London.

Harrington, L. (1995) Sustainability in perspective: strengths and limitations of FSR in contributing to a sustainable agriculture. *Journal of Sustainable Agriculture*, 5 (1/2), 41–59.

Harwood, R. R. (1990) The History of Sustainable Agriculture, in *Sustainable Farming Systems*, (eds Edwards *et al.*), pp.3–19.

Heinen, J. T. (1994) Emerging, diverging and converging paradigms on sustainable development. *Sustainable Development*, 1(1), 22–3.

HMSO (1994) *Sustainable Development: the UK Strategy*, CM 2426, HMSO, London.

Holdgate, Sir Martin (1993) *The Biological Basis for Sustainable Development*, Institute of Biology Charter Award Lecture, 28 Oct. 1993, London.

Holdgate, Sir Martin (1994) Environment and agricultural policies, in *Priorities for a New Century – Agriculture, Food and the Rural Policies in the European Union*, (eds B. J. Marshall and F. A. Miller). CAS Paper 31, April 1995, Centre for Agricultural Strategy, University of Reading, 47–62.

IUCN/UNEP/WWF (1991) *Caring for the Earth: the Second World Conservation Strategy*, IUCN, UNEP and WWF.

O'Riordan, T. (1985) What does sustainability really mean?, in *Sustainable Development in an Industrial Economy*, CEED Report 1985.

Park, J. (1993) *Towards a sustainable systems framework: the assessment of silvoarable agroforestry as an innovative cropping practice*, PhD Thesis, Cranfield Institute of Technology, UK.

Pearce, D. (1993) *Blueprint 3: Measuring Sustainable Development*, Earthscan Publications, London.

Pearce, D., Markandya, A. and Barbier, E. B. (1989) *Blueprint for a Green Economy*, Earthscan Publications, London.

Riley, Sir Ralph (1992) The Challenge to Science in Food and Agriculture. Ninth RASE Symposium *Towards Sustainable Crop Production Systems*, RASE Monograph Series No. 11, 13–15.

Spedding, C. R. W. (1994) *Sustainability in Animal Production Systems*, Hammond Memorial Lecture, BSAS Winter Meeting.

Thompson, Dick (1995) Soil, financial asset or environmental resource? *RSA Journal*, July, 56–67.

Wilson, E. O. (1992) *The Diversity of Life*, Penguin Books, Harmondsworth.

World Bank (1988) *Renewable Resource Management in Agriculture*, Operations Evaluation Department, World Bank, Washington, DC, iv.

Chapter 12

Anderson, D., Reiss, M. and Campbell, D. (1993) *Ethical Issues in Biomedical Sciences: animals in research and education*, Institute of Biology, London.

Ansell D. J. and Done, J. T. (1988) *Veterinary Research and Development: Cost-Benefit Studies on Products for the Control of Animal Diseases*, CAS Joint Publication No. 3, Centre for Agricultural Strategy, University of Reading.

Barber, Sir Derek (1991) *The State of Agriculture in the United Kingdom*. A Report to the Royal Agricultural Society of England, November 1991, RASE.

Bateson, P. (1992) Do animals feel pain? *New Scientist*, **25 April**, 30–3.

Brambell, F. W. R. (1965) Report of the Technical Committee to Enquire into the Welfare of Animals kept under Intensive Livestock Husbandry Systems. Cmnd 2836, London, HMSO.

Campbell, A. V. (1995) The animal rights debate. *ANZCCART (Australian and New Zealand Council for the Care of Animals in Research and Teaching) NEWS*, 8(1), 1–3.

Carruthers, S. P. (ed.) (1991) *Farm Animals: It Pays to be Humane*, CAS Paper 22, Centre for Agricultural Strategy, University of Reading.

FAO (1985) *State of Food and Agriculture*. 1984, FAO, Rome.

FAWC (1981) *Advice to the Agriculture Ministers of Great Britain on The Need to Control Certain Mutilations on Farm Animals*, MAFF.

FAWC (1982) *Report on the Welfare of Poultry at the Time of Slaughter*, Farm Animal Welfare Council, Surbiton.

FAWC (1984) *Report on the Welfare of Livestock (red meat animals) at the Time of Slaughter*, Reference Book 248, HMSO, London.

FAWC (1986) *Report on the Welfare of Livestock at Markets*, Reference Book 265, HMSO, London.

FAWC (1988) *Advice to Agriculture Ministers on Transportation of Unfit Animals*, Farm Animal Welfare Council, Surbiton.

FAWC (1990) *Advice to Ministers on the Handling and Transport of Poultry*, MAFF PB0125.

FAWC (1991a) *Report on the Welfare of Laying Hens in Colony Systems*, MAFF PB0734.

FAWC (1991b) *Report on the European Commission Proposals on the Transport of Animals*, Farm Animal Welfare Council, Surbiton.

FAWC (1993) *Report on Priorities for Animal Welfare Research and Development*, MAFF PB1310.

Guise, J. (1991) Humane animal management – the benefits of improved systems for pig production, transport and slaughter, in *Farm Animals: it pays to be humane*, (ed. S. P. Carruthers), CAS Paper 22, Centre for Agricultural Strategy, University of Reading, 50–8.

Harrison, R. (1964) *Animal Machines*, Watkins, London.

HMSO (1990) *Guidance of the Operation of the Animals (Scientific Procedures) Act 1986*, HMSO, London.

IUDZG (1993) *The World Zoo Conservation Strategy*, Chicago Zoological Society, USA.

MAFF (1994) *Poultry Litter Management*, MAFF Publications, London.

McInerney, J. (1991) Assessing the benefits of farm animal welfare, in *Farm Animals: it pays to be humane*, (ed. S. P. Carruthers), CAS Paper 22, Centre for Agricultural Strategy, University of Reading, 15–31.

Pickering, D. C. (1992) World Agriculture, in *Fream's Principles of Food and Agriculture*, (ed. C. R. W. Spedding), Blackwell, Oxford, pp. 45–93.

POST (1992) *The Use of Animals in Research, Development and Testing*. Report from the Parliamentary Office of Science and Technology.

Smith, J. A. and Boyd, K. M. (eds) (1991) *Lives in the Balance – the ethics of using animals in biomedical research*. Report of a working party of the Institute of Medical Ethics.

Spedding, C. R. W. (1982) *Agricultural Developments and Consumer Choice*, SCI Conference.

Spedding, C. R. W. (1988) *Introduction to Agricultural Systems*, 2nd edn, Elsevier Applied Science, London.

Spedding, C. R. W. (1994) Animal Welfare in Europe. *JAVMA*, **204**(3), 384–7.

Spencer, C. (1995) The meatless tradition and its significance. *RSA Journal*, **June**, 51–60.

Stevenson, P. (1994) *For their Own Good: A Study of Farm Animal Mutilations*. A Compassion in World Farming Trust Report, CIWF, Petersfield.

Webster, John (1995) *Animal Welfare: A Cool Eye towards Eden*, Blackwell, Oxford.

Wilson, E. O. (1992) *The Diversity of Life*, Penguin Books, Harmondsworth.

WSPA/The Born Free Foundation (1994) *The Zoo Inquiry*, World Society for the Protection of Animals, London.

Chapter 13

Boardman, R. (1986) *Pesticides in World Agriculture*, Macmillan Press, London.

Conway, G. R. and Pretty, J. N. (1991) *Unwelcome Harvest*, Earthscan Publications, London.

Harris, P. (1992) The principles of crop production, in *Fream's Principles of Food and Agriculture*, (ed. C. R. W. Spedding), 17th edn, Blackwell, Oxford, pp. 94–145.

HEC (1983) *Action to Prevent Coronary Heart Disease*, Workshop Papers for a meeting at the University of Kent. Health Education Council.

Jollans, J. L. (1984) Health in the agricultural population of England and Wales. *Agricultural Administration*,

Elsevier Applied Science, London, pp. 31–44.

Jollans, J. L. (1985) *Fertilisers in UK Farming*, CAS Report No. 9, Centre for Agricultural Strategy, University of Reading.

Malik, M. and West, A. (1989) The educational needs of the catering industry with special reference to social services catering, in *The Human Food Chain*, Elsevier Science Publishers, London, 125–42.

Mellor, J. W., Delgado, C. L. and Blackie, M. J. (eds) (1987) *Accelerating Food Production in Sub-Saharan Africa*, Johns Hopkins University Press, Baltimore and London

Miller, F. A. (ed.) (1990) *Food Safety in the Human Food Chain*, CAS Paper 20, Centre for Agricultural Strategy, University of Reading.

Moseley, B. E. B. (1990) Ionizing radiation, in *Foods for the '90s*, (eds G. G. Birch and G. Campbell-Platt), Elsevier Science Publishers, London, pp. 25–38.

Paul, A. A. and Southgate, D. A. T. (1978) *McCane and Widdowson's 'The Composition of Foods'*, HMSO, London.

Phillips, J. D. (1992) Medicinal Plants. *Biologist*, **39**(5), 187–91.

Proctor, F. J., Goodliffe, J. P. and Coursey, D. G. (1981) Post-harvest losses of vegetables and their control in the tropics, in *Vegetable Productivity*, (ed. C. R. W. Spedding), Macmillan, London.

Spedding, C. R. W. (1988) *Diet and Agriculture*, RSA Journal, **135**(5382), 388–97.

Spedding, C. R. W. (ed.) (1989) *The Human Food Chain*, Elsevier Science Publishers, London.

Spedding, C. R. W. (1993) Global food safety and the use of agrochemicals, in *Food Safety: The Challenge Ahead*, (eds G. G. Birch and G. Campbell-Platt), Intercept, pp. 29–37.

Waites, W. M. (1990) The magnitude of the problems, in *Food Safety in the Human Food Chain*, (ed F. A. Miller), CAS Paper 20, Centre for Agricultural Strategy, University of Reading, 27–38.

WHO (1979) *DDT and its Derivatives*. Environmental Health Criteria 9, WHO, Geneva.

Chapter 14

Clarke, M. (1989) *Biological and Health Sciences*. A Project 2061 Panel Report for the American Association for the Advancement of Science.

HMSO (1990) *Encouraging Citizenship*, HMSO, London.

Lopez, C. (1992) Scientists seek the quantum leap. Leonardo: the age of discoveries. *Independent*, Supplement, 86–8.

Miles, C. M., Winstanley, M. A., Gunning, J. G. and Durant, J. (1994) Towards a public consensus of biotechnology. *Science in Parliament*, **51**(2).

Pimentel, D. (ed.) (1993) *World Soil Erosion and Conservation*, Cambridge University Press, Cambridge.

Spedding, C. R. W. (1979) *An Introduction to Agricultural Systems*, 1st edn, Elsevier Applied Science Publishers, London.

Chapter 15

Bunting, A. H. (1979) *Science and Technology for Human Needs, Rural Developments, and the Relief of Poverty*. IADS Occasional Paper, New York.

Campbell-Platt, G. (1989) Educational transfer in food technology, in *The Human Food Chain*, (ed. C. R. W. Spedding), Elsevier Science Publishers, London, pp. 113–24.

Clarke, P. (1989) Education about infant feeding, in *The Human Food Chain*, (ed. C. R. W. Spedding), Elsevier Science Publishers, London, pp. 161–72.

Corcoran, K. and Dent, B. (1994) Education and extension: a perpetuating paradigm for success, in *Rural and Farming Systems Analysis*, (eds J. B. Dent and M. J. McGregor), CAB International, pp. 255–66.

Gartner, J. A. (1990) Extension education: top(s) down, bottom(s) up and other things, in *Systems Theory Applied to Agriculture and the Food Chain*, (eds J. G. W. Jones and P. R. Street), Elsevier Science Publishers, London.

Hills, Sir Graham (1995) The knowledge mountain. *Chemistry in Britain*, **March**, 201.

Kuhn, T. (1962) *The Structure of Scientific Revolutions*, University of Chicago Press, Chicago, USA.

Lakatos, I. and Musgrave, A. (eds) (1970) *Criticism and the Growth of Knowledge*. Proceedings of the International Colloquium in the Philosophy of Science, London, 1965, Vol. 4. Cambridge University Press, Cambridge.

Malik, M. and West, A. (1989) The educational needs of the catering industry with special reference to social services catering, in *The Human Food Chain*, (ed. C. R. W. Spedding), Elsevier Science Publishers, London, pp. 125–42.

Newton, I. (1675) Letter to Robert Hooke. 5 Feb., 1675.

Popper, K. (1963) *Conjectures and Refutations*, Routledge, London.

Ravetz, J. R. (1971) *Scientific Knowledge and its Social Problems*, Oxford University Press, Oxford.

Senge, P. M. (1990) *The Fifth Discipline*, Doubleday Currency, New York.

Snow, C. P. (1964) *The Two Cultures: And a second look*, Cambridge University Press, Cambridge.

Spedding, C. R. W. and Wagner, M. A. (1978) Unpublished report.

Spedding, C. R. W. (ed.) (1983) *Fream's Agriculture*, 16th edn, John Murray, London.

Turner, S. A. (1989) 'Where do carrots grow?' – ideas for teaching and learning about food and diet in schools, in *The Human Food Chain*, (ed. C. R. W. Spedding), Elsevier Science Publishers, London, pp. 143–59.

Youngs, A. J. (1989) Education for the Food Industry, in *The Human Food Chain*, (ed. C. R. W. Spedding), Elsevier Science Publishers, London, pp. 93–111.

Chapter 16

Bunting, A. H. (1979) *Science and Technology for Human Needs, Rural Developments, and the Relief of Poverty*. IADS Occasional paper, New York.

Duckworth, W. E., Gear, A. E. and Lockett, A. G. (1977) *A Guide to Operational Research*, 3rd edn, Chapman & Hall, London.

Ford, Brian, J. (1994) *The process of scientific innovation* (presentation), Corporate Affiliates' Forum, London: Institute of Biology, 27 October 1994.

Ford, Brian J. (1995) The secret story of scientific discovery. *Biologist*, **42**(4), 137.

Heap, R. B. (1995) *Agriculture and Bioethics – Harmony or Discord?* Fifth Annual Lecture of The Royal Agricultural Society of England, 18 February 1995, RASE, England.

Hird, S. and Peeters, M. (1991) Whose molecule is it anyway? *New Scientist*, **27 July**, 8.

HMSO (1993) *Realising our Potential: Strategy for Science, Engineering and Technology*. Government White Paper, HMSO, London.

HMSO (1995) *Report of the Committee to Consider the Ethical Implications of Emerging Technologies in the Breeding of Farm Animals*. Chairman The Rev. Prof. Michael Banner, HMSO, London.

Institute of Biology (1991) *The Public Funding of Research and Development in Biology*. A policy study prepared by the Natural Resources Policy Group, Institute of Biology, London.

Koestler, A. and Smythies, J. R. (eds) (1969) *Beyond Reductionism*, Hutchinson, London.

Kuhn, T. (1962) *The Structure of Scientific Revolutions*, University of Chicago Press, Chicago.

Lakatos, I. and Musgrave, A. (eds) (1970) *Criticism and the Growth of Knowledge*. Proceedings of the International Colloquium in the Philosophy of Science, London, 1965, Vol. 4, Cambridge University Press, Cambridge.

Martin, B. and Irvine, J. (1989) *Research Foresight: Priority Setting in Science*, Pinter, London.

Popper, K. (1959) *Logic of Scientific Discovery*, Hutchinson, London.

Ravetz, J. R. (1971) *Scientific Knowledge and its Social Problems*, Oxford University Press, Oxford.

Schuh, G. E. (1988) *Agricultural Research; Still a Good Investment?* CGIAR Annual Report, 1986–87, Washington, DC.

Science Museum (1994) *UK National Consensus Conference on Plant Biotechnology: Final Report*, 2–4 November 1994, Regent's College London, Science Museum, London.

Spedding, C. R. W. (1980) Prospects and limitations of operations research application in agriculture – agrobiological systems, in *Operations Research in Agriculture and Water Resources*, (eds D. Yaron and C. S. Tapiero), North Holland Publishing Co., Amsterdam, New York and Oxford pp. 67–77.

Spedding, C. R. W. (1995) Bioethical aspects of research policy in the agricultural and food sciences, in *Issues in Agricultural Bioethics*, (eds T. B. Mepham, G. A. Tucker, J. Wiseman), Nottingham University Press, Nottingham, pp. 19–29.

Tribe, D. (1994) *Feeding and Greening the World*, CAB International.

Waddington, C. H. (ed.) (1968–70) *Towards a Theoretical Biology*, Vols 1–3, Edinburgh University Press, Edinburgh.

Williams, P. (1995) *The UK Innovation Lecture*, 16 February 1995, The Queen Elizabeth II Conference Centre, London.

Chapter 17

Ashdown, Paddy (1989) *Citizens' Britain: A Radical Agenda for the 1990s*, Fourth Estate, London.

Banner Committee (1995) Report of the Committee to Consider the Ethical Implications of Emerging Technologies in the Breeding of Farm Animals. HMSO, London.

Barnoud, F. and Rinaudo, M. (1986) New perspectives in large-scale procurement of pulps and fibres from European agricultural products, in *Alternative Uses for Agricultural Surpluses*, (eds. W. F. Raymond and P. Larvor), CEC, Elsevier Applied Science, London, pp. 74–6.

Bennett, R. (1989) Personal communication.

Carruthers, S. P. and Jones, M. R. (1983) *Biofuel production strategies for UK agriculture*. CAS Paper 13, Centre for Agricultural Strategy, University of Reading.

CSO (1988) *Annual Abstract of Statistics*, HMSO, London.

Coghlan, A. (1993) Engineering the therapies of tomorrow. *New Scientist*, **24 April**, 26–31.

Conway, G. R. and Barbier, E. B. (1990) *After the Green Revolution*, Earthscan Publications, London.

Conway, G. R. and Pretty, J. N. (1991) *Unwelcome Harvest*, Earthscan Publications, London.

Debach, P. and Rosen, D. (1991) *Biological Control by Natural Enemies*, 2nd edn, Cambridge University Press, Cambridge.

FAO (1985) *State of Food and Agriculture, 1984*, FAO, Rome.

Guggenheim, K., Weiss, Y. and Fostick, M. (1962) Composition and nutritive value of diets consumed by strict vegetarians. *British Journal of Nutrition*, **16**, 467–74.

Heap, R. B. (1995) *Agriculture and Bioethics – Harmony or Discord?* Fifth Annual Lecture of The Royal Agricultural Society of England, 28 Feb. 1995, RASE.

HMSO (1990) *This Common Inheritance: Britain's Environmental Strategy*, Cmd 1200, HMSO, London.

Hodges, R. D. and Scofield, A. M. (1988) Biological husbandry: an introduction to its scientific foundation. *Modern Organic Farming and Horticulture*, **1**(1), 9–15.

Holden, P. (1989) *The Case for Organic Agriculture*. Proceedings 1989 National Conference on Organic Food Production, 4–6.

House of Lords Select Committee (1993) Report of the Committee on Science and Technology on Regulation of the United Kingdom Biotechnology Industry and Global Competitiveness. Paper HL80 of 1992/93 Session.

Jollans, J. L. (1985) *Fertilisers in UK Farming*. CAS Rep No. 9, Centre for Agricultural Strategy, University of Reading.

Kurtha, A. N. and Ellis, F. R. (1970) The nutritional, clinical and economic aspects of vegan diets. *P1 Fds Hum Nutr*, **2**, 13–22.

Lampkin, N. (1990) *Organic Farming*, Farming Press, Ipswich.

MAFF (1988) *Annual Review of Agriculture 1988*, CM299, HMSO, London.

MAFF (1989) *Agriculture in the UK; 1988*, HMSO, London.

Martin, S. and Tait, J. (1992) Attitudes of selected public groups in the UK to biotechnology, in *Biotechnology in Public* (ed. J. Durant), The Science Museum for the European Federation of Biotechnology.

Miflin, B. J. (1992) Progress in Plant Breeding: its role in regulating the need for inputs. Proceedings of 9th Royal Show International Symposium *Towards Sustainable Crop Production Systems*, 1–7 July 1992, Cambridge. Royal Agricultural Society of England, 51–53.

Neville-Rolfe, E. (1990) British agricultural policy and the EC, in *Agriculture in Britain: changing pressures and policies*, (ed. D. Britton), CAB.

North, J. (1990) Future agricultural land use patterns, in *Agriculture in Britain: changing pressures and policies*, (ed. D. Britton), CAB.

Oliver-Bellasis, H. (1991) What if? A vision of the future of our industry – a farmer's view. *Farmland Market*, **36**, 14–15.

Polkinghorne Committee (1993) Report of the Committee on the Ethics of Genetic Modification and Food Use, HMSO, London.

POST (1994) Regulating biotechnology. *Post note*, 55,

November 1994. Parliamentary Office of Science and Technology.

Postgate, J. (1995) Eugenics returns. *Biologist*, **42**(2), 96.

Rexen, F. and Munck, L. (1984) Cereal crops for industrial use in Europe. Report prepared for the Commission of the EC, EUR 9617 ER.

Spedding, C. R. W., Walsingham, J. M. and Hoxey, A. M. (1981) *Biological Efficiency in Agriculture*, Academic Press, London.

Spedding, C. R. W. (1989) The impact of organic foods and vegetarianism, in *Foods for the '90s*, (eds G. G. Birch, G. Campbell-Platt and M. G. Lindley), Elsevier Applied Science, London, pp. 231–41.

Straughan, R. (1989) *The Genetic Manipulation of Plants, Animals and Microbes*, National Consumer Council, London.

Straughan, R. (1992) *Ethics, Morality and Crop Biotechnology*, Department of Arts and Humanities in Education, University of Reading.

Taylor, J. P. and Dixon, J. B. (1990) *Agriculture and the Environment: Towards Integration*, RSPB, Sandy.

Thompson, S. and Braun, E. (1978) Cropping for plant-based agriculture. *Fd Policy*, **3**(2), 147–9.

USDA (1980) *Report and Recommendations on Organic Farming*, USDA, Washington.

van Emden, H. F. (1987) Paper to: Biological control of pests, pathogens and weeds: development and prospects. Royal Society Discussion meeting London, 18–19 February, 1987.

Vogtmann, H. (1995) Organic farming – trend of the nineties? *Elm Farm Research Centre Bulletin*, No. 16, April 1995, 2–5.

Whitby, M. (1990) Multiple land use and the market for countryside goods. *Journal of the Royal Agricultural Society of England*, **151**, 32–43.

Index